Powder Mixing

Powder Technology Series

EDITED BY

BRIAN SCARLETT and GENJI JIMBO

Delft University of Technology
The Netherlands

Chubu PowTech Plaza Lab
Japan

Many materials exist in the form of a disperse system, for example powders, pastes, slurries, emulsions and aerosols. The study of such systems necessarily arises in many technologies but may alternatively be regarded as a separate subject which is concerned with the manufacture, characterization and manipulation of such systems. Chapman & Hall were one of the first publishers to recognize the basic importance of the subject, going on to instigate this series of books. The series does not aspire to define and confine the subject without duplication, but rather to provide a good home for any book which has a contribution to make to the record of both the theory and the application of the subject. We hope that all engineers and scientists who concern themselves with disperse systems will use these books and that those who become expert will contribute further to the series.

Particle Size Measurement
Terence Allen
5th edn, 2 vols, hardback
(0 412 75350 2),
552 and 272 pages

Chemistry of Powder Production
Yasuo Arai
Hardback (0 412 39540 1),
292 pages

Particle Size Analysis
Claus Bernhardt
Translated by H. Finken
Hardback (0 412 55880 7),
428 pages

Particle Classification
K. Heiskanen
Hardback (0 412 49300 4),
330 pages

Pneumatic Conveying of Solids – A theoretical and practical approach
G.E. Klinzing, R.D. Marcus, F. Rizk and L.S. Leung
2nd edn, hardback (0 412 72440 5),
624 pages.

Powder Surface Area and Porosity
S. Lowell and Joan E. Shields
3rd edn, hardback (0 412 39690 4),
256 pages

Principles of Flow in Disperse Systems
O. Molerus
Hardback (0 412 40630 6),
314 pages

Processing of Particulate Solids
J.P.K. Seville, U. Tüzan and R. Clift
Hardback (0 751 40376 8),
384 pages

Powder Mixing

Brian H. Kaye

Professor of Physics
Laurentian University
Ontario
Canada

CHAPMAN & HALL

London · Weinheim · New York · Tokyo · Melbourne · Madras

Published by Chapman & Hall, 2–6 Boundary Row, London SE1 8HN, UK

Chapman & Hall, 2–6 Boundary Row, London SE1 8HN, UK

Chapman & Hall, GmbH, Pappelallee 3, 69469 Weinheim, Germany

Chapman & Hall USA, 115 Fifth Avenue, New York, NY 10003, USA

Chapman & Hall Japan, ITP-Japan, Kyowa Building, 3F, 2-2-1 Hirakawacho, Chiyoda-ku, Tokyo 102, Japan

Chapman & Hall Australia, 102 Dodds Street, South Melbourne, Victoria 3205, Australia

Chapman & Hall India, R. Seshadri, 32 Second Main Road, CIT East, Madras 600 035, India

First edition 1997

Typeset in 10/12 Times by Blackpool Typesetting Services Limited, UK
Printed in Great Britain by The University Press, Cambridge

ISBN 0 412 40340 4

A catalogue record for this book is available from the British Library

♾ Printed on permanent acid-free text paper, manufactured in accordance with ANSI/NISO Z39.48-1992 and ANSI/NISO Z39.48-1984 (Permanence of Paper).

Contents

Wordfinder index

This index can be used to find the first occurence and definition of technical terms used throughout the book.

1 Mixing technology

1.1 AN HISTORICAL OVERVIEW OF POWDER MIXING SCIENCE AND TECHNOLOGY

The art of mixing powders is an ancient one. Wall paintings from Egyptian tombs show technicians mixing powders using a pestle and mortar to create cosmetic mixtures and medicines. The mixing of the ingredients of gunpowder has a long history reaching back to at least around 700 BC. A manuscript dating from that time warns Chinese technicians in a discussion of the art of gunpowder making that

> Many people when mixing gunpowder ingredients have caused fires which singe the beards of those engaged in the manufacture of such powders and burns down the building.

The making of cakes and pastries is another ancient art involving the mixing of powders. Today the cosmetics, pharmaceuticals, explosives and food industries are major users of mixing equipment.

One of the problems that the student soon encounters when studying the scientific literature on powder mixing science and technology is a confusing terminology. This confusion arises from the fact that powder mixing is an activity of many industrial technologies and there has been little or no attempt to co-ordinate the evolution of a systematic and carefully defined vocabulary to be used by scientists from many different technical backgrounds. Throughout this text we will discuss historical terminology usage in powder mixing and suggest a consistent set of terms, which will be used throughout the book. To provide ready reference for the reader to the way in which words are used in this text, the wordfinder approach has been adopted. In this technique a word is defined the first time that it is encountered in the text. The first occurrence is highlighted using a bold font and listed in the wordfinder index placed at the beginning of the book. When a word is encountered later on in the text, the reader may wish to refresh his or her memory with regard to the technical term and the wordfinder index will direct the reader to the definition location. Another reason for discussing vocabulary used in powder mixing is the need to organize a consistent set of key words to be used to retrieve useful research papers from the world's rapidly expanding scientific literature. I once did a computer search for literature on the behavior of suspensions of solids in liquids and received from the computer a listing of a paper entitled 'The Theoretical Behaviour of Ideal Suspensions'. At some expense a copy of the paper was obtained, only for me to discover that the investigations described in the scientific publication were concerned with the performance of springs in automobile suspension systems!

Unless scientists pay more attention to the organization of their literature in such a way that the computer can be used to store and retrieve the information, they will find themselves floundering in a sea of irrelevant literature. One of the projects underway at Laurentian University is to take the listing of scientific publications presented in this book and annotate it with key words so that an information retrieval system can be organized on the computer system being used to print the reference list. This is obviously an ongoing project, and in the future it may be possible to issue the literature list with annotated key words, as a separate entity from the list of references used, to make this book useful to the working industrial scientists.

The first thing to note about the vocabulary used in this book is that the term **blend** will not be used, except when it occurs in the name of a piece of equipment. Two main contributory sources to the development of the English language are root words from Latin and Greek, referred to as 'classical' languages, and root words from the Anglo-Saxon which is the parent language of English. As a consequence of the way in which the language is developed, there are many doublets in the English language which have virtually the same meaning, with one of the pair being from Latin root words and the other from Anglo-Saxon roots. This is the situation with regard to the words 'mix' and 'blend'. In the development of the vocabulary used in the book, we will often refer to the meaning of the root words of a technical term. Some readers may be impatient with the discussion of the origin of vocabulary, since they regard themselves as practical people who are anxious to get on with the job. However, a great deal of time and money is wasted by a failure to communicate precisely in a technology such as powder mixing. It is essential that words used in discussing or posing technical problems should be precisely defined and used consistently. For example, as we shall discuss later in this section, the word 'random' is used in a very sloppy way by various authors, and a failure to define the concept can result in two scientists talking at cross purposes and never focusing on the real problem.

The Latin root word *miscere* means 'to mix'. This root word has given us the word **mixture**, to describe the product of the physical intermingling of more than one finely divided component when those components retain their physical identity. A **mixer** is a piece of equipment in which a mixture is achieved. The word 'blend' comes from an old English root word *blanden* meaning exactly the same as the Latin root word. Therefore, blend and blender are doublets of mixture and mixer. Some investigators have used the word 'blend' to indicate the gradual addition of a minor ingredient to a mixture, but the usage is not universal, nor useful, and is not recommended or used in this text.

When discussing the problems of mixing technology it is useful to differentiate between two main classes of powders. One type, described as **free flowing powder**, is typified by such systems as coarse dry sand and coarse metal powders. In general, free flowing powders are relatively easy to mix, but the resultant mixture is prone to segregation of the components during subsequent handling procedures. **Cohesive powders** are ones in which the constituent fineparticles are aggregated to each other by forces such as electrostatic forces and liquid bridges caused by humidity. (The powder does not have to look moist to contain aggregates formed by internal moisture; this aspect of cohesive powder structure is discussed in greater detail in Chapter 3 on powder rheology. **Rheology** is the scientific term for the study of the flow of substances, a word derived from the Greek word *rheos* meaning 'to flow'.) The subject of moisture control of cohesive powder properties is very complex. In some situations one must reduce the environmental humidity to make it possible for powders to flow. In other situations, one must increase the humidity to avoid problems due to electrostatic charge. For example the environmental conditions in the laboratory at Laurentian University can vary widely, reaching 100% humidity in the summer when the temperature outside the building is close to 100° Fahrenheit, but in the winter, when the external temperature can be −20° Fahrenheit, internal humidity values can be so low that dry powder can climb up the walls of a plastic container because of electrostatic forces. Additives are available which can convert a cohesive powder, such as fine cement, to one that flows like water. Thus if one adds finely divided silica to the cement powder it can be transformed into a free flowing powder. These agents are known by names such as flowagent or glidants. Flowagents added to powdered ingredients in a mix can promote mixing but can cause problems later in the mixture handling. In general, cohesive powders are more difficult to mix than free flowing powders, but the resultant mixtures are far less susceptible to segregation during subsequent handling than free flowing powders.

Some scientists have used the words **particulate** to describe finely divided material. The term is not used in this book because particle physics to physicists, and many other scientists, is the study of fragments of the atom. Any computer-aided literature search based on the word 'particle' would flood the enquirer with publications on muons, neutrons, electrons etc. The term **fineparticle** is a unique term which can be used to describe the study of the physical behavior and properties of finely divided materials including sprays. The term 'fineparticle' will be used throughout this text. The kind of confusion that can arise when using the word 'particulate' is illustrated by Beddow in his textbook [1] when he makes the statement

> The most immediate savings which can be released in the operation of particular mixes is that of mixing time.

In this sentence he does not mean special mixtures but particles being mixed. The confusion is obvious.

Even though the art of mixing powders is an ancient tradition, the science of powder mixing is relatively new. Thus, as pointed out by Karl Sommer in an introduction to the history of powder mixing,

> Despite the long history of dry solids mixing, or perhaps because of it, comparatively little is known of the mechanisms involved. All the experience that has been gained ever since solids were first mixed by man some 30,000 to 40,000 years ago has been handed down through the ages formerly from medicine man to medicine man, now a days from foreman to foreman. By continuous trial and error a degree of perfection has been achieved that could hardly been improved upon by scientific approaches [2].

Although there is some truth to these comments on the history of powder mixing, there are many industrial scientists today who would not regard the mixing technology available as constituting a body of knowledge which is a state of perfection that could hardly be improved upon. This is because industry is involved in a never ceasing effort to lower costs, improve efficiency and evolve specialist mixtures. J.K. Beddow, in another review of the history of powder mixing, comments that

> Prior to 1940 there has been little incentive to develop efficient powder mixing equipment for the powder processing industries as a whole. Since the power consumed by mixing equipment was not large and the economic gains that could be achieved by improved mixer design were not a sufficient stimulus to initiate research, effort focused on better design of mixing equipment [1].

The beginning of the scientific study of mixing equipment and mixture structure can be traced back to a pioneering paper published in 1943 by P.M.C. Lacey [3], on the structure of mixtures when the ingredients have had their positions within the mixture randomized. Beddow points out that Lacey's paper had only one reference, to a book on the theory of statistics [1]. The word **stochastic** is used by mathematicians who study randomly varying systems. The word comes from a Greek word meaning 'I guess'. Thus if we are throwing pairs of dice in a game of chance, the number that appears when the dice are thrown is known as a **stochastic variable** because we can only guess which number will be generated by the throw of the dice. After the pioneering work of Lacey, there was a very rapid growth in the literature on the stochastic aspects of powder mixing. Beddows has commented on the growth of this literature by pointing out that in the first major work on powder science published in 1948, Dalla Valle's pioneering work called *Micromeritics*, the subject of mixing was not discussed [4]. (Incidentally, it should be noted that Dalla Valle advocated the use of the term 'Micromeritics', which he coined as an overall term for the subject of powder science and technology, including spray technology. The name is still to be found in some scientific publications but has not been widely adopted.) In 1966 the text *Particulate Technology*, by Clyde Orr, devoted 30 pages to the subject of powder mixing [5].

The rapid growth of the literature on powder mixing is evident by the fact that a bibliography published in 1972 contained 350 references and one prepared in 1976 contained 650 references [6]. As one contemplates such a wealth of literature, one would anticipate that all of the mixing problems of industry could be solved if only the busy technologist could find time to read the accumulated wisdom represented by such a list of references. In fact after the 1970s, interest in the scientific aspects of powder mixing appeared to wane, and a scientist from the United States reviewing a book published in 1986 described powder mixing as

> An important but academically unfashionable subject in the United States [7, 8].

However, academics who have apparently shied away from powder mixing for 10 to 15 years may now be poised for another major set of investigations into powder mixing because of the current interest in chaos and catastrophe theory [9–12].

The fact that very little changed in the art of powder mixing during the 1970s and early 1980s is emphasized by some comments made by Weidenbaum at the commencement of his chapter on the mixing of powders in a textbook edited by Fayed and Otten [13]. In these comments he pointed out that the chapter in the book was the direct reproduction of a chapter written some 12 years earlier for Perry's handbook of chemical engineering [14]. He said that it was being published unchanged because nothing noteworthy had happened in the intervening period. The apparent failure to use the information generated by 30 years of academic investigation is commented on in a chapter on the mixing of powders written by J.C. Williams of Bradford University (one of the acknowledged international experts on powder mixing) in a book published in 1986. Dr Williams states

> During the past 30 years there has been much work done at the universities in the study of solids mixing but the results of this effort are not yet widely applied in industrial practice [15].

At first sight it is surprising that such information is not being used by the industrial community. After conducting workshops for industrial scientists concerned with powder mixing for several years, I think that one of the major problems is that the scientists carrying out the investigations into powder mixing processes in the 30-year period leading up to 1986 tended to use complex mathematical symbolism and vocabulary remote from the background of those who had to use powder mixers. For this reason, a great deal of the information generated in the academic world was inaccessible to the practical scientist. This is one of the reasons why excessive use of mathematical symbolism is avoided in this book. When I began to plan this book, I started to look again at some of the high-powered scientific papers on the statistics of powder mixtures that I had collected over the last 40 years. As I re-read them I remembered some comments made by Dr J.E. Gordon in one of his delightful books on material science and structures. These books are *The New Science of Strong Materials, or Why You Don't Fall Through the Floor* [16], and *Structures, or Why Things Don't Fall Down* [17]. The titles are a good intimation of the flavor of the books. I personally gained a better appreciation of the science of materials through reading these two paperback books than from a university course at the second year honors level. When commenting on textbooks of elasticity, Dr Gordon made the comments

> Since the subject became popular with mathematicians about 150 years ago, I am afraid that a really formidable number of unreadable incomprehensible books have been written about elasticity. Generations of students have endured agonizing boredom in lectures about materials and structures. In my opinion, the mystic and mumbo-jumbo is overdone and often beside the point. It is true that, the higher flights of elasticity are mathematical and very difficult but then, this sort of theory is probably only rarely used by successful engineering designers. What is actually needed for a great many ordinary purposes can be understood quite easily by an intelligent person who will give his or her mind to the matter. ... The engineering professor is apt to pretend that to get anywhere worthwhile without higher mathematics is not only impossible but that it would be vaguely immoral if you could. It seems to me that ordinary mortals like you and me can get along surprisingly well with some intermediate and I hope more interesting state of knowledge [16].

If one substitutes 'powder mixing' for the word 'elasticity' in the above quote, you have a statement of how I feel about some of the complicated theories and mixing indices of published work on powder mixing.

In all fairness however, it should be pointed out that after Dr Gordon has severely criticized the boring mathematical complexity of textbooks, he goes on to add that one does, however, need a basic mathematics survival kit to be able to proceed in a quantitative manner. He writes:

> What we find difficult about mathematics is the formal, symbolic presentation of the subject by pedagogues with a taste for dogma, sadism and incomprehensible squiggles. For the most part, wherever mathematical argument is really needed, I shall try to use graphs and diagrams of the simplest kind [16].

If I can do this for powder mixing science and technology as successfully as Dr Gordon did it for structures and materials, I will be satisfied.

A more serious criticism of the academic literature on powder mixing than its use of mystifying language and mathematics is that it may not have any relevance to the real world. Thus Dr Hersey, who carried out many investigations into the powder mixing problems encountered in the pharmaceutical industry, made the following comments on the powder mixing scientific literature:

> Mixing indices based upon the theoretically randomized state are useless for following pharmaceutical mixing operations. Mixing theory has mostly been evolved by consideration of the mixing of coarse (above 100 micron) noncohesive ideal particles such as glass beads or sand. As a result, the importance of the theory developed is questionable (quoted by Beddow [1]).

Other comments made by Hersey are as follows:

> Most systems (described in scientific papers on mixing) dealt with experimental work on idealized binary systems. Examination of the more practical multicomponent systems showed they do not behave as theory predicts them. Individual components behave distinct from each other and whilst one ingredient may become more well mixed others will be segregating. Hence, the use of tracers to follow the mixing of one ingredient as an indicator for another is not practical. Few attempts are made to use current powder mixing theory to predict results (in industrial practice) suggesting that either the theory is inapplicable or too difficult, or too time consuming, to be applied in practice (quoted by Beddow [1]).

When reviewing some of Hersey's comments, Beddow stated:

> This type of comment is very strong when presented in learned journals where the current style is to be polite to those with whom you disagree [1].

Again as pointed out by Beddow,

> 150 years ago, scientists did not pull their critical punches quite so much and would use adjectives which today would be considered ungentlemanly [1].

A free translation of Dr Hersey's comments in shop floor vocabulary may well be unprintable. The premature death of Dr Hersey was a great loss to powder mixing technology. If he had been able to continue his work, the science of powder mixing would probably be much further along the road to the successful resolution of many problems than it is today [18, 19].

From the perspective of the early 1990s, it is beginning to appear that the fundamental problem with the research into powder mechanisms described in so many of the academic papers out of the 650 listed in the review paper of 1976 [6] is that scientists had a false anticipation of the role that deterministic science has to play in powder mixing technology. In the philosophy of science the theory of **determinism** concerns itself with the basic idea that all events in the universe are connected by a cause-and-effect chain. This idea, that observed phenomena had comprehensible causes underlying their pattern of occurrence, was one of the major ideas responsible for the development of science from Newton onwards [9]. The predictability of the movements of the universe as a cascade of cause-and-effect relationships is often known as **Laplacian determinism** from the fact that one of the most confident expounders of physical determinism was the French scientist Pierre Simon Laplace (1749–1827). Prior to 1970 the belief was widespread among scientists that such problems as weather forecasting and the predicting of the behavior of complex systems only needed better and bigger computers to be solvable by deterministic modeling and reasoning. However, in the early 1970s, experiments carried out by pioneers such as Lorenz and May indicated that in practice many systems which were generated by the interaction of many causes behaved in such a complex manner that, for all useful purposes, the behavior of a system in the future was unpredictable, that is chaotic, within broad limitations. [19, 20].

The term chaos originally was a philosophical-theological term applied to the state of the universe as envisaged by philosophers before the creator or creators, known as gods, structured the universe. Thus a dictionary definition of **chaos** is

> The state of matter before it was reduced to order by the creator. Chaos according to Greek mythology was the most ancient of gods, Night was his daughter [21].

In the 1970s and 1980s a new discipline concerned with the study of the properties of complex systems became known as 'deterministic chaos'. Inevitably the name has been shortened by popular usage to 'chaos' [21].

It has always been an implicit assumption in powder mixing research that the problems of efficient design of powder mixing equipment were solvable in a deterministic manner, provided that we gained more understanding of the causes which contributed to the performance of the mixer. Therefore many of the scientific papers studied the action of contributory mechanisms to powder mixing in the hope that later such knowledge could be synthesized into a grand theory of powder mixing. However, it is now becoming evident that a powder mixer is a chaotic universe and that the predictability of the performance of a powder mixer from fundamental principles may be so limited, that in essence, the powder mixing specialist has to be guided by empirical data generated for the particular powders of interest in any given situation. In other words the systematic study of the dynamics of powder mixing equipment is properly a branch of deterministic chaos and the failure of 30 years of academic study to significantly improve the performance of powder mixing equipment is due to fundamental philosophical problems and not the inadequacies of the research investigations. It is interesting to note that in the area of fluid mixers, a subject closely allied to powder mixing, experts have begun to realize that fluid mixing is a branch of deterministic chaos. Thus Oldshue, the author of an authoritative textbook on fluid mixing, recently stated:

> Mixing processes are so complex that it is not possible to define process requirements using parameters that involve fluid mechanics (personal communication).

The shift from Laplacian determinism to the perspective of chaos theory in liquid mixing is demonstrated by the fact that Ottino has recently started to apply the theories of deterministic chaos to fluid mixing processes [23, 24].

In a recent review of powder mixing equipment, one of the leading experts of powder mixing theory and practice, L.T. Fan, makes several statements which are indicative of a fundamental shift in the attitudes and expectations of the future focus of powder mixing investigations. Thus he states:

> Various mathematical models for [powder mixing] mechanism have been proposed and numerous mathematical expressions for the rates of powder mixing based on these mechanisms have been developed. While many of the models and expressions are deterministic or microscopic, some resort to stochastic approaches. This may be attributed to the difficulties in delineating the inherently complex nature of solids mixing processes by means of the deterministic approaches. Our understanding of solids mixing processes and the design of mixers for powders has mainly been carried out heuristically [24].

Again,

> Due to the complexity of powder mixing behaviour, describable only by a large number of parameters the experience gained with a pilot scale mixer may not be reliable for scale-up. Therefore an effective design procedure employing both heuristics and algorithms needs to be developed [24].

Some readers may be unfamiliar with the terms heuristic and algorithm. The term **algorithm** is used by mathematicians to describe the various steps required to complete a calculation or to carry out a mathematical experiment. Thus the mathematician would say that he or she had an algorithm for deriving the solution to an equation. The term **heuristic** is defined in a disctionary as

> Serving to discover: *method*, system of education under which the pupil is trained to find out things for himself [irreg. f. Gk *heurisko*, find] [21].

The term *eureka* meant in Greek, 'I have found it'. History tells us that Archimedes was so pleased when he discovered his principle that he became the first streaker, running naked down the main street yelling '*Eureka, eureka*'. When Dr Fan uses the term heuristic with regard to the evolution of powder mixers, he is using it in the same sense as computer programmers use the term. A dictionary of computer terms has the following comment on **heuristic progamming**:

> Heuristic programming, an approach to computer programming based on trial and error methods for the solution of problems [25].

A heuristic is a rule of thumb, strategy, method or trick used to improve the efficiency of a system which tries to discover the solutions of complex problems.

There is a growing appreciation amongst powder technologists that powder mixing is on a par with weather forecasting or aircraft design. Understanding contributory mechanisms and basic interactions of various causes can create a broad picture of probable events, but the actual performance of given devices with specified powders is so complex that detailed prediction lies outside the realm of real science. One must always test mixing performance with real systems in the same way that the aeronautical engineer must go to the wind tunnel to check out prediction and fine tune his or her models.

When one adopts the philosophy of deterministic chaos to deal with powder mixing problems, one soon realizes that one of the difficulties to be faced in the evolving technology is the efficient management of information collected over many different mixing situations. Fan suggests that a partial solution to such information handling problems will be the development of expert systems [24].

Dr Fan and his colleagues define an expert system as

> A generalized inference engine and a rule base that takes input data and assumptions, explores the inferences derivable from the rule base, yields conclusions and advice and offers to explain its results by retracing its reason for the user [24].

In everyday language an expert system in its simplest form is an intelligent robot assistant that stores all the information available and can display information when the right buttons are pressed. At Laurentian University we have begun work on the design of an expert system for powder mixing studies which we have named POMM [26]. The name is an acronym derived from the phrase 'Powder Mixing Monitor Expert System'. At its present stage of evolution, POMM is a robot or as one of my colleagues has described it 'a statistician in a box'. However, as POMM evolves and we gain more expertise in programming the system, we are hopeful that it might develop into a cyborg. **Cyborg** is short for cybernetic organism. **Cybernetics** is the name for a study of self-controlled, self-mobile systems capable of achieving goals. The word comes from a Greek root word *Kybernetes*, which was the name given by Greeks to the sailor in charge of steering a ship toward a distant destination. Thus a heuristically programmed POMM system will discover for itself significant patterns in its accumulated data and possibly suggest new line of activity to a technologist. For example, if POMM were to be automaitcally recording temperature and humidity in the working environment of a powder mixing system it may, by checking through its data banks, discover that problems were always experienced with a powder mixing system when the humidity and temperature were low, a situation which could be caused by high condensation problems either on the walls of the mixer or on pipes being used to feed ingredients into the mixer. Thus it could print out the following message:

> I would like to advise you that my data banks suggest that we are having production problems on the days when we have high humidity and low temperature. Perhaps a dehumidification procedure for the ingredient feed system may lead to better performance of the mixing equipment.

(The human expert could probably discover this correlation of bad performance with temperature and humidity for themselves. The above illustration is given here to explain what is meant by heuristic programming of an expert system.) Display systems available with POMM at its current state of development are discussed later in this chapter [26, 27].

1.2 A HOLISTIC APPROACH TO POWDER MIXING

It is the perspective of this book that it is useful to adopt what can be described as a **holistic** approach to powder mixing. In 1963, I attended a scientific meeting at the Warren Spring Laboratories in Stevenage, England, a laboratory which over the years has been active in powder mixing research [28–30]. At that meeting Dr Valentin, the group leader, was asked the question

> How do you move a powder mixture?

The answer that Dr Valentin gave was

> You don't.

In thc shocked silence that developed as a reaction to this answer, Valentin went on to explain that moving a mixture can cause segregation. The technologist who is involved in powder mixing must be concerned for the fact that the mixture he or she generates must be delivered to the process for which it is intended. Sometimes such a delivery of a mixture can be achieved by using a system which mixes as it delivers using passive means mixers as connectors (Chapter 9). In other situations it is necessary to consolidate the mixture in some way before delivering it to a subsequent process. Thus in the pharmaceutical industry, wet granulation of the mixture with subsequent drying and milling is standard practice to stabilize a mixture which is subject to segregation. To demonstrate the problems of delivery of mixed systems, Dr J.C. Williams used to use the system shown in Figure 1.1(a) to illustrate segregation occurring when powder was poured out of a container [31]. As the pile of powder builds up, it is subject to intermittent avalanches, the cascading of which tends to segregate the large fineparticles from the small fineparticles in the heap, as shown in the figure. This example demonstrates that the technologist concerned with mixing a system must know the size distribution of the ingredients and the rheology of the powders into and out of the mixer. The science of characterizing fineparticle distribution is a topic in itself, but in Chapter 2 a brief review of the necessary information which has to be acquired when characterizing a powder to be used in a mixer is presented.

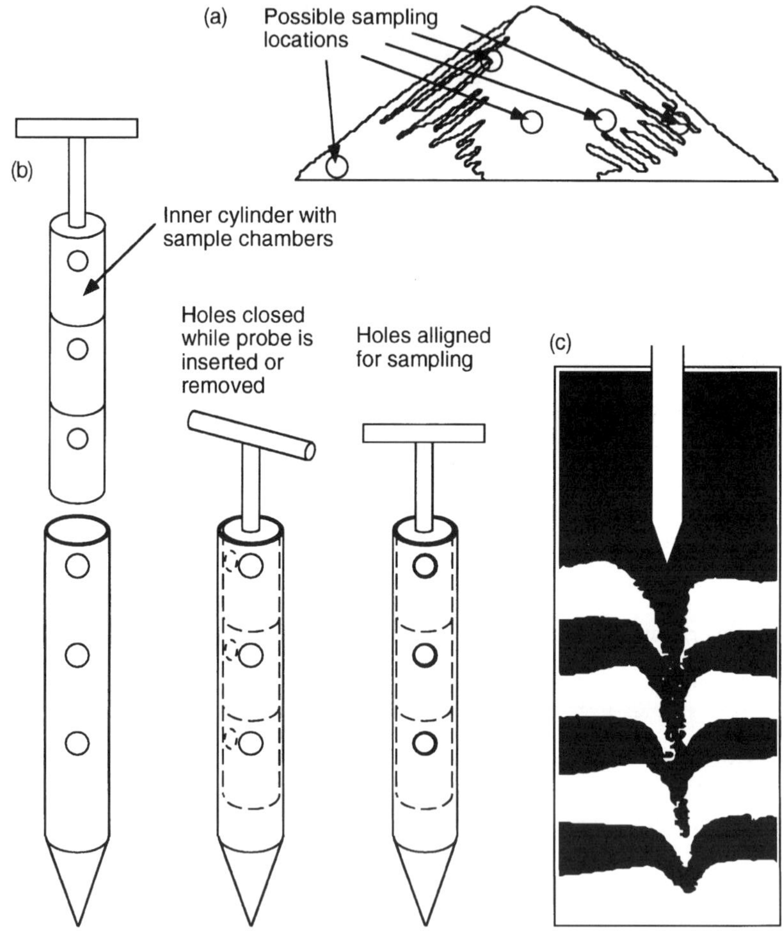

Figure 1.1 Pouring a mixture into a heap can cause separation of the ingredients. The sampling of the powder mixture can be carried out with a thief sampler. The act of sampling can disturb the mixture. (a) Christmas tree demonstration of the type of segregation that can occur within a mixture of free flowing powders poured into a heap. Black glass spheres 1 mm in diameter, white beads 0.2 mm [31]. (b) The thief sampler can be used to extract a sample from a packed bed of powder mixture [32]. (c) The effect of a thief sampler on the powder being sampled. (d) Sand will not flow through an hourglass if the internal air is evacuated, showing that the usual flow of powder through the orifice is air lubricated. (e, f) Powder flow through an orifice can be complex. ((a) used with the permission of J.C. Williams, Bradford, UK; (b) used with the permission of C. Schofield.)

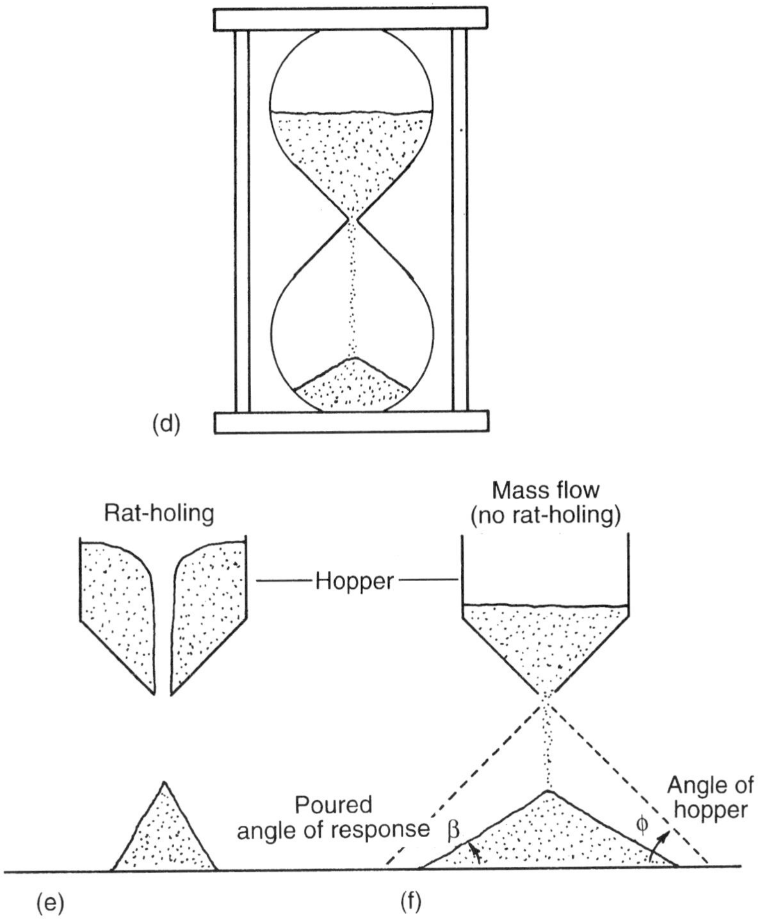

Figure 1.1 *Continued.*

All too often, technologists concerned with powder mixing overlook the fact that a major ingredient of any powder is air or other gas. An aerated powder behaves very differently from one that has sat for a long time. Thus if one wishes to insert any type of probe into a powder mixture, such as a thief probe shown in Figure 1.1(b) [32], the physical resistance of the powder mixture to the passage of the probe increases rapidly as the air inside the powder bleeds out under the vibration experienced by the powder mixture in a factory location. If one takes a well-packed mixture of sand and colored material, the insertion of a probe into the static settle powder requires relatively high force, with subsequent distortion of the layers of the mixture, as illustrated in Figure 1.1(c). But in fact, the sampling of an aerated powder with a thin probe may not cause such a severe distortion to the mixture as that shown in the system of Figure 1.1(c). It may be that a mixture in a mixer should be given a very low level of aeration at all times to prevent consolidation of the mixture by vibration and to facilitate subsequent emptying of the mixer and/or the probing of the mixture to characterize the performance of the mixer.

All too often there is a failure to recognize that when powder is poured into a vessel, the air already in the vessel must leave the vessel. The consequent flow of gas can segregate ingredients of the mixture by creating fluidized bed conditions in the mixture. Over the years, I have made studies of the problems of creating objects from metal and ceramic powders. One of the unsuspected causes of problems in those processes is the fact that when pouring the powder into the mold, the fines of the powder are blown upwards by the exiting air. Thus the fines migrate to the top part of the mixture, leaving the lower part with a coarse open texture and the top part of a more dense structure with the fines located in the spaces between the coarser metal grains. Unfortunately, such segregation is not detected until after the part has been created by fusing the parts together using heat.

A simple experiment which can demonstrate the role of air in determining the behavior of a powder is as follows. Sandflow timers (hourglasses) of the type shown in Figure 1.1(d) were common in the average household before the advent of cheap digital clocks. If one connects the upper chamber of such a sandflow timer to a vacuum pump, one can take out the air. If one attempts to operate the timer without any internal air one finds that the powder will not flow through the neck of the device. For this reason, vacuum filling of molds with metal or ceramic powders is far more difficult than imagined by the uninitiated. The properties of powder in vacuum became an important subject of discussion in the early 1960s when space scientists were concerned that the first astronauts to the moon might discover that the moon's surface was covered with a 2 mile layer of dust. They were concerned with the possibility that the space capsules would sink down into the dust and never be seen again. Such fears turned out to be groundless, however. Research work initiated at the time showed that in the hard vacuum that exists on the surface of the moon a deep bed of powder would have been impenetrable by anyone walking over its surface because of the very high friction between the powder grains. This was something that perhaps scientists should have been able to predict, but which they had to discover by empirical experiments because the full implications of their existing knowledge were not immediately apparent. It should be noted that if one were to be working with a metal powder in a hard vacuum, any movement of one grain past another would result in the destruction of the oxide film on the surface. The two freshly generated virgin surfaces in contact with each other would then immediately weld together to form a virtually unbreakable bond.

The structure of the sand timer of Figure 1.1(d) can also be used to illustrate an important aspect of the feed systems used to feed a powder mixer. The upper part of the sand timer can be regarded as a storage bin feeding an ingredient into a powder mixer. It will be noted that the powder accumulating in the bottom half of the sand timer forms a conical heap. In such conical heaps the angle made by the side of the cone with the horizontal is called the **poured angle of repose**. Historically, industry had many problems with storage bins used to feed powder into industrial processes. It took a long time for industry to realize that the flow of powder out of a storage bin was governed by the angle of the bin bottom and the relative size of the opening through which the powder was flowing, expressed as a function of the largest fineparticle present in the powder flowing out of the bin. In general terms the **orifice** (exit opening at the bottom of the bin) will always block with a bridge of fineparticles forming over the orifice if it is smaller than five diameters of the largest fineparticle in the powder flow. In general, to be sure of steady flow through the orifice it should be at least 20 diameters across. The angle made by the slope of the bottom of the bin, called the **bin bottom angle**, should be greater than the poured angle of repose for the material in the bin to move down in the uniform manner without differential flows in various regions of the bin, which, when developed, can be a source of segregation of the material in the bin. A bin which has a bin bottom angle greater than the poured angle of repose is called a **mass flow bin**. Note again, in powder handling technology sometimes the word **hopper** is used a a synonym for bin. In this book the word **bin** is used to describe storage systems for powder handling. In a badly designed bin, flow can develop into a preferential movement down the core of the bin, a process known in the industry as **rat-holing**. Rat-holing can cause the powder to stop flowing out of the hopper once a core has left the bin, as illustrated in Figure 1.1(e), a condition known as **hangup**. More disastrously the sides of the rat-hole can, in some circumstances, start to collapse into the hole, with the consequent trapping of air in the powder. The turbulence caused by the falling powder then leads to aeration of the powder, with further instabilities of the rat-hole sides. If the process cascades in an uninterrupted manner, the collapse of the rat-hole can lead to the mixture of air and powder behaving likc a low viscosity fluid and the powder flows out of the orifice at the bottom of the bin like water; a situation known as **flooding**. I once visited a factory where they made carbon electrodes for the aluminum industry, shortly after a large bin full of carbonblack had flooded. The entire factory had been given a carbonblack dusting!

I remember hearing Dr J.C. Williams (who is a very vigorous advocator of the proper design of mass storage bins) lecture on one occasion in which, after a company had paid a substantial fee to a consultant to have him determine the angles of repose of a powder and the proper angles for the bin bottom angle for a new bin design, he visited the factory and discovered they were using flat bottom bins instead of the mass flow bins he had designed for them. These square bottom bins had to be vibrated vigorously to shake the powder out into the subsequent processing in an uneven stream. He was told that his designs had been ignored because the headroom in the plant was such that they could not get bins of enough volume into the storage area if they had used the bins with the proper bin bottom angle. One wonders if the company eventually paid for such an oversight with a flooding episode. It also reminds me of the definition of an industrial consultant as someone who is paid well enough that he does mind his recommendations being ignored. It is interesting to note that the designers of sand timers learnt a long time ago to make the bottom bin angle larger than the poured angle of repose (Figure 1.1(f)).

It is suggested that before beginning any mixing study, the technologist prepares a flow sheet based on the diagram presented in Figure 1.2. When constructing such a flow sheet the following questions should be asked:

- Is the flow of any or all of the ingredients into the mixer inadequate? Should one add a flow modifying ingredient to the mixture?

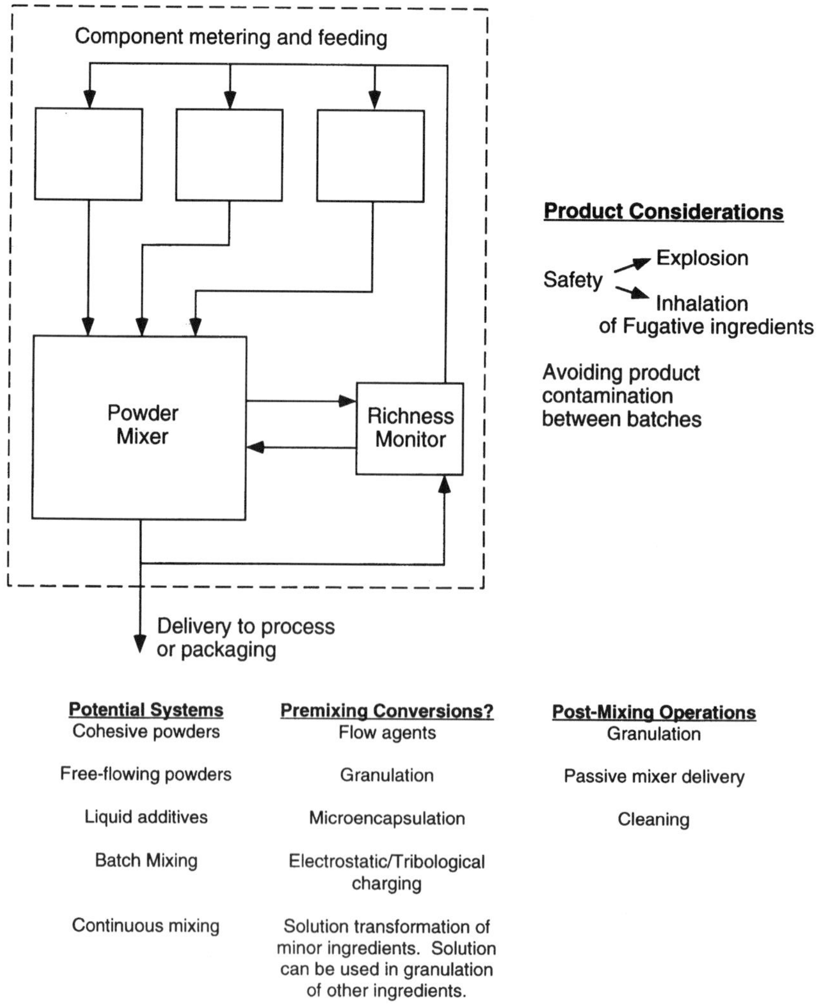

Figure 1.2 Flow chart for planning a holistic approach to the powder mixing process.

- Should one pretreat an ingredient to facilitate the mixing process? Thus should one consider agglomerating one or more of the ingredients, or use a microencapsulation technique to combine two or three minor ingredients to make a coarser grain composite powder, which then becomes easier to mix into the overall system?

When selecting a particular mixer for a process one should also ask the question:

- What are the economics of cleaning the mixer if it is necessary to change the product produced with the same mixing equipment?

This question is very important since, for example, in the pharmaceutical industry it is absolutely necessary to avoid cross-contamination of the product from one batch to another. Another question is:

- Is there any danger of fugitive powders from the mixer?

As the name implies, a fugitive component in a mixture has a tendency to escape from the mixture when it is handled. Thus anyone who has opened a packet of soft-drink mix containing an artificial sweetener has probably tasted the sweetener before using the drink when some of it becomes airborne from the act of pouring the powder into a container, creating a fugitive emission from the package. Artificial sweetener powders are usually very fine in order that the sweetener may dissolve quickly in the added water. In the case of the soft-drink powder mix, the consequences of the fugitive emission are negligible, but in other powder handling industries the problem can cause severe difficulties. A few years ago, some workers in an explosives factory inhaled small quantities of very fine dynamite which caused the workers to suffer from vascular constriction when not working with the powder. This in turn caused severe headaches over the weekend. In another situation, it was rumored that male workers handling estrogen powders inhaled fine powder with subsequent changes in their hormone balance and their sexual characteristics.

One should also always consider the potential problem of explosion hazards when using powders which can burn. A full understanding of the problems which can be encountered when mixing powders can probably only be gained by considering many case histories. For example, I was once consulted on the problems involved in mixing a pharmaceutical powder which normally had operated very well over a long period of time. One day a whole batch had to be destroyed because it would not come out of the mixer in a free flowing state. Subsequent investigation showed that the mixture had had silica flowagents added to it to improve flow. On this particular day, a power cut had resulted in the mixture standing in the mixer for a period of several hours, during which it was subjected to environmental vibration. As a consequence, vibratory consolidation had altered the entire flow properties of the system in such a way that the mixture could not be induced to leave the mixer by any reasonable applied pressure or other technique. (Studies of the effect of silica flowagents on powders compacted by vibration are discussed in Chapter 3.) In a complex industrial situation, changes are sometimes made to a mixing technology which causes subsequent problems in another branch of the manufacturing process. Communication between the groups can be so poor that the group having problems find it difficult to trace their problems back to the fact that the group preparing the mixture added a flowagent to the grinding process. For example, if one adds magnesium stearate flowagent to a pharmaceutical product one may alter the bio-availability and/or disintegratability of the tablets being manufactured from that mixture. The group having trouble with the measurments of the bio-availability of a set of tablets may be unaware of the change in manufacturing process (i.e. the addition of a flowagent) which has altered the surface characteristics of the ingredients of the mixture.

When reporting on a powder mixing operation the technologist should always record the following information:

1. the volume fraction of the various ingredients;
2. the presence of any moisture in the powder;
3. the surface condition of the powder, i.e. if it was oxidized or coated with stearate (very often a metal powder is coated with stearate as part of the manufacturing process and a failure to know about this coating can create havoc in the efficiency of a mixing process);
4. any tendency of the powders to disintegrate during handling and to create dust;
5. the shape and size of aggregates and primary fineparticles;
6. environmental moisture (humidity);
7. any observation of electrostatic phenomena in the powders being handled;
8. a sampling protocol used to explore a mixture;
9. the size of powder samples used to assess the mixture structure (for reasons that will become apparent after the discussion given in section 1.5, the sample of powder used to investigate mixture structure should always be quoted as the ratio of the volume of the largest fineparticle present in the mixture to the volume of the sample size or vice versa);
10. mixing time and handling procedures to deliver the mixtures to container and/or storage bins;
11. any change of powder supplier.

This overall report should always accompany any batch of mixture leaving the mixing station and being used in subsequent processes. Any property tests such as optical reflectance, tap density, permeability, percolation structure etc. should also be reported. If size distribution data are given for the ingredients, the methods used to characterize the size distribution and shape should be listed, since size information for all fineparticles other than hard dense spheres are a function of the method used to characterize the size and shape of the powder [32, 33].

1.3 MIXERS AND POWDER MIXING MECHANISMS

The simplest type of powder mixer which has been employed in industry is the horizontal rotating drum [34]. In such a device the powders to be mixed are added to the drum, which is then set into a rolling motion with the powder rising up the wall and cascading down, as illustrated in Figure 1.3(a). The volume percentage of the drum filled with powder is called the **powder charge**, a quantity sometimes expressed as a fraction of the internal volume of the powder mixer. The powder charge must not be too large or there will be insufficient freedom of motion within the mixer for the powders to mix. The speed of

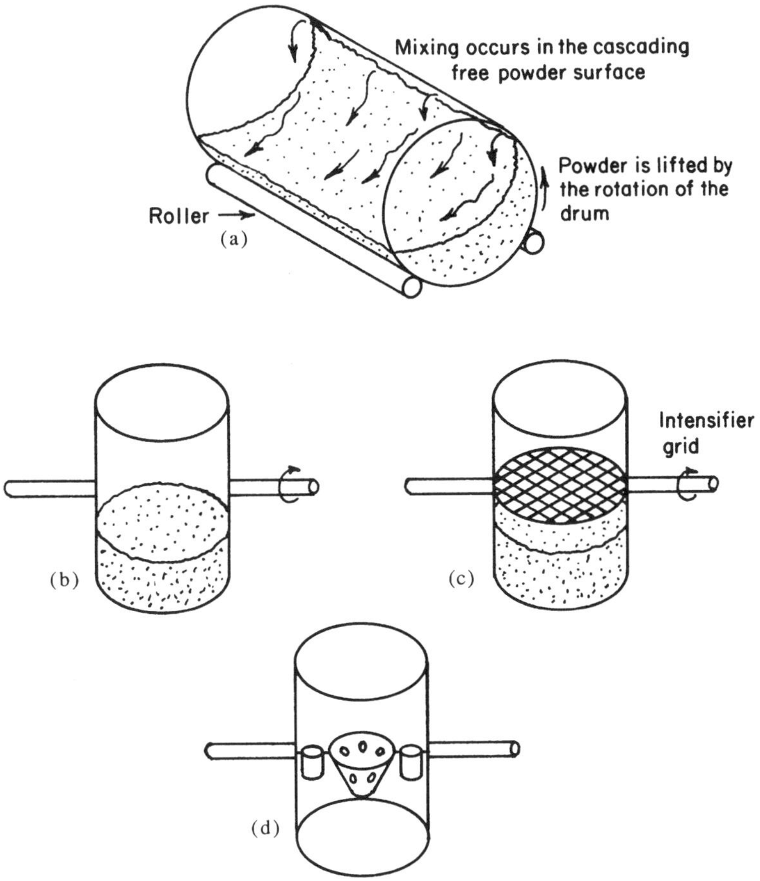

Figure 1.3 Drum mixers are inexpensive to construct and can perform efficiently. (a) Rolling drum mixer. (b) Transient fluidized bed drum mixer. (c) Mixing intensifier grid. (d) Fluidized bed drum mixer equipped with chaos inducing inserts and sampling cups.

rotation cannot be too high because at high speeds of rotation the powder will be centrifuged to the walls and held in position. The only real mixing mechanism in such a drum is the random motion of the fineparticles of the powder as they cascade down the moving surface of the powder [34]. This fact is indicated in Figure 1.3(a). For reasons that will become more apparent as our discussion proceeds, a horizontal rotating drum mixer is classified as an active, captive, mixer. This name indicates that parts of the mixer move, but that the mixing container remains in a fixed position relative to the general geometry of the mixer and to the driving mechanism activating the mixer. From a technological perspective one would not normally expect the rotating drum mixer to be particularly efficient since there are relatively few randomizing actions within the movement of the mixer.

Another way to use a drum to create a mixing device is illustrated in Figure 1.3(b). I first encountered this type of mixer (**transient fluidized bed drum mixer**) on a consulting visit to a company where a practical engineer had created the mixer only to find that his manager could not believe that anything so simple was capable of achieving a good mixture. We were able to show by experiments that indeed the mixer was functioning efficiently. However, I cautioned the practical engineer that this time he had been very lucky and that the functioning of the device was somewhat critically dependent on the volume charge of powder added to the mixer and also the physical properties of the particular ingredient he was using in that mixer. It could be that if he attempted to use the same mixer with a different product he could find himself in serious trouble with severe segregation in the same mixer, with the fines being continually located at the top of the mixture. The mixer was being operated by tipping the drum over and back. When inverted, the powder charge was falling to the bottom on the drum in a highly irregular manner, forcing the air at the bottom upwards and creating a transient fluidized bed. To understand what is meant by a transient fluidized bed, consider the basic system employed in a fluidized bed as shown in Figure 1.4. **Fluidized beds** generally can be created either with a gas or a liquid. Since our concern in this book is basically with powder systems, we will confine our discussion to systems in which the powder is fluidized by means of a gas (D. Boland, chapter in Harnby *et al.* [35]). In the system shown in the figure, compressed air is used as the fluidizing material. If one is using work environment air (defined as **ambient air**) it should be realized that the air always contains moisture and that if a compressed gas is allowed to expand it will usually cool, creating moisture changes that can be a source of undesired agglomeration in the powder. On the other hand, very dry air being used to fluidize a potentially combustible powder always carries with it the danger of explosions. Before using fluidized beds as powder mixing systems, one should always check on any potential explosion hazards. If there is any danger of explosion, one can sometimes use an inert gas as a fluidizing medium. Gases leaving a combustion process usually are a mixture of carbon dioxide and nitrogen, but may have to be dried to remove moisture produced by the combustion process when the exhaust gas is cooled. Alternatively a recirculating system using purified nitrogen can be used.

The fluidizing air is pumped into the air chamber underneath the porous plate and allowed to pass up slowly through the settled powder bed. Successful fludization of a powder requires that the plate have holes smaller than the smallest fineparticles in the powder to be fluidized and be of uniform porosity over the base of the powder bed. As the air pressure is increased, the velocity of air moving up through the powder will increase and, at a certain critical speed, the air will begin to levitate the grains of powder so that the powder bed

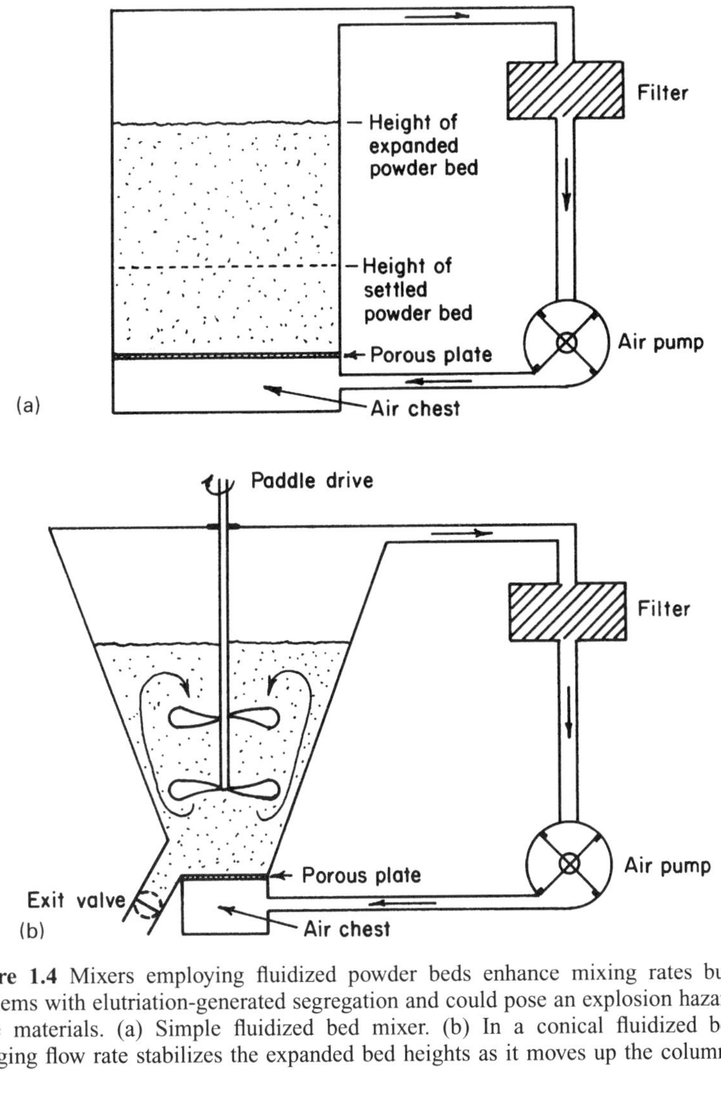

Figure 1.4 Mixers employing fluidized powder beds enhance mixing rates but have problems with elutriation-generated segregation and could pose an explosion hazard with some materials. (a) Simple fluidized bed mixer. (b) In a conical fluidized bed, the changing flow rate stabilizes the expanded bed heights as it moves up the column.

occupies a higher volume. The height of the expanded powder bed depends upon the velocity of the air being moved upwards through the bed. In a simple fluidizing chamber of the type shown in Figure 1.4(a), one only obtains a stable fluidized bed if the fineparticles of the powder are monosized and of uniform density. There is a critical size of fineparticle at which the fluidizing gas is able to transport the fineparticles out of the bed in the exiting gas stream. In such a situation, the fines of the powder bed are said to be elutriated out of the bed. In fact elutriation of powders in a cylindrical column such as that of Figure 1.4(a) is an important technique for fractionating powders into two different size groups [32, 33]. The word **elutriate** means 'to wash out'. It is related to the term **pollute** which originally meant 'the dirt washed out' of an article into the water used to carry out the cleansing process. It should be noted that sometimes a cohesive powder can be turned into a free flowing powder by elutriating the fines out of the powder, a process which will be discussed in more detail in Chapter 3. It is usual to provide a fluidized bed system with a filter since, even if all of the fineparticles start off at the same size, there is usually considerable turbulence in the powder bed, leading to intergrain collision. If the material is at all **friable** (able to break down into smaller parts under relatively low forces) fines are created in the fluidized beds.

A modified fluidized bed system having a recirculating capacity along with a conical fluidizing chamber is shown in Figure 1.4(b). The conical chamber creates a more stable fluidized bed because as the air moves up the chamber its velocity falls, so that it is no longer able to carry the fineparticles upwards. However, the conical fluidized bed is more prone to separation of the ingredients of the mixture into fines and coarse, since the coarse fineparticles, when there is a range of fineparticle sizes and densities in the mixture, or dense fineparticles, may not be able to make it up into the higher regions of the fluidized bed. Note that in some literature on fluidized bed behavior the terms **flotsam** and **jetsam** are used to denote the material that floats to the top of the bed and other material that sinks to the base. Although this usage is deeply entrenched amongst the chemical engineers studying fluidized beds it is an unfortunate terminology since in maritime law flotsam and jetsam are both items that float in the sea. Jetsam differs from flotsam in that it is cargo that has been deliberately thrown over from the ship in an attempt to lighten the ship when it is traveling through a severe storm. The legal ownership of jetsam is different from flotsam and hence the need for separate names. Engineers lacking knowledge of the English language gave new meaning to perfectly prevailing legal term and created ambiguity where none was needed. The terms will not be used further in this book. If one needs names for such different material simple **floaters** and **sinkers** seems to be an adequate vocabulary based upon root words existing in the living language. To avoid the problems caused by fractionation of the ingredients of the powder mixture by the conical fluidized bed, whilst exploiting the advantages of the very high degree of freedom enjoyed by the individual powder grains in such a system, it is possible in some cases to stir the conical bed as indicated in

Figure 1.4(b). It should be remembered that whenever one is constructing a powder handling system one has to be careful how many close fitting shafts, valves etc. one places in the system, since any powder that ends up inside the closely fitting system can create havoc with the operation of moving parts. It may be that a better way of creating turbulence in the conical fluidized bed is to have narrow high-speed jets of air fitted at certain locations along the side of the conical wall to create turbulence in the fluidized bed.

Doctor Fan, in his discussion of mixing technology [24], points out that it is difficult to fluidize cohesive powders, because channeling of the air streams occurs through the powder bed. His comments are very appropriate if one is interested in creating a fluidized bed reactor; however, for the purposes of mixing, one is not too worried about uniform fluidization, particularly as one can often use baffles and moving parts which tend to break up any channels that start to track through the powder bed. In powder mixing, the aim of fluidization is to create freedom of motion for the grains of the powder and one is not overly concerned with the uniformity of the fluidized bed.

The practical engineer who devised the mixer of Figure 1.3(b) was also unknowingly exploiting tribological phenomena. **Tribology** is the study of friction and wear, a name coined from the Greek word *tribos*, meaning 'to rub'. When electrostatic charge is generated by a frictional encounter between two surfaces, the electrostatic charge generated is referred to as a **triboelectric charge**. Many successful mixing operations are achieved inadvertently because the components being mixed become electrostatically charged as they rub against each other in the operation of the mixer. In such situations the mixing process is more properly described as **electrostatic coating**.

In fact the engineer using the drum mixer of Figure 1.3(b) was electrostatically coating one component with another in a transient fluidized bed system. If he had attempted to mix free flowing glass spheres of two different sizes, the experiment would have been a dismal failure. This fact is another indication of how experiments with idealized systems are often irrelevant to the performance of real systems. It should also be remembered that the same triboelectric charge that can enhance mixing can also generate sparks which, under certain circumstances, can trigger an explosion. If one were to be building one's own version of the fluidized bed drum mixer, one could enhance performance by placing a mixing intensifier grid across the drum, as shown in Figure 1.3(c). In general, the more that one can use inserts to randomize the position of any particular powder grain within a mixer by creating chaotic conditions, the better. However, the use of such rods, grids etc. must be balanced against the difficulty of fabrication and cleaning. In general, mixer designers have overlooked the provision of sampling devices within the mixer; the design of the drum mixer shown in Figure 1.3(d) incorporates sampling cups on the intensifier bar placed across the drum.

In Figure 1.5 a series of mixer configurations related to the drum mixer, variously referred to as Y-mixers, V-mixers etc., are shown. Although the literature rarely states it explicitly, all of these mixers rely on transient fluidized beds to achieve mixing of ingredients when the mixer is rotated quickly [36, 37]. Thus in the Y-mixer, during the first rotation the powder falls down into the two arms of the mixer. Part of the mixing action arises from the diversion of the original amount of powder into the two arms which then converge turbulently on the second inversion of the mixer as the powder pours back into the bottom of the system. The mixing action of the V-mixer is essentially the same as that of the Y-mixer. The V- and Y-mixers are also widely used to add liquid to a powder mixture via a central bar with spray nozzles, as suggested by Figure 1.5(b). It was probably discovered empirically that the insertion of a spray injection

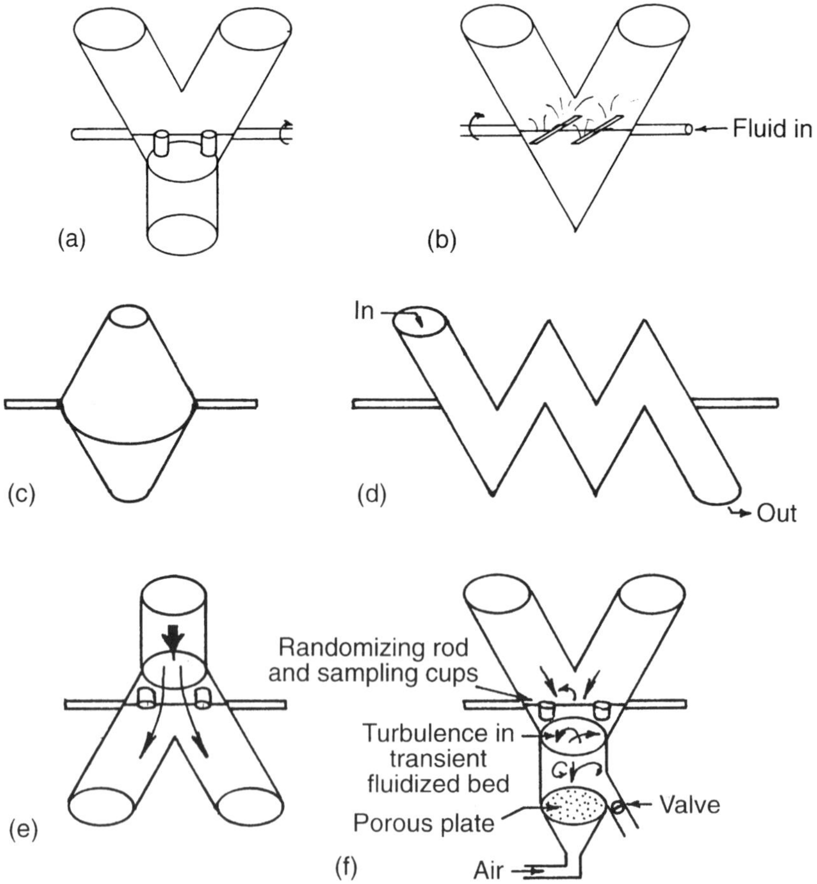

Figure 1.5 Y- and V-mixers, and related designs, employ transient fluidized beds to achieve powder mixing. (a) Y-mixer with sampler attachment [36]. (b) V-mixer with spray attachment. (c) Dual cone blender [37]. (d) Zig-zag mixer [36]. (e) Lambda position return flow of a Y-mixer. (f) Suggested modification to a Y-mixer to facilitate sampling and monitoring of the mixer.

horizontal bar improved the performance of the mixer. For this reason these bars are sometimes described as **intensifier bars**. In this book the term **chaos enhancers** will be used to describe grids and bars added to mixer systems to increase the randomization of the ingredients. Improved mixing from such inserts arises from the fact that as the powder moves past the spray bar it diverts the powder flows and increases the turbulence within the mixer. As we shall discuss in a later section, the question of location sampling systems within a mixer has been virtually ignored in the technical literature. It would appear to be a good modification of a mixer of the Y and V type to always place a chaos enhancing bar across the middle of the system with the provision of sampling cups on the bar, as suggested by Figure 1.5(a).

Thus far in our discussion of mixers, all the mixers have been active, captive, batch mixing systems. In Figure 1.5(d) a variation on the V-mixer, known as the **zigzag mixer**, is shown [36]. This is a chain of V-mixers. The system can function as a continuous mixer with ingredients being placed in the first leg of the mixer and a batch delivery of mixed ingredients at the other end of the system.

When one examines the mixing mechanisms operative in a Y-mixer, it would appear that there are advantages to be gained by making one end of the mixer porous, from sintered metal or glass frit. This frit could be connected at the end of the mixing process to an air chest to provide a low level of aeration to keep the internal friction of the powder mixture at a low level. This would aid the emptying of the mixer to a subsequent progress. The design of such a system is shown in Figure 1.5(f).

In Figure 1.5(e) the inverted Y-mixer is shown to illustrate the fact that some workers describe the Y-mixer in such a position as being in the **lambda position**. There are some discussions in the scientific literature as to whether it is better to load components to be mixed in the two arms of the mixer in the lambda position or as layers in the base of the Y in the other position of the mixer. In my opinion such discussion is not relevant to the performance of the mixer, and surely it is better not to expect the mixer to do everything. If one can introduce the components as parallel concurrent streams, a coarse preliminary mixing would be achieved in the assembly of the system even prior to the operation of the mixer.

In Figure 1.6 a new type of mixer developed at Laurentian University, called a **free-fall tumbling mixer**, is shown. This type of mixer differs from the captive mixers discussed earlier in this section in that the mixing chamber is not linked to the driving mechanisms. The free-fall tumbling mixer was originally developed for the mixing of small quantities of explosive powders and also to improve sampling technology when setting out to characterize a powder system. The mixing chambers are usually half filled with the ingredients to be mixed. The chamber is then placed inside the slowly rotating tumbler drum, which is coated with foam rubber lining. The type of foam rubber lining is similar to that used in packing crates with large dimples, such as those placed in egg packing crates.

This foam rubber lining serves three purposes. First, the friction between the mixing chamber and the tumble drum lifts the chamber a certain height during the rotation before it freely tumbles down to the bottom of the drum. During its fall, the cups in the foam lining tend to divert the falling chamber in an irregular manner while at the same time protecting the chamber from physical shock if the internal ingredients are sensitive to shock damage. Furthermore, the foam rubber lining ensures very quiet operation of the equipment. The gentle tumbling action can be achieved with a very low power driving source and the whole system is very economical and efficient. The performance of the system will be discussed in a later chapter. Investigations have shown that all of the mixing chambers shown in Figure 1.6 generate efficient mixing conditions. The mixing is achieved by creating a chaotic fluidized powder bed inside the tumbling mixing chamber; hence the need to limit the powder charge in the mixer to 0.5 of the volume of the mixing chamber. If necessary the mixing action can be increased by placing randomizing rods across the mixing chamber, which also serve to carry sampling cups. Although the mixing has only been tested for small volume mixing chambers, there is no reason why the system cannot be used for relatively large mixing chambers. The operational fabrication costs are so low that one can afford to use several mixing chambers to achieve mixing of relatively large batches of powder. Furthermore, all cleaning costs are avoided since in essence

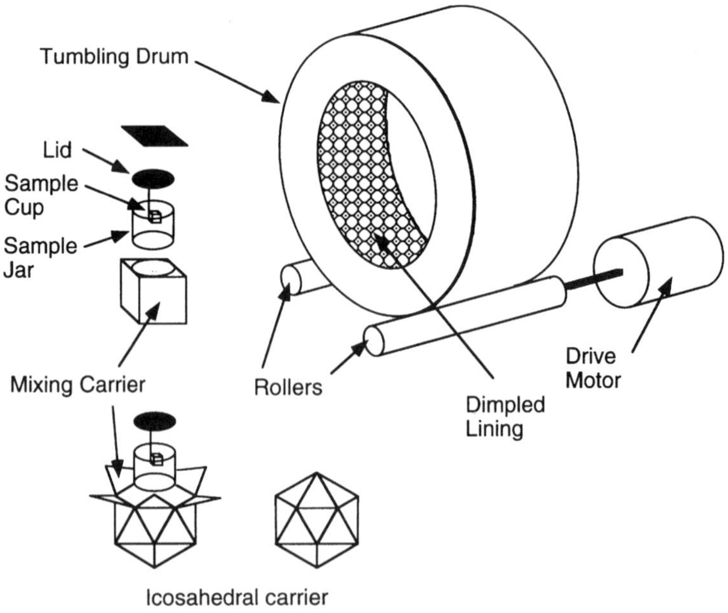

Figure 1.6 The free-fall tumbling mixer creates chaotic conditions using mixing chambers of various configurations, which tumble at random in the foam-lined drum. A commercial version of this mixer/sampler is known as the AeroKaye [38].

the square mixing chamber can be a disposable box, or certainly a cleanable box, for different batches of powder [38].

In Figure 1.7 the basic system of a different type of mixer, known variously as a **motionless** or **Static Mixer**® is shown. I prefer to call these systems **passive mixers** since Static Mixer® is a registered trade name of a specific company and motionless does not convey the idea that in this type of mixer the powder is passed through an array of passive baffles which act as chaos creating elements, to generate the required mixture. Figure 1.7 is based on a specific commercial instrument, the **Luwa mixer**, which is known by the trade name of Blendex® [39]. In a passive mixer, mixing is achieved by passing the ingredients over chaos inducing structures, assembled along the axis of the mixer. Passive mixers

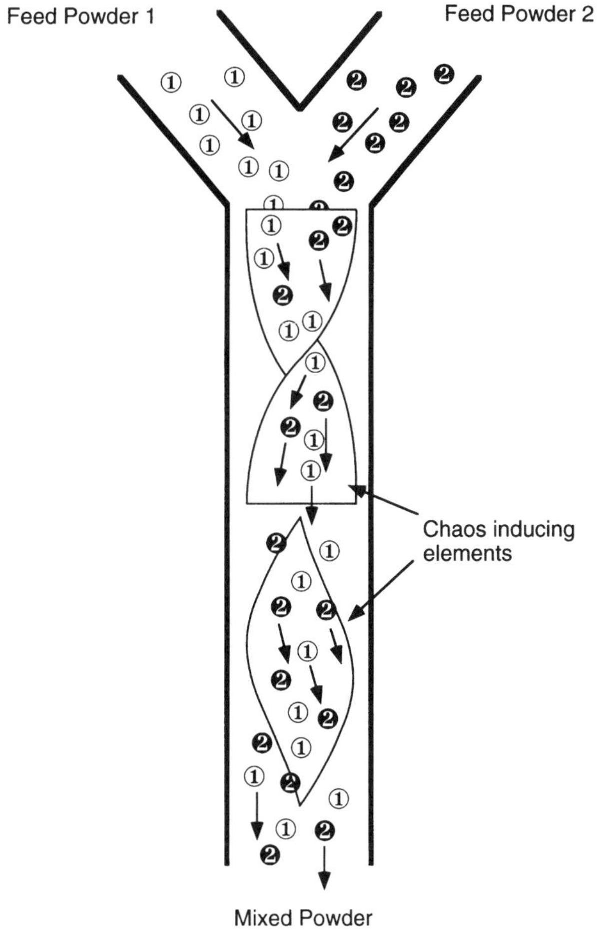

Figure 1.7 A passive mixer uses chaos inducing structures assembled along its length to achieve mixing.

have not been extensively used for powder mixing, but there is a potential for the development of more efficient chaos creating elements. Also, wherever possible, a passive mixer should be used to deliver a mixture, created in a primary device, to a process in order to maintain mixture structure during handling. This aspect of passive mixture utilization will be discussed in more detail in Chapter 8.

Current research at Laurentian University is investigating variations in the chaos creating structures employed in a passive mixer to improve its efficiency both for free flowing and cohesive powders. Figure 1.8 shows a hybrid passive mixer which could simulate the behavior of an infinitely long passive mixer. For the sake of clarity, only one line of very simple chaos creating elements are shown along the length of the cylinder. Powder to be mixed would be allowed to pass through the vanes and would encounter a second line of vanes part way along the periphery of the cylinder. Slow rotation of the chamber containing the lines of chaos creating elements could simulate an infinetly long passive mixer. (The fact that the container would have to rotate is the reason for the term hybrid in the description of the mixer, since it is not a true passive mixer. The term **hybrid** is defined in the dictionary as the offspring of parents of two different species, a **mongrel**. Thus a hybrid mixer is a combination of two different types of mixers which cannot be classified in a simple manner.) The hybrid passive mixer of Figure 1.8 could be used in a batch mode or continuous mode. To

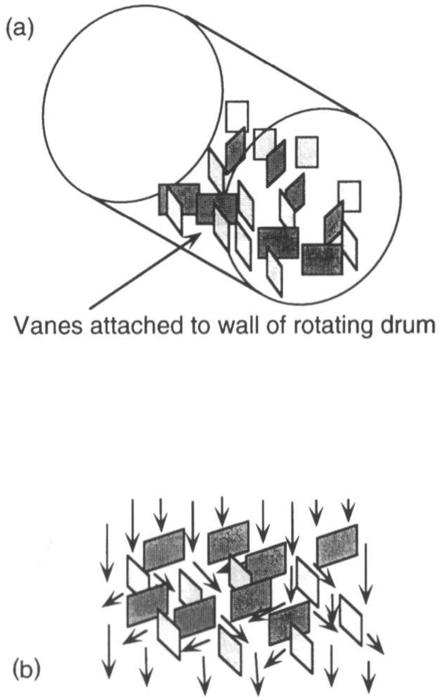

Figure 1.8 (a) Hybrid mixers can approximate the action of an infinitely long passive mixer. (b) Powder is mixed randomly as it flows over the vanes as the drum rotates.

operate in a continuous mode it is necessary for the cylinder axis to slope slightly downwards so that powder would not only be chaotically mixed around the periphery of the cylinder but could also migrate slowly to the exit at the end of the cylinder. Alternatively, a small number of small holes could be placed along a line drawn on the periphery of the cylinder to allow small amounts of the mixed powder to drop through into a process position located underneath. The same system could be used to deliver a mixture into the cylinder, which is then rotated slowly to give uniform delivery through a series of holes onto a process line spanning the whole length of the cylinder. It should be noted that continuous mixing is one of the goals of many industrial processes and that, at the current stage of industrial development, the major limitation on the performance of continuous mixers is our inability to meter and feed powdered ingredients in a smooth and consistent manner into the continuous mixer. Furthermore, industry lacks the online sensors to ensure that a continuous mixer is operating satisfactorily. In Chapter 5 the use of what are known as Poisson trackers for monitoring the performance of continuous mixing equipment will be discussed.

In most discussions of powder mixing presented in chemical engineering textbooks, the mechanisms contributing to the creation of a mixture are usually described as **convective mixing**, **diffusional mixing** and **shear dispersion**. The use of these terms originally arose from engineers moving into powder mixing from a background in liquid mixing.

Dr J.C. Williams has pointed out that the intellectual perspective created by experience with mixing liquids is not necessarily valid for studies of powder mixing [31]. In a liquid, the molecules of the liquid are in constant random motion from intermolecular bombardment, a phenomenon known as **Brownian motion** [40]. If a powder is moved around without creating any space between the powder grains, the only movement of the powder that can exist is the percolation of fines down through the powder. To give any individual grains of powder movement within a mixture it is necessary to dilate the powder bed. The word **dilate** means 'to expand'. Sometimes the dilation of the powder bed is not obvious to an external observer and may be being generated by vibration or low level aeration. Therefore an essential difference between a liquid and a powder is that the equivalent of Brownian motion does not exist within a powder unless the powder is dilated by vibration or by deliberate aeration of the powder, such as aeration that occurs in fluidized beds. It has already been pointed out that in many traditional mixer designs such as rotating drums and V-mixers, aeration of the powder is an important aspect of the operation of the mixing system. Again in fluid mixing, the term convection is used to describe gross fluid currents within the liquid either caused by differential heating of the container or by mechanical agitation such as stirring. In the fluidized bed system of Figure 1.4(b) the engineer would say that paddles are creating convective mixing currents concurrent with the random motion of the individual powder grains caused by the turbulent fluidization.

In Figure 1.9 the concepts of diffusional dispersal (Brownian motion type) of powder grains and convective movement of powder grains are illustrated [41]. The display of Figure 1.9(a) represents a distribution of aggregated monosized black powder in a white powder. The small black square labeled α is considered to be typical of the monosized black elements distributed in the white background. The profile β represents two single grains touching each other, γ three grains touching each other etc. The position of every black grain in the white background is stored in a computer. To simulate diffusion dispersion of the grains it is imagined that, by some physical means, the powder bed is expanded to give the individual grains freedom of motion on a small scale. It is then further assumed that each grain moves in a random direction. The computer program used to simulate such a representation of diffusional mixing inspects each grain sequentially, selects a direction at random from a set of tables and then moves the grain according to this program. Thus in Figure 1.9(b)(i) the movements given to individual grains leading to the breakup of the aggregates are illustrated by means of arrows and the resultant dispersion of the aggregate in mixture by small-scale diffusional motion is illustrated in Figure 1.9(b)(ii). The concepts of convectional mixing are illustrated in Figure 1.9(c). It is

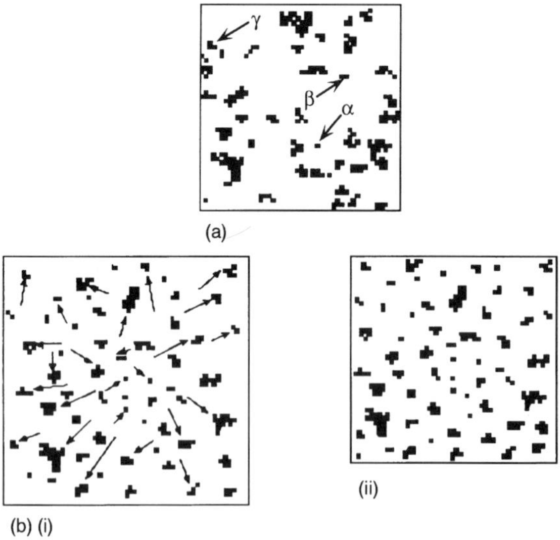

Figure 1.9 Chemical engineers active in powder mixing technology usually describe three different mechanisms active in a powder mixing process: diffusion, convection and shear dispersion. (a) A simulated, loosely agglomerated mixture. (b) (i) Dilated mixture, achieved by shaking or fluidization, allows powder grains to change position by undergoing randomwalks in a process known as diffusion. (ii) Recompacted bed after dilation–diffusion. (c) In convective mixing, bulk movement of the powder followed by random tumbling achieves mixing. (d) Shear creates dispersion by moving layers relative to each other within the powder bed.

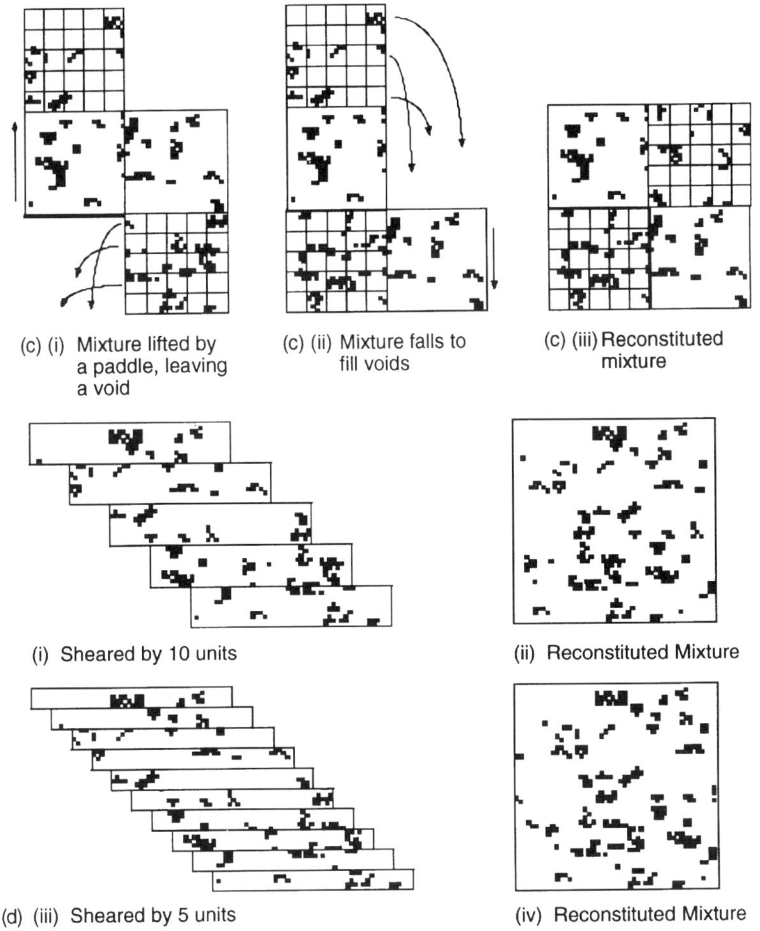

Figure 1.9 *Continued.*

assumed that the four blocks of the mixture shown in the figure were originally forming a square and that a gross displacement of the powder mixture along the direction of the arrow causes the displacement shown in part (i) of the diagram. It is then assumed that the bottom left-hand corner portion of the simulated mixture collapses into the vacant space with the consequent tumbling of the top right-hand corner into the space created as the original top left-hand corner moves down. The final situation for the convectively displaced mixture is shown in Figure 1.9(c)(iii). The reader might like to note that POMM, our **expert system** being developed at Laurentian University, not only acts as an advisor during powder mixing studies but can act as a tutor in the educational mode. In this mode, it can simulate variations in diffusional dispersion of a mixture and convective patterns of the type shown in Figure 1.9.

A major difference between liquid mixing and powder mixing arises from the coarse-grained nature of the universe of items contained in a powder mixer. The reader will recall that a major intellectual revolution occurred in physics when physicists found that they had to realize that energy and matter were not infinitely divisible and that there was a basic structural element to the universe, which was the smallest amount of material that could be created. This was known as the **quantum** of the universe. There is some difference of opinion on the origin of the word quantum. I prefer the story which says that it was the word used to describe the load that could be carried by a cart. Thus a whole supply of material was made up of several cart loads, just as an energy level in physics was made up of several basic units. To a physicist a powder mixer is a small universe in which the quantum of existence is not only variable (the fineparticle sizes of the ingredient) but the size of the smallest quantum is relatively large with respect to the container. Many of the basic theories of quantum physics can in fact be used to describe the behavior of powder systems. It is beyond the scope of this textbook, but one could, for example, use the concepts of entropy to describe the disorganized structure of powder mixtures. When considering fluid mixing, the quantum of existence (the molecules of the liquid) is so much smaller than the container in that, in general, the diffusional dispersion of the liquids is relatively small compared to the convection displacement in the system. In other words, the ratio of the average displacement from Brownian motion is many times smaller than the displacement caused by convection. On the other hand, in a powder mixer, true continuous diffusion does not exist, and small-scale chaotic dispersal from agitation of a dilated powder bed is only one or two orders of magnitude different from gross displacements created by the action of items such as rotating paddles and diversionary baffles. From a deterministic chaos perspective, diffusional movement and convective displacement can be integrated to a randomwalk theory of dispersion in which the distribution of probable steps during a dispersion excercise are a mixture of many small probable steps (diffusional dispersion) and large rare event leaps (convectional displacement). Such a randomwalk approach to combined diffusion–convection displacements can be integrated into a theory known as Levy flights. In Chapter 6 we will discuss the modeling of powder mixing using the concepts of randomwalk and Levy flights.

The third term used to describe dispersion mechanisms in the classical terminology of the chemical engineer is known as shear dispersion. In shear dispersion, what is known as a shear stress is created in a system by moving one surface over another to cause displacements in the direction of the moving surface. Experienced cake makers know of the important role of shear in the dispersion of aggregates, and their chosen mixing device is a wooden spoon with a large flat underside which when moved over the surface of the bowl creates shear forces to disperse lumps (aggregates) in their cake mixture. Over the centuries, pharmaceutical specialists have always used shear dispersion with their pestle and mortar, and this is the reason why the curvature of the bottom of the pestle is often of the same radius of curvature as the base of the mortar. In the grinding action used with the pestle of mortar, shear dispersion is used to improve the mixture texture. The dispersive action of shear can be modeled on the computer as illustrated in the sequence of sketches of Figure 1.9(d). Under the influence of shear stress, the mixture can be considered as consisting of a series of thin plates that slide over each other in the same way that a pack of cards would slip when pushed along the top surface when the base is stuck to the table. In the first two parts of the sketch, the slippage caused by the shearing of the simulated field of view of part (a) is shown and then reassembled to show what the mixture would look like if it had formed a continuous surface around a cylinder. The next part of Figure 1.9(d) shows the shearing that would take place if the slippage occurred in thinner sheets, i.e. under higher shear stress. The two sketches illustrate (in a rather artificial way it must be confessed) how shearing dispersion only works when the effective shearing takes place over a distance of the same order of magnitude as the size of the aggregates that one wishes to disperse. This leads to two further observations. First of all, as Dr Fan points out in his comments [24], a notorious aspect of powder mixing research is that pilot plants often give little guidance to the function of larger systems. One of the reasons for this failure of powder mixing equipment to scale upwards is that the space available to shear a mixture does not scale the same way in which the volume of a mixer increases. Thus if the shearing action is created by an internal part of the mixer moving over the surface, when one makes a bigger mixer, the volume of the mixer increases by the radius cubed, whereas the surface available for shearing only increases by radius squared. In other words, if in the domestic situation one attempts to use a bigger bowl, the volume increases much more rapidly than the surface area of the spoon that one has available to shear the mixture between the spoon and the surface of the bowl.

An examination of all of the mixing systems described in the previous part of this section will establish that there are no shearing mechanisms in the operation of the mixer to create dispersion forces to break down aggregates in the mixture. Early in my involvement with powder mixing, I was involved in a study of the preparation of colored powder to incorporate into plastic to make telephone headpieces. The problem was that even if one manufactured the colored batch powder, a mixture of white powder and pigment, in exactly the same way, the human eye could still tell the slightly different shades of color in the finished product. To overcome this problem, the manufacturers of powder-filled plastics, and those involved in the paint industry and cosmetic industry, very often prepare relatively large batches of master mix supplies of colored pigments. These master mixes are used in small amounts to color products, but once the supply of master mix runs out, the next batch of material may not be color compatible with the previous product. This problem arises because of the difficulty of dispersing pigments of the order of one micron in other systems with a systematic and reproducible efficiency, since most powder mixers do not have shearing forces acting over very small physical distances. At that time I felt that the problem was, and still is, that people ask too much of a powder mixer. To intermingle components such as cocoa, flour and sugar in a cake mix is a relatively easy task, but to disperse a color generating pigment in a matrix is a very difficult problem. Perhaps what one should do is intermingle the ingredients in a powder mixer and then pass the intermingled product through a shear intensive dispersing device. For example, with the master mix supply used to color plastics, perhaps the mixture should be intermingled as much as possible in a V-mixer and then passed through a pinmill of the type shown in Figure 1.10(a). As the powder mixture moves through the small spaces between the rapidly rotating pins and the static pins, very high shear forces are generated over short regions, and this is sufficient to disperse the mixture ingredients. In

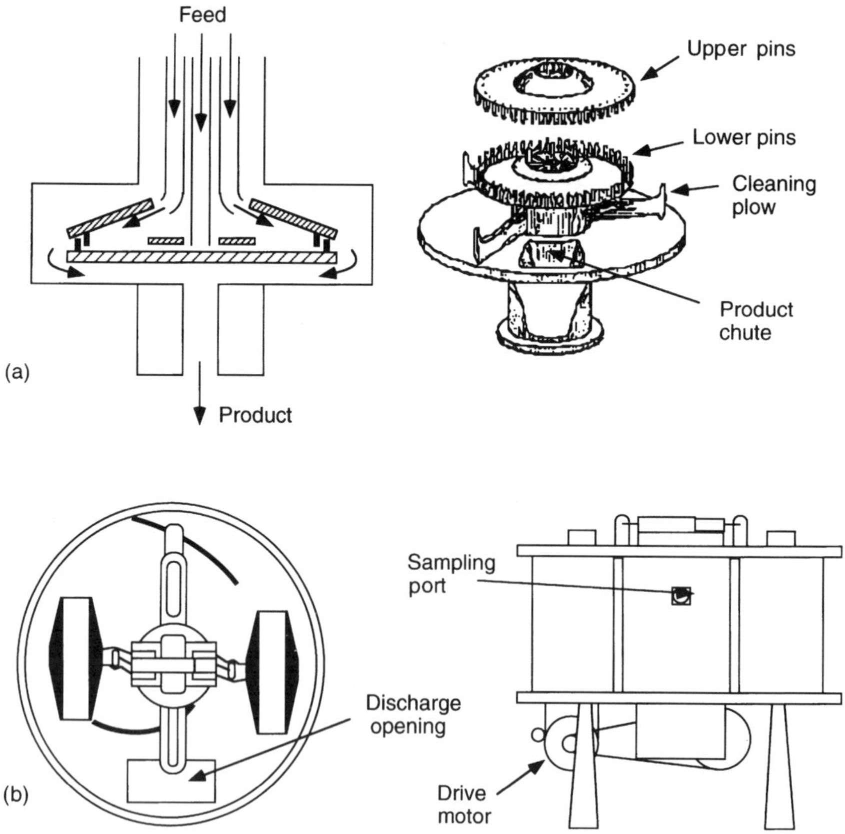

Figure 1.10 Ancillary devices external to the main powder mixer may be the optimum strategy to intimize the internal structure of a powder mixture by means of high shear rates in pinmills or mulling devices. (a) A pinmill can be used as a powder mixer [42,43]. (b) A mulling device can apply high shear forces to a powder mixture [44]. ((b) Courtesy of Bruce W. Dienst, President, National Engineering Company, 751 Shoreline Drive, Aurora, IL 60504, USA.)

my opinion it helps to focus on the absence of shear forces in many mixers and to actually use the vocabularly 'chaos creating mechanisms' and 'shear dispersing mechanisms' as two different functions required to achieve intimate powder mixtures. The pinmill is a relatively expensive machine and can be difficult to clean. A simpler and lower cost procedure would be to take the horizontal drum mixer of Figure 1.3(a) and add steel rods lying along the length of the mixer, which, as they tumble over each other, can create high shear zones in the mill. One can also use balls to increase the shearing forces in a drum mixer. But even then it may be necessary to go to a device such as a triple roll mill to create shearing forces to disperse the powders. Technologists in the various industries have solved the problem of the need to shear a mixture to create a higer level of intimacy than one can achieve in a fluidized bed mixer by going to automated pestle and mortar technology or to mulling techniques. In a **mulling technique** the material to be dispersed is placed in the bottom of a relatively shallow pan, and large wheels are rolled over and skidded on top of the powder mixture lying at the bottom of the mulling wheel [43]. Such devices require auxiliary plows to lift up the crushed mixture and place it back in the path of the oncoming mulling wheel. A plan of a mulling device is shown in Figure 1.10(b).

New procedures variously called hybridization and mechanofusion also provide high shear forces to bring about a specialized form of aggregate dispersion (see discussion of the production of structured mixes in section 1.4 and in a discussion of the evaluation of mixture structure by optical inspection in Chapter 5).

1.4 WHAT IS AN IDEAL MIXTURE? TECHNIQUES FOR DESCRIBING THE STRUCTURE OF MIXTURE

Over the years, various workers have used terms such as ideal mixture, random mixture, perfect mixture etc. to describe the system they generate, or which are the goal of their efforts to create mixtures. To help visualize what the various workers intended to convey by such terminology it is useful to simulate the appearance of powder mixtures by transforming the random number table shown in Figure 1.11(a) using what is known as a Monte Carlo routine.

The term **Monte Carlo routine** is used to describe any mathematical experiment in which the bahavior of a system is simulated and which incorporates stochastic behavior modeled using a randomness generator to vary the behavior of the system. The term was first used in science in the publication on a stochastic process by Metropolis and Ulam in 1949 [45]. Historically the term Monte Carlo was used as a code name during the Second World War for top secret calculations being carried out to predict the flux of neutrons in an atomic bomb. The flow of millions of neutrons following random paths through a massive array of uranium molecules could only be modeled on a computer and not predicted from theory [45, 46].

Since the paths of the neutrons varied at random and since the building of the atomic bomb was a big gamble, the calculations were given the code name Monte Carlo because of the fact that Monte Carlo (the capital of the tiny principality of Monaco) was the gambling center of the world. The fact that

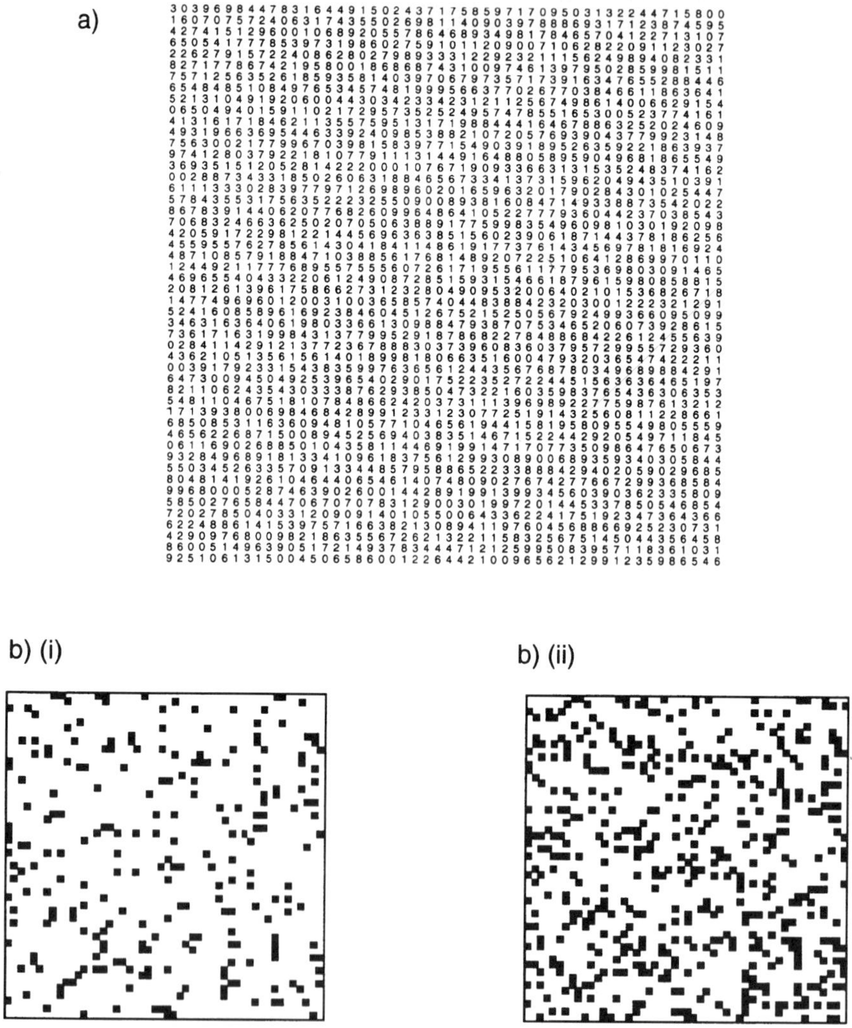

Figure 1.11 A random number table can be used to simulate the appearance and structure of various types of powder mixtures. (a) Typical random number table used to simulate the structure of chaotic powder mixtures. (i) A simulated 10% mixture created by changing all 4's to black pixels. (ii) A simulated 20% mixture created by changing all 4's and 5's to black pixels.

computers are becoming more and more powerful and less and less expensive is making it possible to simulate the behavior of many systems using Monte Carlo routines. Powder mixing is an area where mathematical experiments using Monte Carlo routines are generating useful information to guide experimental studies.

The concept of randomness is very complex and the term random tends to be used in a very loose way to mean 'non-structured'. A dictionary of everyday English gives the definition of random as

> Haphazard, without aim or purpose, heedlessly [ME & OF *random* etc., great speed (*randir*, gallop); -m as in *ransom*].

This definition hints that a **random variable** is one which behaves unpredictably, in the same way that a wild horseman will gallop in many different directions apparently without reason. In a **random number table**, such as that of Figure 1.11(a), the digits 0–9 appear at any location in the table with the same probability and completely independent of the value of any number in any neighboring squares. Thus to print out a table, the computer is instructed to print the digits between 0 and 9 chosen 'completely at random', without reference to any numbers previously printed out and with the probability of each digit between 0 and 9 having the same probability of 0.1 with all possibilities being 1.0. In fact, the generation of such a sequence of random numbers is a very difficult task and should not be attempted by amateurs in the computer world [41]. To convert the table of Figure 1.11(a) to a simulated mixture, one imagines that each digit is occupying a small square of a mosaic representing the whole array of numbers. One then uses a conversion routine to change the digits into tiny squares representing elements of the mixture. In computer graphics an array of tiny squares which becomes a picture is referred to as a **mosaic**. The individual squares of the mosaic are known as **pixels**. This term is a shortened form of the term picture element. The various constituents of a mixture can be shown by representing a component of the mixture by different representations of the unit pixel. The simplest mixture to represent is a **binary mixture** with only two components. One component is represented by a white square and the other by a black square. The richness of a given component in the mixture is usually described as the **volume fraction** of that ingredient. All too often some scientists report components of a mixture as a **weight fraction**. Because the structure of a mixture is determined by the way in which the various components of the mixture compete for space in the volume of the mixture, it is more useful to work with the volume fraction than the weight fraciton of a component in the mixture. When discussing the structure of a mixture in two-dimensional space such as many of the mixtures simulated in this book, the volume fraction of the component being studied is identical to the **area fraction** of the simulated mixture. The random number table of Figure 1.11(a) can be converted into simulated mixtures of 10, 20, 30% etc. by the simple strategy of converting a given digit into a black square to give a mixture of volume fraction 0.1, 0.2 etc.

Thus in Figure 1.11(b)(i) a 0.1 volume fraction mixture (10% of one ingredient) has been generated by turning every 4 in a random number table into a black square and all other digits to white squares.

In Figure 1.11(b)(ii) a binary mixture of 0.2 volume fraction of the ingredient represented by the black square was generated by turning every 4 and 5 into black squares.

The first simulation of a binary mixture in two-dimensional space similar to the system shown in Figure 1.11(b) was published by Lacey in 1943 [3]. At that time, computers did not exist and the construction of diagrams such as Figure 1.11(b) would involve a very tedious and labor intensive construction routine. Today diagrams of this type can be generated on personal computers and can be used very effectively to train new technologists in the physical appearance and properties of various powder mixtures. Mixtures such as those of Figure 1.11(b) were described by workers in the field as **random mixtures** because every discrete element of the mixture was located at random in the occupied space. As mentioned earlier, the term random is often used very loosely and it is preferable, in my opinion, to call mixtures such as those in the diagram **chaotic mixtures**.

When discussing the performance of powder mixing equipment with technologists, it becomes apparent that one of the problems faced by some workers is that they have no experience in judging what can be expected from the chaotic arrangement of ingredients to produce a chaotic mixture. Thus in a powder mixing workshop, participants were shown the simulated 20% powder mixture of Figure 1.11(b)(ii). They were asked if, in their judgement, such a simulated mixture could be considered to be well mixed. Many of the participants in the workshop, who were told that the black ingredient was a monosized powder component dispersed in a matrix of white powder felt that the mixture still contained agglomerates which could be dispersed by a more vigorous randomization of the ingredients. If they had proceeded to try to improve upon the system of Figure 1.11(b)(ii) in a chaos creating mix they would be asking the impossible of their mixer. Any attempt to disperse the groupings of monosized pigments in such a simulated mixture would only result in the breakup of some agglomerates with the formation of others. In this book the association of fineparticles by chance in a chaotic structure will be decribed as stochastic clusters [47]. The development of stochastic clusters in a dispersion of powders in a mixture or a composite material is an important property of the mixture. When discussing such clusters it is important to differentiate between agglomerates and aggregates. The past usage of these two terms in the scientific literature is ambiguous. Whenever one is reading a scientific paper or a technical report one should carefully determine the meaning intended by the writer for either word. In this text, the term **aggregate** will be used to describe a system which is a loose association of visible subunits which disintegrate in the process in which it is used. An **agglomerate** is a conglomerate of visible subunits which maintains its structure throughout any subsequent handling of material. One of the major mistakes that is made when carrying out a study to characterize a powder is to subject the powder sample being used to arrive at the size distribution data with an inappropriate dispersion technique. For example, if someone were to be looking at the size distribution of a titanium dioxide pigment for use in a paint, it would not be appropriate to disperse the powder gently in liquid since aggregates of titanium dioxide which will disperse in the severe shear conditions of the triple roll mill being used to disperse the pigment into paint or plastic will not be matched by the forces created by gentle stirring. Under the stress of 'gentle stirring' the loose aggregates of pigment will maintain their existence throughout the size analysis procedure, with the consequence that size distribution data for the material will be much coarser than the operative existence of the pigment in the coating. On the other hand, if one is working with a pharmaceutical powder, then one should not disperse it with ultrasonics for studying the size distribution of the powder with respect to sedimentation dynamics if the most severe treatment that the powder is going to receive during later use is a gentle stirring in water.

When making composite materials in which a powdered ingredient is dispersed in a matrix, the clustering of the powder fineparticles added to the system can be an important aspect of the structure of the system. For example, if the black pixels of the simulated mixture represented a white pigment being dispersed in a paint, then the clustering of the pigment would represent a loss of efficiency in the scattering power of the white pigment in the paint. An appreciation of the fact that a chaotic mixture structure created by the randomizing mechanisms employed in powder mixing results in stochastic clustering focuses the attention of the technologist on the fact that, to improve on his or her composite material, it is necessary to develop procedures which interfere with the natural stochastic clustering in a chaotic structure. Sometimes technology has solved such a problem **empirically** (by experiment and trial and error) without even knowing the reason for its technological innovation. For many years technologists knew that one could add a component to a paint mixture which was known as an **extender**. Originally, extender pigments were added to lower the cost of the paint without any intention of modifying the optical properties of the paint. Thus calcium carbonate, which was much cheaper than titanium dioxide, a widely used powder, was added to a paint mixture containing titanium dioxide as an extender pigment. The calcium carbonate was basically of the same refractive index as the supporting matrix and theoretically should not contribute to the whiteness of the paint by optical scattering. However, it was well known by paint technologists that such extender pigments improved the light scattering power of the more expensive titanium dioxide pigments and that for unknown reasons the amount of titanium dioxide placed in a film of paint could be reduced if calcium carbonate was added to the paint. It is now apparent that the extender pigment competed for space within the structure of the paint film and prevented the buildup of large stochastic clusters of expensive pigment. In Figure 1.12(a) the size distribution of the stochastic clusters in a simulated

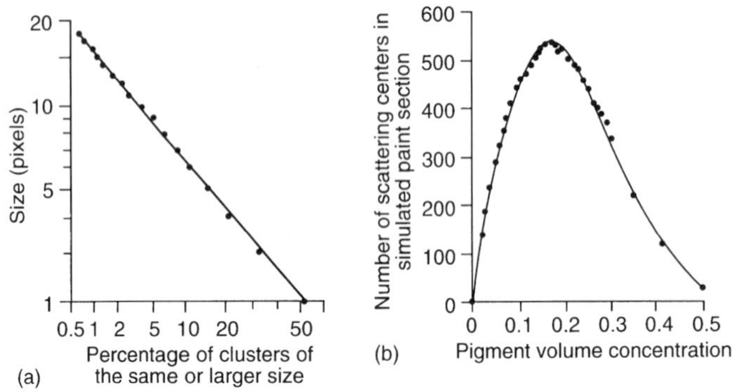

Figure 1.12 Cluster formation by 'random chance' in a powder mixture can be studied by using Monte Carlo routines [41,45]. (a) Size distribution of clusters formed in a simulated 20% pigment by volume, chaotic dispersion of monosized pigments. (b) As the pigment loading of a mixture increases the number of scattering centers, single particles or clusters, peaks at 17% pigment by volume.

20% mixture of monosized black pixels is shown. (Remember this is for a two-dimensional system; in the three-dimensional square the cluster distribution would be different.) Figure 1.12(b) is a graph of the number of independent pigments and/or stochastic clusters in a pigmented coating as a function of the volume percentage of the monosized pigment. This shows that if one is making a mixture in which one is concerned with the existence of independent action centers such as light scattering in a pigment or crack stopping in a dispersion strengthened plastic, the number of independent centers of the pigment in the material starts to decline after a volume percentage of 17%. The physical significance of this figure is that, after a powder dispersion loading greater than 17%, if one were to imagine another fineparticle to be added to the simulated matrix it would have a greater chance of joining a cluster than of forming an independent center of activity [41].

In some situations, the technologist is interested in discovering when a continuous path via one component exists in a mixture structure. For example, in the pharmaceutical industry it is common practice to add a passive powder to an active drug with the subsequent processing of the mixture to form a tablet or capsule. The inert passive carrier of the active drug, known as the **excipient** in the drug mixture, can be substances such as potato starch or lactose (dried milk powder). For example, if one looks at the simulated mixture for 40% black pigment in Figure 1.13, one can regard the white background as constituting a matrix having continuous paths through the mixture. Scientists creating continuous release drug systems are interested in creating mixtures through which continuous paths exist for the slow delivery of the drug system. Also the disintegration of the tablet may be expedited by the existence of a continuous path in the excipient which will disintegrate relatively easily when the tablet is placed in the fluid. The existence of continuous pathways through heterogeneous

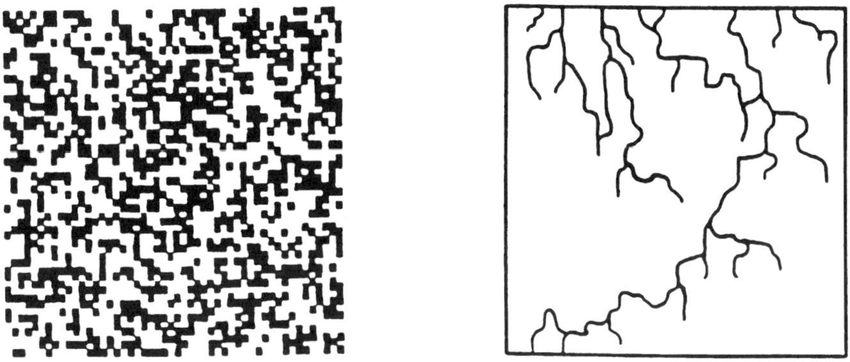

0.4 PVC

Figure 1.13 A continuous percolating pathway can be found through the background (white) of this 40% pigment by volume simulated mixture [41,47].

bodies such as a powder mixture has received a great deal of study in recent years because of the interest in conducting paths in solid state systems etc. Physicists, when developing the theorems associated with the growth of stochastic clusters as the volume fraction of the species of interest increases, until they reach a continuous path, have named the subject area **percolation theory** [48]. Unfortunately, the term percolation was already widely used in powder mixing technology to describe the physical movement and subsequent segregation of a small-sized ingredient in a mixture being subjected to vibration [49, 50]. Thus one might find the following statement in the literature:

> If one adds finely divided *X* to coarse *Y*, the mixture will segregate on standing due to an environmental vibration creating percolation of the fines down through the matrix of the coarse material.

Because of the confusion created by the different usage of the words in physics and in mixing technology, a percolating mixture may either be a solid mixture with a continuous path through the mixture created by one of the components or may actually be a mixture undergoing physical segregation because of the bodily movement of one of the finer constituents. Whenever the term percolation is used in this text its intended meaning will be clearly stated.

Determining when a continuous path will exist through a mixture depends upon how effective the pathway has to be for a given physical property and also on the treatment that the material receives after assembly. Thus as one looks at the cluster shown in Figure 1.14 which is a stochastic cluster taken from the simulated 20% pigmented coating such as those of Figure 1.11(c)(ii). If the touching of one black pixel to another is regarded as constituting a possible pathway for some physical purpose, then theory shows that a continuous pathway will exist whenever the monosized pigment loading exceeds 0.5 volume fraction. The 'just touching' type of percolating path is described by the physicist as **bond percolation**. If the mixture is compressed after it is assembled, the just touching bonds of the stochastic cluster could be squashed together to make a more viable path for such properties as electrical or heat conductivity. Another approach to the study of continuous paths through a heterogeneous system such as a powder mixture is known as **site percolation**. In site percolation adjoining pixels in a simulated mixture are only considered to be forming a union if they butt on to each other **orthogonally** (at right angles). Thus from the perspective of site percolation, the cluster of Figure 1.14 is not a single cluster but constitutes seven 'just touching' smaller percolating clusters. Site percolation is usually assumed as being necessary when one is modeling pathways for the movement of liquids through a body. Although modified (by applied pressure), site percolation might be appropriate in some cases for gaseous diffusion through a porous body. Studies have shown that a mixture becomes a percolating structure for some physical property from the perspective of site percolation theory at a volume fraction of just over 59% by volume. Thus if one looks at the simulated field of view of Figure 1.13, one can trace a continuous path through the matrix from one side to another. Such theory would indicate that if one relies only on stochastic clustering created by chaotic dispersion of the percolating

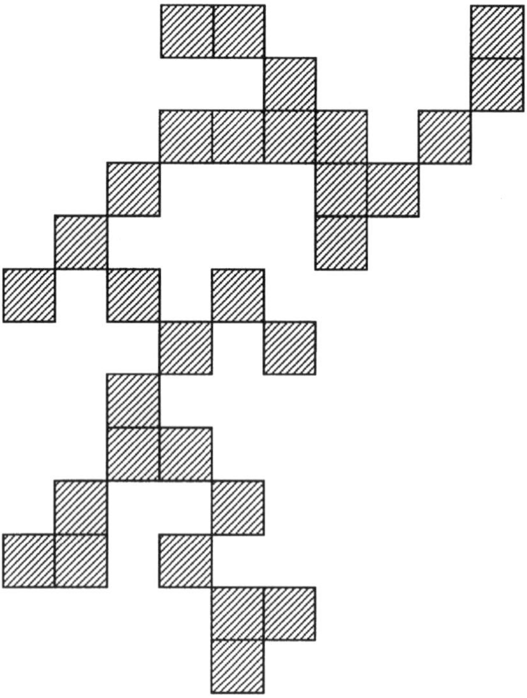

Figure 1.14 The appearance of the largest cluster found in a series of simulated 20% pigment by volume, chaotically dispersed, monosized fineparticles.

component then effective percolating paths exist in the mixture when the value of that component increases above 60% solids in a completely dense mixture.

In our simulation of mixture structure so far, conversion algorithms for going from the random number table to the simulated mixture are simple and straightforward. Thus to stimulate a 0.3 component volume mixture one would convert any three digits in the table into black pixels, say 3, 6 and 9. To construct simulated mixtures of intermediate component fraction involves a more complicated algorithm. Thus to create a simulated mixture of 0.01 volume fraction, one would only convert one tenth of a given digit into a pixel. To operate a conversion routine one would use an auxiliary random number table in the following way. To convert a random number table into a simulated 0.01 volume fraction number only one in ten of the specific digit would be turned into a black pixel. Therefore one would have to have a routine for converting a given digit every time the prime search for conversion resulted in, say, 5 being located. Thus consider searching the primary random number table according to the following search strategy.

To locate each 5 to become a black pixel down column 1, back up column 2, down column 3 etc.: the first stage of conversion would create a sequence as follows:

50000 00000 00050 50000 50000 00500 00000

To select only one in ten of these digits for conversion each time a 5 is encountered, one checks the auxiliary table to see if that 5 should be converted into a black pixel or left as a blank in the simulated matrix. Thus in a probability of conversion search using the auxiliary table and running down column 1, up column 2, down column 3 etc., we could decide that only if we encountered a 6 in the second table would we convert the 5 of the above sequence into a black pixel. Thus our search of that auxiliary table would generate the sequence

3 1 4 6 2

Therefore only the fourth number 5 would become a black pixel in our simulated mixture.

Lacey called the chaotic mixtures such as those of Figure 1.11(c) random mixtures. In the scientific literature such random mixtures were sometiomes called **ideal mixes**, perfect mixes or **natural mixes** in the sense that they were the natural consequences of randomizing forces built into mixing systems. The terms ideal and perfect should be avoided since they have no objective meaning in the discussion of mixture structure. The terms **operationally achievable mixture** and **satisfactory mixture** are two practical terms that can be used to describe mixtures. The first indicates that the structure being achieved in the powder mixing process is the best intermingling of the components that can be achieved by the chosen process. Satisfactory mixture is an operationally determined evaluation of the structure of the mixture being achieved with respect to the intended use of the mixture in a subsequent process. What constitutes a satisfactory mixture will be a function of the penalty incurred for an 'out of limits' sample in a given situation.

In this book we will use the term **legal variation** to describe the range of mixture richness lying within the limits of statistical fluctuation that can occur by chance when sampling a mixture to determine its richness. To understand the range of legal variations that can occur by chance when sampling a chaotic powder mixture we will consider the simple system of the binary mixtures.

In Figure 1.15 a series of simulated chaotic mixtures is shown, in which the black monosized components range in volume fractions from nominal 0.02 to 0.10 in increments of 0.01. When such a sequence of simulated mixtures is shown to beginners in powder mixing technology, they are totally unprepared for the extent that stochastic clustering contributes to the loss of individual fine-particle centers in the system. Thus if one were to be dispersion hardening a composite material by dispersing powder in the material, the individual fine-particle centers act as crack stoppers when the material is subject to stress [17]. As the series of simulated structures of Figure 1.15 shows there is considerable loss of independent centers due to stochastic clustering as the volume of dispersed fineparticles increases in the structure. The same considerations apply to the efficiency of colorant pigment fineparticles in paint and cosmetic powders where aggregates are not as efficient at coloring the system as the individual dispersed systems. Another aspect of Figure 1.15 that newcomers to the subject find surprising is the variation in the actual concentration of a system of given nominal concentration. Thus most people find it hard to believe that a system which is nominally 0.02 by volume can actually, with a fairly high probability, manifest a structure which is 24% away from a nominal structure, as shown by the simulated systems in Figure 1.15. The reader may also like to carry out another interesting experiment and ask people to rate the percentage of the field of view covered by black squares when shown the fields of view of Figure 1.15 in random order. Operators will generate an amazingly wide range of estimates for the systems.

In Figure 1.16, the loss of primary centers in a simulated matrix as the volume fraction of dispersed material is increased is summarized. Figure 1.17 shows the variation to be expected in a set of simulated mixture structures of the same nominal black fineparticle content. The departure from nominal richness of this set indicated in the display of the nine views given in Figure 1.15 demonstrates the need to have a firm understanding of the inherent variation in mixture structure when the mixtures are created by the chaotic dispersal of ingredients. This inherent variation in richness is an important consideration when one comes to consider how to sample the contents of a mixer for richness evaluation (see discussion of sampling protocol in the next section of this chapter). It should be noted that, in the language of the mathematician, the set of simulated mixture systems of Figure 1.17, which should be identical if stochastic variables were not at work, are described as being **statistically self-similar**. Thus the mathematician would say that the set of simulated mixtures of Figure 1.17

represents the legal variations in a statistically self-similar set representative of a system with the nominal value of 0.05 volume fraction.

Deciding when one can treat a mixture as being virtually homogeneous rather than heterogeneous is a matter of the scale of scrutiny with which one examines the mixture. Figure 1.18 depicts the appearance of a 0.1 mixture (10% by

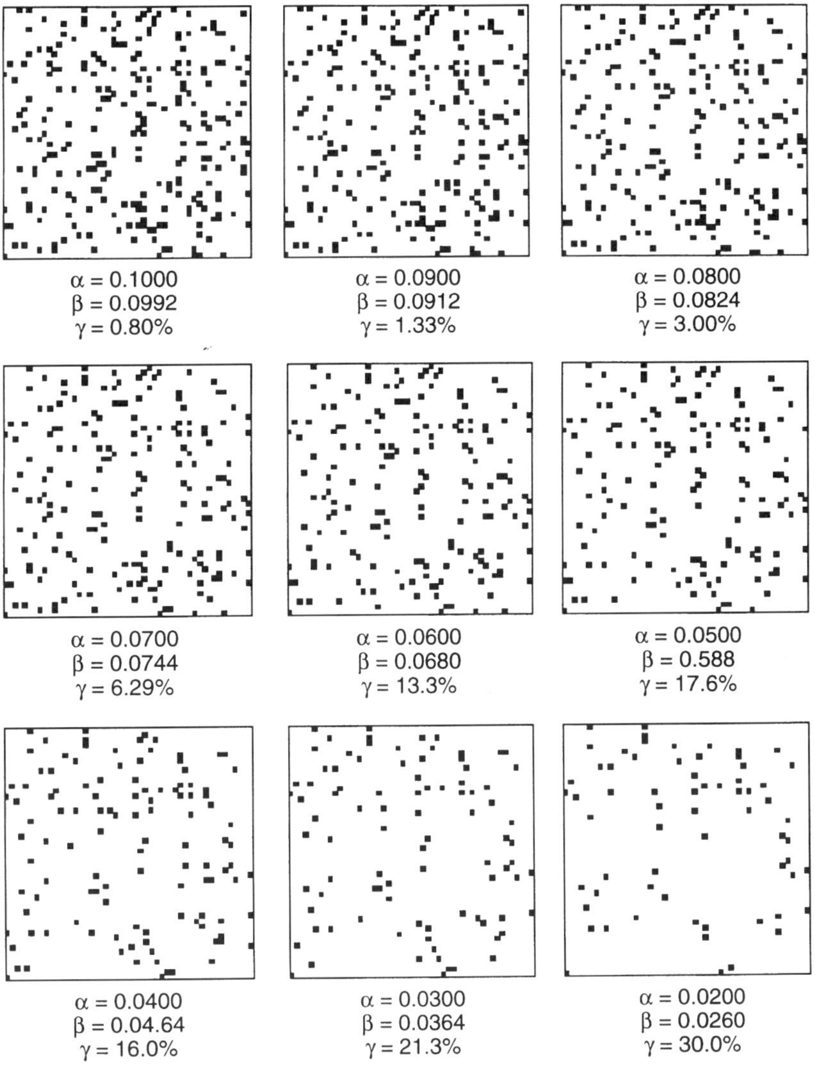

Figure 1.15 Simulated mixtures showing pigment volume concentrations of nominal 10% down to 2% (0.10 to 0.02) in 1% increments, shows the loss of primary scattering centers as stochastic clustering increases at higher concentrations. α = nominal pigment concentration; β = actual pigment concentration obtained by the simulation; γ = percent deviation of the simulated mixture from nominal.

volume) at various scales of scrutiny. The scale of scrutiny of the various fields of view is expressed as a ratio of the pixel size to the field of view pixel size. It can be seen that as this ratio tends to zero the appearance of the field of view it becomes homogeneous as far as the eye can tell. The reader is, however, cautioned that the acceptability of the heterogeneous mixture as a virtually homogeneous structure depends also upon the operational unit into which the mixture is to be fabricated. Thus if the last of the simulated fields of view represented a cattle food biscuit, with the dispersed species represented by vitamin and mineral additives, the cow is going to be munching on a homogeneous biscuit as far as her digestive system is concerned. If, however, the coarsest level of inspection represents a tablet of powerful potentially toxic drug (one which is beneficial at low levels of dosage but which at a relatively low threshold of concentration can create toxic side effects) then one should certainly not regard such a tablet as being homogeneous, especially if the patient is going to be taking half-tablets. I remember reading one particular story about the treatment of chickens with a powerful antibiotic added to their feed. The description of the results reads

> Some chickens did not get enough protective drug and died; the others got too much and died anyway [51].

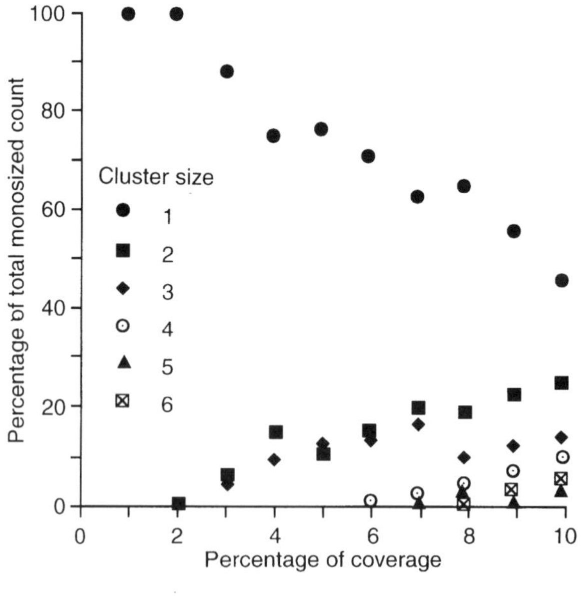

Figure 1.16 Plot showing the effect on primary count loss as the percentage coverage on a simulated field of view is increased for fields of monomized fineparticles. In such fields, clusters form by random juxtaposition of monomised fineparticles as they arrive, and are not true agglomerates.

Such a result with chickens is not going to stir the population into a froth, but the same results with an antibiotic on human patients would certainly lead at least to a demand for better mixture control! Incidently, the sequence of illustrations of Figure 1.17 illustrates the movement from a quantized universe with a grain size which has to be taken into account to a large-scale universe in which quantum effects are lost in the pseudohomogeneity of the large sample. The word pseudo comes from a Greek word meaning 'false'. A pseudonym is a

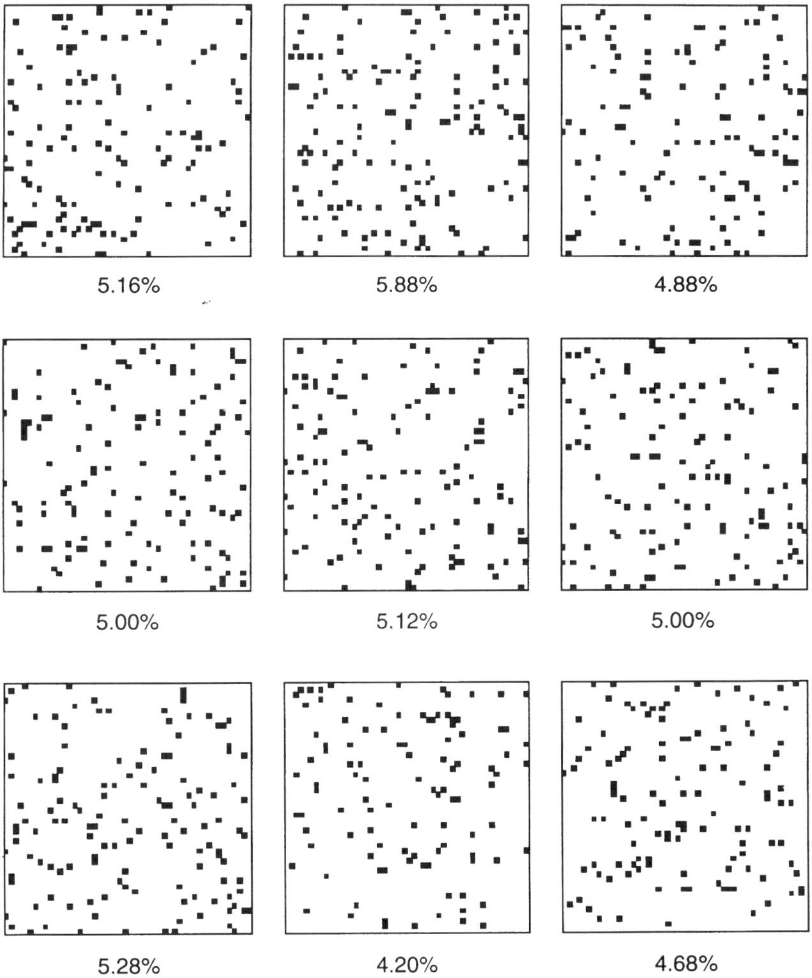

Figure 1.17 The range in actual richness of a region of a powder mixture created by chaotic dispersion can legally vary over a wide range of values; far more than the range anticipated by the newcomer to powder mixture technology. The variation can usefully be expressed in normalized format by dividing the manifest percentage volume fraction by the nominal.

false name. However, a pseudohomogeneous body is not a falsely homogeneous body; the term is intended to describe the fact that for many 'intents and purposes' the body can be treated as if it were homogeneous, although in the strictest sense it always remains heterogeneous. This is the reason why a physicist must consider quantum effect in solid state detectors, but can treat copper as being a continuous material capable of conducting electricity. The wire is so much bigger than the quantum of the Universe.

In Chapter 6 we will describe a dispersal mechanism in a model of mixing behavior as being a pseudo-Levy flight. In this context the term means that, although strictly speaking the distribution of probabilities of various step sizes in a randomwalk do not fit the hyperbolic function of the true Levy flight distribution, in many ways the system behaves as if it were a Levy flight. Thus the use of 'pseudo' in this context warns the reader that the dispersal model is not actually a Levy flight as defined by the pure mathematician, but for many purposes one can treat the system as if it were a Levy flight.

The concept that the best possible mixture system that one could create in a practical situation was one in which the components had had their positions

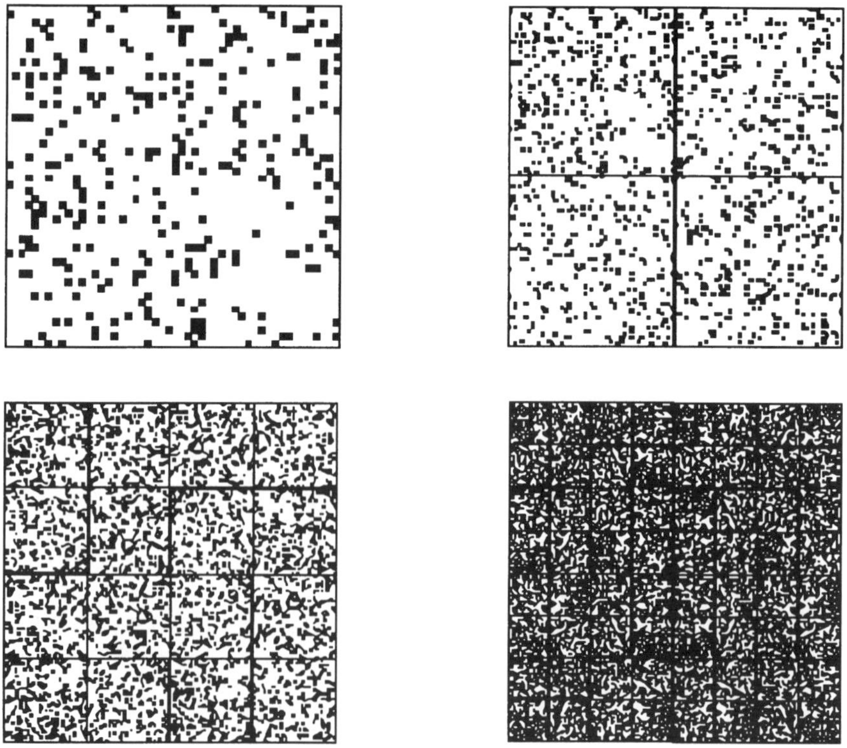

Figure 1.18 A heterogeneous mixture becomes effectively homogeneous when large samples are inspected or used in a process.

completely randomized by creating chaotic conditions in the mixer, dominated powder mixing research for approximately 30 years. However, beginning in the early 1970s, two different groups of workers started to move toward a concept that by exploiting or modifying the physical properties of the ingredients one could achieve better than chaotic mixtures. These structured systems have been variously called, amongst other terms, **ordered**, **regimented** and **structured mixtures**. The first group of people moving towards the use of structured mixtures were Dr Hersey and colleagues in Australia [18, 19, 52–59]. They were concerned for the need to develop mixtures for drug delivery in which the variation in active drug components was minimized. Essentially this group set out to exploit tribological forces to coat one component with another. However, as the technology developed the vocabulary proliferated, as illustrated by the information summarized in Figure 1.19. Thus Orr, in an excellent article on powder mixing in the pharmaceutical industry, used the term regimented mix to differentiate between the random mix, which is our chaotic mix, and ordered mixtures, as shown in Fig 1.19(a) [60–62]. His regimented mix is obviously a kind of structured mix in which the dispersed component is organized into a pattern. This type of mix has also been called a **checkerboard mix** from its resemblance to the board used in chess and checkers. I prefer to use the term structured mix to describe any mixtures which is better than chaotic and then to add some descriptive term describing how the mixture was assembled. Thus the regimented mix of Orr becomes a **layered ordered mix**. The type of mix that Orr calls an ordered mix with freedom of movement, as shown in Fig 1.19(b), I would describe as a heterogeneously coated powder created by electrostatic forces. Many of the food powder–flowagent mixtures reported by Peleg, discussed in Chapter 3, are mixtures of this kind. The term structured ordered unit as used by Orr indicates a coated powder which is agglomerated by the same material with which the coating has been achieved. Orr also discusses incomplete ordered mixing when dispersion of agglomerates of the powder used to coat the larger grains has not been achieved, as indicated in Figure 1.19(c). Staniforth has introduced terms such as total mixing, pseudorandom, imperfect ordered etc. [63–68]. Fan has given an excellent summary of all the various terms that have been used in the pharmaceutical industry. If the reader encounters a term on mixing that he or she has not met before that is not in this book, they will probably find it described in Fan's reference article [24].

The other group of workers that moved towards 'better than chaos' mixtures were a group of materials investigators at the Illinois Institute of Technology Research Institute (known by the initial letters IITRI) who were interested in designing optimum paint films for painted structures. The loss of scattering centers from stochastic clusters in a paint film represented loss in pigment efficiency. If one could avoid stochastic clustering in a paint film, one could increase the reflectance of the paint being used for thermal control of spacecraft. This meant that one could lower the weight of paint on the spacecraft, which in turn meant that less fuel was needed to launch the spacecraft. Research into preventing stochastic clusters in paint films and in composite reinforced materials is an important area of mixing research, and one of the major techniques being used to develop structured paint films is microencapsulation. In this process the pigment or other fineparticles are either completely or partially covered with a thin film of material different from the support matrix of the paint film and the pigment [69]. Figure 1.20 shows data from an early study on the improved scattering power of paint films when extender pigment competed for space in the chaotically organized film. The original work on structured paint

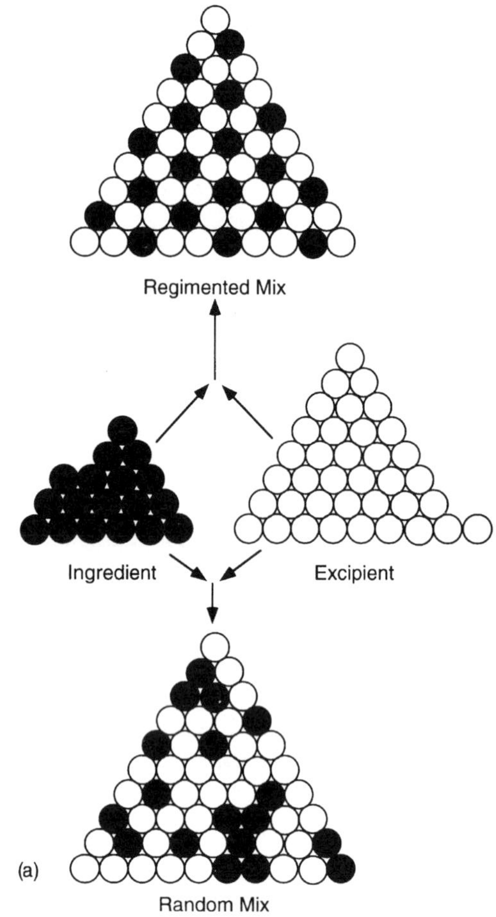

Figure 1.19 As pharmaceutical specialists attempted to develop 'better than random' mixing systems the terminology they used proliferated. Several mixes are described by Orr. (a) Comparison of a 'regimented' or 'structured' mix with a random mix. (b) An 'ordered' mix involves the coating of the larger ingredient by the finger ingredient. (c) When the fine ingredient is somewhat agglomerated the mixture can become an 'incomplete ordered' mixture. (From Orr [60], with the permission of the author.)

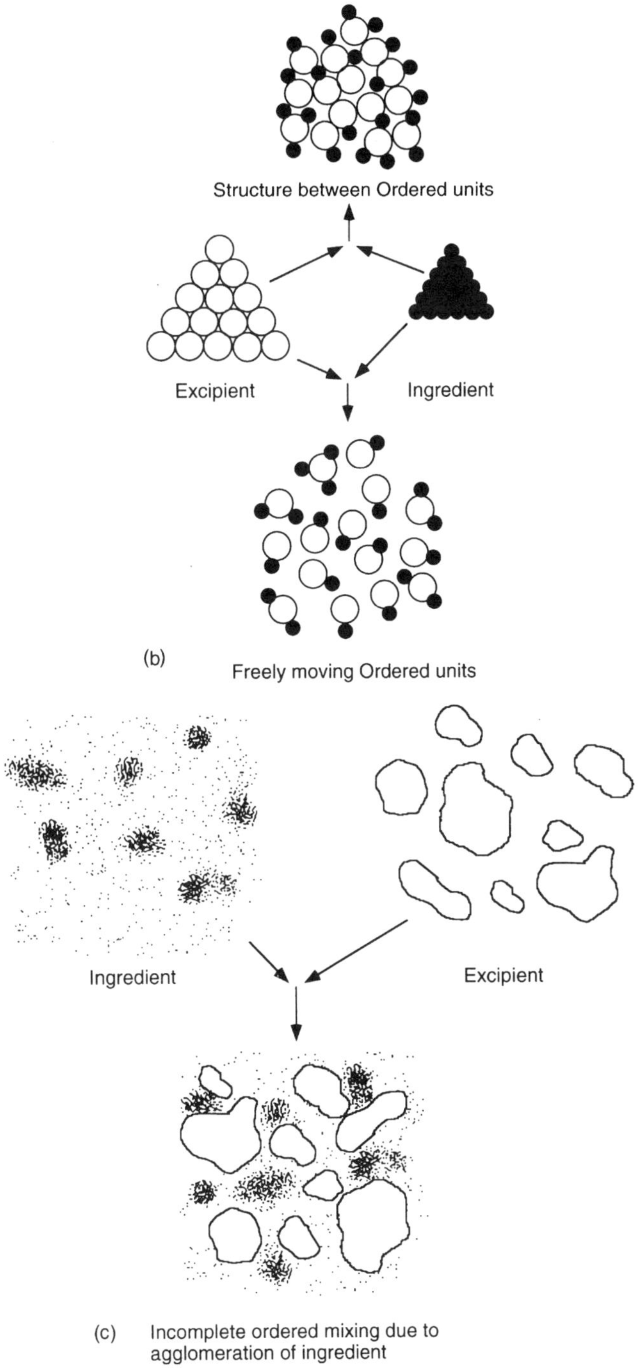

Figure 1.19 *Continued.*

film for spacecraft was reported in a NASA report of 1967, but a more accessible review of the main features of that study is to be found in reference 41.

The basic concepts employed, and the products generated in the new techniques for producing structured mixtures of powder in which one component is used to coat the other are illustrated in Figure 1.21. The Nara Machinery Corporation describes its process for heterogeneous microencapsulation as **Hybridization**® [70]. As already discussed, hybrid means the offspring of two different species; therefore the hybrid powders have complex structure and are

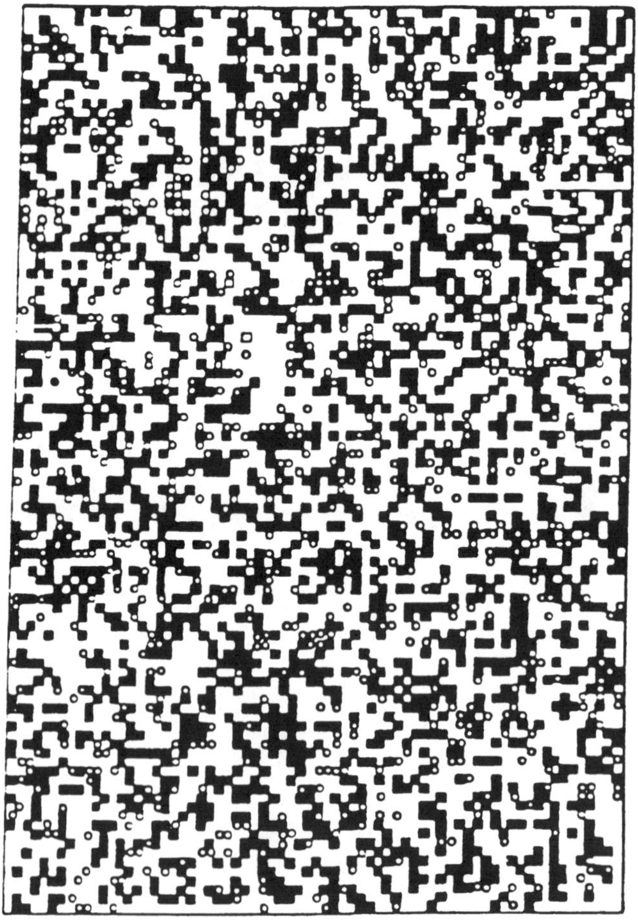

Figure 1.20 The design of structured, 'better than random', paint films is an important area of mixture research. Shown above is the structure of a paint film in which extender pigment competes for space and reduces the size and frequency of stochastic clusters [41].

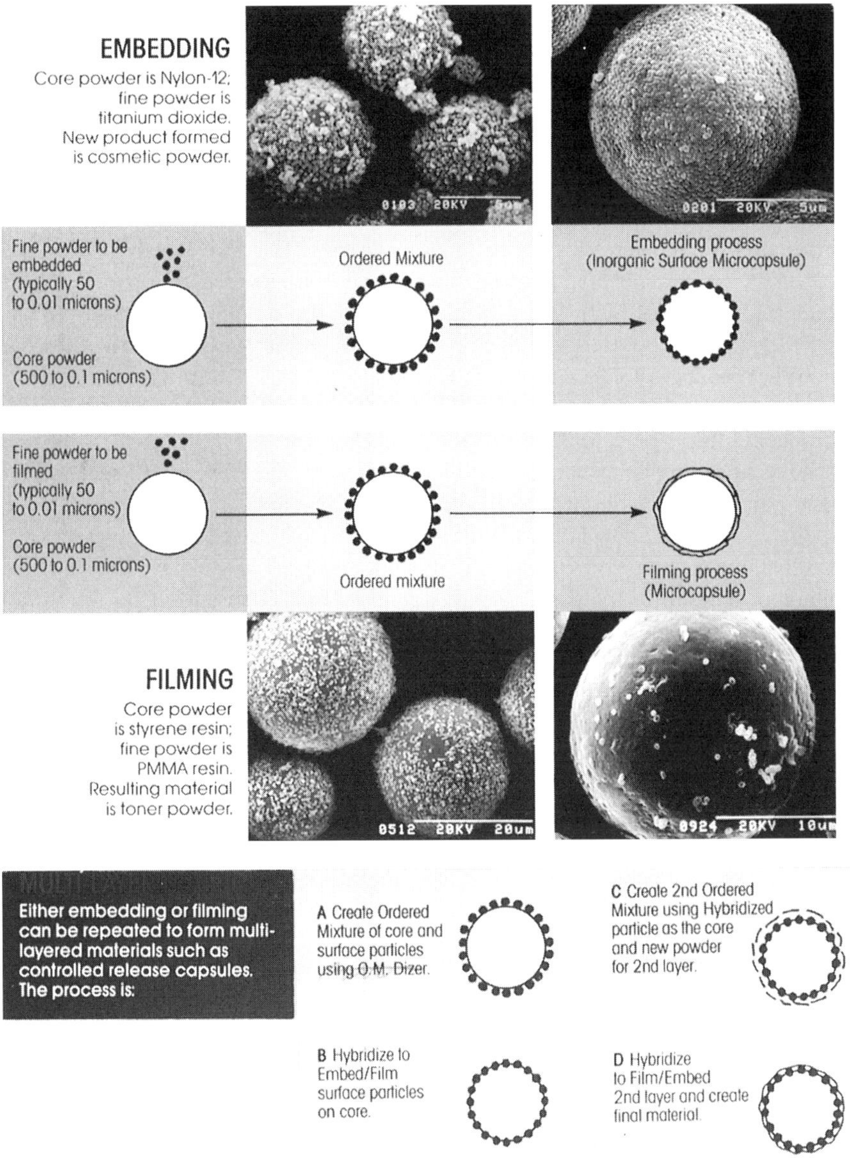

Figure 1.21 Nara Machinery Corporation's Hybridization® process for heterogeneous microencapsulation [69]. (The hybridization system (NHS) is a piece of joint patent equipment of Prof. & Dr Masumi Koishi (Faculty of Industrial Science and Technology Science, University of Tokyo) and Nara Machinery Co., Ltd. (Mr Yorioki Nara, Japan.))

the combination of two different powders which would normally not readily form a coherent mixture. In this book the term **heterogeneous microencapsulation** is used to describe the coating of one powder with another when the added coat does not form a continuous coherent film, as is the case of **coherent microencapsulation** with a layer of gelatin etc. The techniques for producing heterogeneously coated fineparticles such as those shown in Figures 1.21–1.23 have been given various names such as **dry impact blending** and Mechanofusion® (a trade name for a process invented by the Hosokawa Company) [71]. In this book the term **dry grinding** will be used to describe the process for generating such heterogeneous coated fineparticles. The process can be used in a series of stages to develop complex fineparticles with layered outer structures, as indicated in Figure 1.21. In dry grinding usually one employs a relatively coarse powder as core powder and then the added ingredient used to coat the core powder produces a structured mixture of the type illustrated in Figures 1.21–1.23. The **hybrid fineparticles** of these figures illustrate several reasons for producing this type of mixture. First of all, the powder of Figure 1.22, which is used as an ingredient in cosmetic powders, is obviously a very efficient way of using an expensive pigment which is likely to become even more expensive in the next decade. (The world's supply of titanium dioxide, the white pigment used to coat the Lucite core powder is limited and one can expect the price of titanium dioxide to increase.) In the studies which I carried out on spacecraft paint, our concept was to encapsulate the titanium dioxide to separate it from other titanium dioxide fineparticles and to increase the number of light diffracting surfaces.

It is probable that for hundreds of years the powder technologist has been unknowingly producing some type of heterogeneous encapsulated fineparticle when using a dry pestle and mortar grinding to achieve an intimate mix. It may be that, when trying to assess the structure of a powder mixture from light reflectance studies, heterogeneous encapsulation will make the results of gross light scattering signals difficult to interpret [72]. Thus a loose mixture of nylon powder and titanium dioxide powder is going to have a very different light scattering power than the heterogeneously coated hybrid powder. The technologist should be aware that when mixing powders using a pestle and mortar or when employing high shear forces to disperse aggregates one may inadvertently produce hybrid powders. Depending upon the grinding time and the severity of the shear forces, one can create hybrid powders in which the coating layer is adhering to the surface of the core fineparticles or is embedded, as illustrated in Figures 1.21 and 1.22. One can also achieve **filming** of the core fineparticles by using a coating powder which tends to shear-smear under the forces of grinding.

The structure of the coating achieved by Mechanofusion™ is illustrated in Figure 1.22(d). Its heterogeneous nature is likely to amplify the light scattering power of the pigment content of the mixture. Such heterogeneously encapsulated powders will also probably have high surface energies when freshly produced, which will aid in the chemical reaction and/or consolidation dynamics if the hybrid powder is part of a system to be consolidated by scattering.

Another reason for pretreating a powder by a dry grinding process prior to a mixing operation is illustrated by elements (f), (g) and (h) of Figure 1.22. The production of the hybrid powder radically alters the flow properties of the two ingredients, making them much more likely to move freely around the mixer and

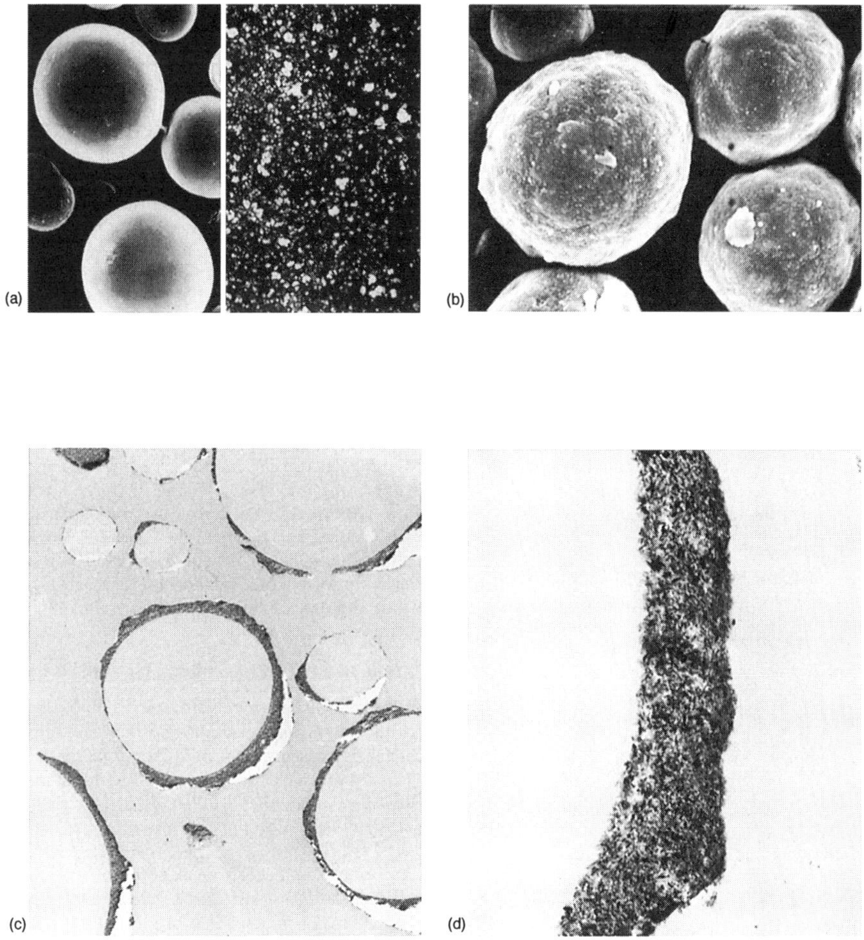

Figure 1.22 Dry powder heterogeneous microencapsulation can be produced by a process known as Mechanofusion®. PPMA = polymethyl methacrylate, a clear plastic which is known as Lucite in the USA and Perspex in Great Britain. (a) Raw materials. (b) Heterogeneous coated material (PPMA + TiO_2). (c) A TEM picture of sliced coated PMMA. (d) Thickness of the TiO_2 layer is about 0.5 mm. (e) Mapping Ti by X-ray microanalyser. (f) Treated powder flows like fluid. (g) PPMA. (h) TiO_2. (From Yokoyama *et al.* [71]. Reproduced by permission of the Council of Powder Technology and the Hosokawa Powder Technology Foundation, Osaka, Japan. © 1995 KONA. All rights reserved).

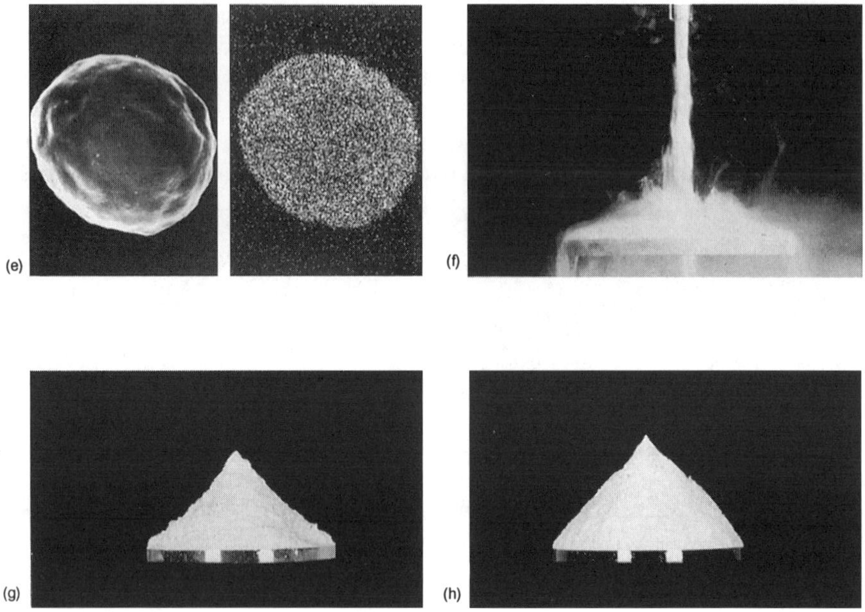

Figure 1.22 *Continued.*

to create a controlled flow of powder into a mixer, if one is interested in continuous mixing.

An interesting application of the use of hybrid powders is reported by Koishi *et al.* [73]. From their data it is apparent that continuous percolating paths for such properties as heat transmission, electrical conductivity etc. can be created at a low volume fraction of the component of the mixture, which is intended to form the percolating pathway by compressing hybrid powders. The creation of such systems from hybrid powders will have many applications in the drug industry in the creation of continuous delivery systems. Furthermore, the pathways outlined in Figure 1.23 will be a method of retrieving information on the voidage and packing characteristics of the powder prior to compression.

Technologists in a hurry like to refer to systems with complex names by convenient acronyms. Some of the technical literature on dry grinding encapsulation uses the term **OM'izers**. It took me a long time to work out that they meant 'ordered-mixture-rizes'. A very clumsy terminology! If one must have an acronym, one can use HEDG, which is formed from 'Heterogeneous Encapsulation by Dry Grinding'. It also conveys the image that one creates a hedge around the core fineparticle with the coating fineparticles. (This acronym is used as a key word in the set of descriptors describing the contents of mixing papers listed in the Laurentian bibliography of mixing literature, which is in preparation.)

Another type of mixture which has not received extensive study in industry is the **assembled mixture**. In the late 1950s, I had some involvement in the creation of **dense concrete** shielding for nuclear reactors. Normally concrete is assembled from the ingredients in a chaotic manner. The four main ingredients of concrete are described by the industry as aggregate, sand, cement powder and water. Note that the building specialists describe the coarse gravel as **concrete aggregate**. This use of 'aggregate' to describe dense chunks of rock is diametrically opposed to the meaning of aggregate as used in this book. We will refer to the builders' aggregate as the coarse-sized fraction of the concrete. The best way to create dense concrete is to form an assembled mixture by first packing the container to be filled with concrete with the coarse fraction. In the case of dense shielding for nuclear reactors this is a special coarse-gravel-type material made from a dense substance such as barium sulfate. When making lightweight concrete for high-rise buildings the coarse fraction of the concrete can be made out of sintered fly ash (the waste material from coal-fired powder stations) or alternatively from foamed glass crushed in a ballmill. The dry coarse fraction when packed together contains a porous network of voids. One then infiltrates this voidage by a much smaller dry powder, in the case of traditional concrete with sand. The final ingredient in an assembled mixture of concrete is provided by filling the fine voids left, by the fine sand within the coarse voids formed by the coarse fraction, with a thin slurry of water and cement powder. The term **slurry** in this book is used to describe suspensions of powdered material in liquid which is of relatively high solids content but which can still flow like a liquid. When the solids content is such that the material does not flow readily, the system is described as a **paste**. Pastes and slurries often have non-Newtonian properties, as will be discussed in Chapter 9. The slurry of cement powder in water contains a wetting agent so that the cement slurry can come into intimate contact with the other ingredients of the concrete. The assembled concrete mixture then sets as the cement powder reacts with the water and binds to the surface of the other ingredients. The study of assembled concrete mixtures was pioneered by C.C. Furnas [74,75].

Investigators who developed the assembled concrete mixtures for nuclear reactors were also interested in teaching the building industry how to develop assembled concrete mixtures for pathways and roads. Unfortunately, the technologists preferred muscle and brawn to efficient strategies in assembling their concrete mixtures. Industry has been slow to exploit the properties of assembled mixtures, although the ceramics industry has developed what is known as **cermets**, in which a premix of ceramic fineparticles is assembled and then a metal is allowed in infiltrate through the voidage in the original material to create conducting percolating paths in the system. The reader should note that in the construction of assembled concrete the fine sand percolates through the coarse grains in the chemical engineer's sense of the word, whereas the percolating path created in the cermet percolates in the physicist's sense.

So far in our discussion of powder mixtures, we have used two-dimensional drawings to describe the structure of a mixture. Three-dimensional description and depiction of the structure of powder mixtures is obviously a complex task, but there are indications that new computer graphic systems are going to enable the specialist to simulate three-dimensional mixtures and to develop mathematical models of three-dimensional chaotically assembled composite structures.

The assembled concrete mixtures that we have discussed are described by the mathematician in terms of a system known as an **Apollonian gasket**. The mathematician's Apollonian gasket in two-dimensional space is shown in Figure 1.23. The Apollonian gasket has the interesting property that one cannot deduce

a)

b)

c)

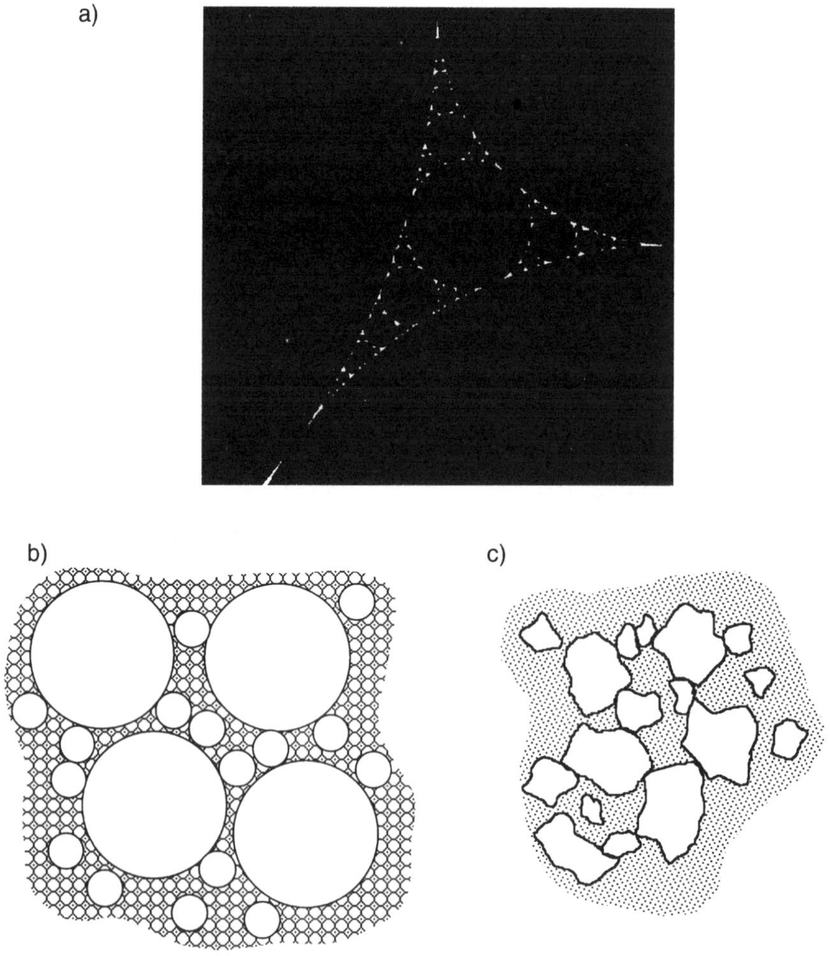

Figure 1.23 The Apollonian gasket is a self-similar packing of spheres which can approximate the structure of some everday mixtures. (a) The Apollonian gasket shows self-similarity when portions are magnified [78]. (b) Pape and Regis's pigeonhole model of sandstone [78]. (c) Buslik's model of concrete [78].

the scale of magnification from the appearance of a depiction of the system [76].

This magnification-independent appearance of a system is described as a **scaling property**. It is an important property of what is known as a **fractal system**. **Fractal geometry** is a new branch of mathematics originally conceived and developed by B.B. Mandelbrot [41, 76]. Amongst other topics, fractal geometry deals with the structural properties and behavior of rugged systems such as irregular surfaces and the structure of porous bodies. The fractal description of chaotically assembled and self-similar composite bodies is proving to be very fruitful and an active area of current material science research. Natural systems are rarely identical with theoretical mathematical models; real sections through concrete are only self-similar over a given range of inspection and tend to be statistically self-similar rather than exactly self-similar. Thus in Figure 1.23(b) a dilated version of the Apollonian gasket, which has been used by Pape and colleagues to model the structure of sandstone, is shown. Sections through the sandstone are a **dilated pseudo-Apollonian gasket** system. Several sections through sandstone at the same magnification would be statistically self-similar and can be described using the concepts of fractal mathematics. Pape and colleagues call such a model of sandstone the **pigeonhole model** [77]. In this name they capture the imagery that the sandstone is made up of a series of different-sized grains and that the coarsest grains form a void system similar to pigeonholes into which the smaller sizes of sand are fitted. Sandstone tends to be an assembled mixture in which the coarse grains of sand are laid down as a porous body, which is then infiltrated with smaller grains of sand during different river flow conditions and eventually silted up with tiny clay fineparticles. From Figure 1.23(c) it can be seen that concrete is also mathematically describable as a dilated pseudo-Apollonian gasket system.

The technique for modeling the structure of a mixture in two-dimensional space using a random number table, described earlier in this chapter, can be extended to describe a chaotically assembled mixture of monosized particles. The methodology is illustrated in Figure 1.24. The model of this diagram was assembled by a student interested in the stochastic structure of pores and pathways in a filter. The model is appropriate, since some of the earlier membrane filters were made by creating a mixture of granular salt (sodium chloride) in a plastic at a sufficient concentration to create contiguous percolating paths through the mixture. The salt was then dissolved out to leave a tortuous path through the filter. Although the particular interest of the model of Figure 1.24 was that of a porous body, the voids of the body could also represent a monosized dispersed ingredient in that body. The physical model of Figure 1.25 was made by modeling the structure of a random number table converted into a two-dimensional representation of the mix by cutting holes in a Styrofoam sheet wherever a pixel in the table had been made to represent a void. Thus the ten fields of view in the two-dimensional sequence of Figure 1.24 represents ten blocks of 100 digits from a random number table. These ten slices of a porous

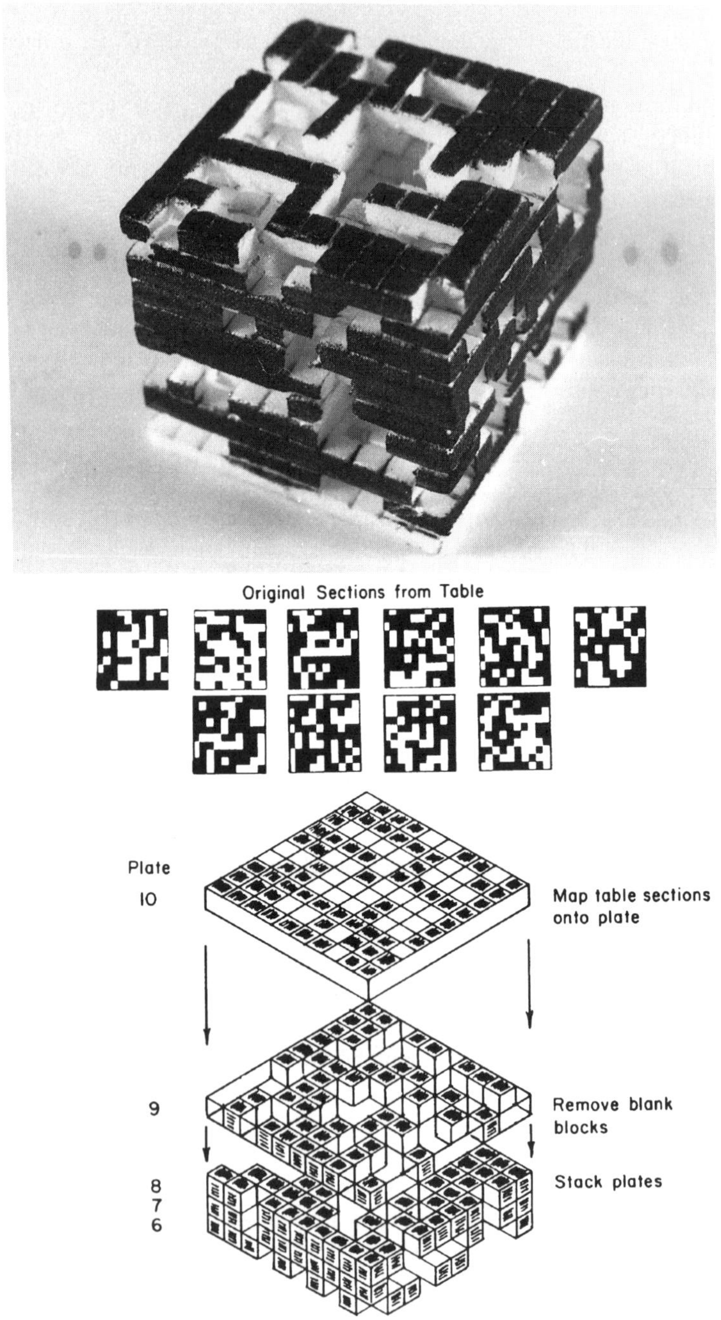

Figure 1.24 An extension of the modeling technique using a random number table to simulate a chaotically assembled mixture of monosized particles can be used to study stochastic cluster size distributions and percolating paths through a chaotic structure.

body were then assembled to make the model as shown. The physical modeling of the chaotically assembled mixture–void system such as that of Figure 1.24 has recently been modeled on a computer by I. Robb of Laurentian University. In Figure 1.25 the buildup of a model of a porous body, as created by Robb, at the

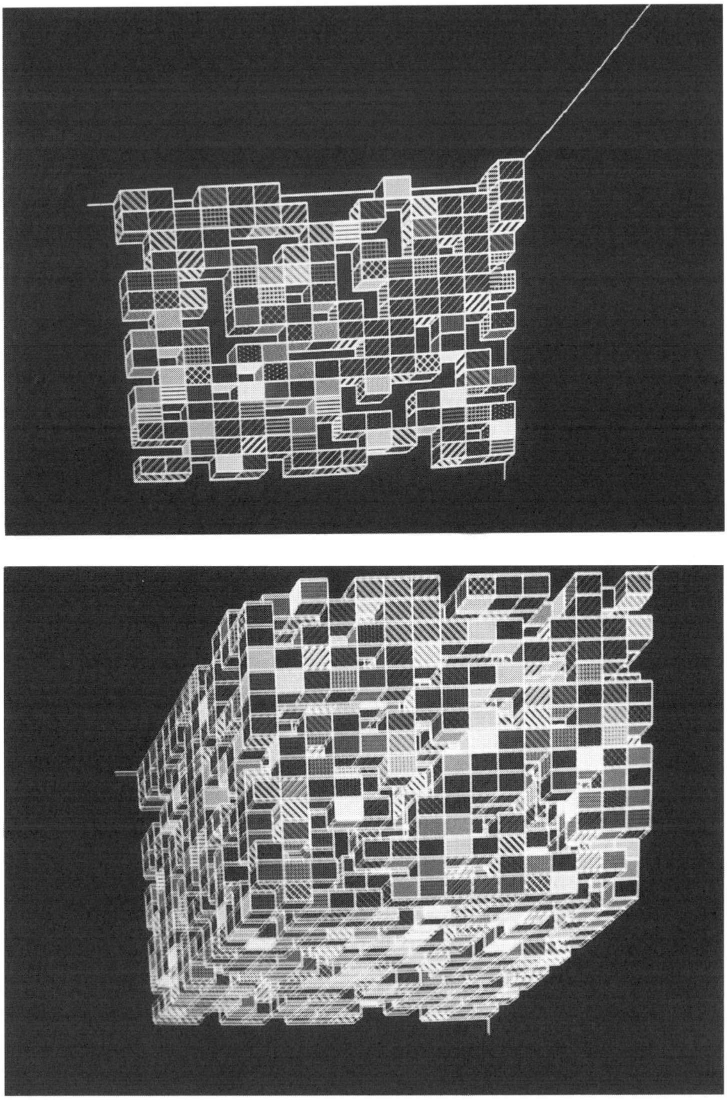

(a) (b)

Figure 1.25 The physical model of Figure 1.24 is being simulated on a computer by I. Robb of Laurentian University. (a) First layer of simulated mixture at 0.3 solids content, the black pixels beinh left as voids in the layer. (b) A 30-layer simulated mixture from which clusters and percolation pathways can be abstracted.

first layer and then at the multilayer level is shown. Robb can extract percolating paths or different clusters out of the assembled model to be displayed separately on the computer screen. His model of a mixture should be very useful for people planning to design continuous delivery drug systems and those concerned with percolation of fluids through a porous body, such as technologists studying secondary oil recovery from sandstones and the pollution of ground water by the movement of pollutants down through soil into wells or other water supplies.

Another system which has been used to create mathematical models of porous bodies and composite materials is the **Menger sponge** [76]. The term sponge is usually used to describe this system because in its original development it was considered to be a set of holes in a three-dimensional body. However, the holes can obviously be interpreted as a different species of material from the support matrix and the Menger sponge can also be used to model powder mixtures. There are many different Menger sponges which can be created mathematically depending upon the size distribution of the phase distributed throughout the matrix. Thus for this system the original solid body is considered to be divided into $9 \times 9 \times 9$ cubes and then the central cube abstracted. Then each of the subsidiary cubes is divided in the same way and the central subcube removed. If this process is carried on an infinite number of times, the resultant sponge has an infinite number of internal cells with an infinite surface area having no volume. This highly abstract model obviously does not match reality, and the abstraction of the central cube process is operative in a natural system over a limited range of iterations [41]. When the subcubes have their positions randomized so that the system is statistically self-similar, the resultant statistically self-similar natural Menger sponge model differs in a very important way from the ideal mathematical system in that, after a certain level of iteration, the randomized subunits become interconnected. The problem of deciding when a self-similar natural Menger sponge becomes interconnected, that is percolating in the physicist's sense of the word, is one of the aspects of composite body structure being studied by I. Robb. The construction of, and the properties of, systems such as the Menger sponge are easier to describe when one considers the two-dimensional analog of the system known as a **Sierpinski carpet**. The construction of an ideal Sierpinski carpet is illustrated in Figure 1.26(a). For the sake of clarity, only the first three stages of the construction of the particular Sierpinski carpet similar to the Menger sponge are shown. If the construction algorithm is carried on an infinite number of times, the ideal Sierpinski carpet becomes a curve with no area and an infinite number of infinitely thin threads. When the position of the holes in the third stage of the Sierpinski carpet are randomized according to the instruction

> Take a square and relocate it with x,y co-ordinates for the center of the square chosen at random from a random number table

then the system of Figure 1.26(b) is generated. Again, at a certain level of volume fraction of dispersed material, that statistically self-similar chaotic version of the Sierpinski carpet becomes interconnected. As will be discussed in section 5.5, a quantity known as the fractal dimension can be used to describe a structure of Sierpinski carpets and hence of dispersed material in a powder mixture.

1.5 GRAPHICAL AND EXPERIMENTAL DESCRIPTION OF MIXTURE STRUCTURE

To determine when a mixture has reached a satisfactory level of intermingling of the components, one must be able to explore the internal stucture of the mixture. The first thing to note about the design of a system for exploring a powder mixture is that one must avoid the trap of seeking to obtain a random sample. A dictionary defines **random sample** as

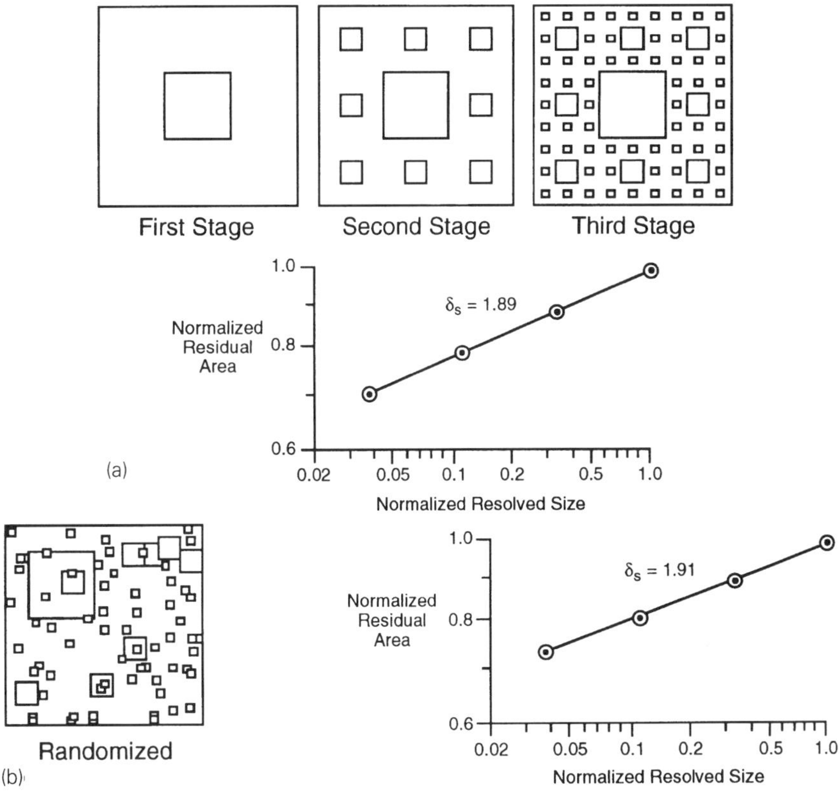

Figure 1.26 Mathematical models which have been used to describe the structure of mixtures and composite materials are the Menger sponge and the Sierpinski carpet. Sierpinski carpets are two-dimensional analogs of Menger sponges [41,76].

> a sample taken in such a way that every individual object or component comprising the group set or mass to be sampled has an equal probability of forming part of the sample [21].

The technologist exploring a mixture does not require a random sample but a **representative sample**. In other words, the material taken from the mixer must represent the state of dispersion in the mixer or the variation from region to region within a mixer. Sometimes people are instructed to attempt to randomize their selection of the location within the mixer at which they take their sample.

If the powder mixture varies chaotically from one point to another and if the mixture is as well dispersed, as is achievable by chaos-created mechanisms, then any sample taken from any point of the mixer is a representative sample. If, on the other hand, the mixture is not as good as it can be, then the last thing one wants to do is to sample the mixture at random. One needs to explore the existence of possible bias by systematically exploring various regions of a mixer. Consider, for example, the cylinder shown in Figure 1.27(a). This is the type of system that one would want to explore for powder mixture structure, which forms the base of a Y-mixer in its rest position. To explore the structure of the mixture, one would wish to take samples from three different planes, taken down through the vertical axis of the mixer, to create a set of 12 samples of the type shown. The map of the locations of the samples could then be used to create a display of the measured richness of the samples taken from the various locations. It will be recalled that we have also suggested that any sampling of this type of mixture should be carried out immediately after the powder settles in the container, and perhaps even while the mixture is being very gently aerated from the bottom.

One should also be aware of the danger of segregation being created in the mixer by vibration before one samples the mixture. Segregation of this type is called **wedge walking**. The physical mechanisms underlying its operation are illustrated in Figure 1.28(a). Random motion of a large fineparticle elevates one end of its ends; subsequently the finer neighborhood fineparticles move into the void created by the temporary displacement of the larger fineparticles. The same technique was used by pioneers to clear large stones off the land by pouring sand under temporarily elevated rocks. The same technique was also probably used in such exercises as building the pyramids, and certainly is the reason for the continuous replacement of flint, stones and boulders on fields, as the largest stones are lifted to the top of the soil by the movement of the finger grains of soil. Wedge walking segregation is not usually a problem over short time periods with cohesive fineparticles, but can be a problem with all mixtures when they are stored. It is the mechanism which ensures that the Brazil nuts are always on top of the peanuts when a tin of mixed nuts is opened [12]. It has been studied extensively by scientists in the food industry where it is important that the larger items in a food mixture do not segregate in the container during storage before the sale of the product [11].

Segregation of the ingredients is a major reason behind the small-scale sale of food mixtures for such items as ready-mix cakes, since if one were to supply enough for five cakes, wedge walking segregation might create disasters in the kitchen. Ideally any powder mixture which has been stored for any length of time should be remixed before being used or sampled. Having decided on the location of a set of samples to be used to examine the structure of the mixture, the next question to answer is how much material should be sampled? Normally

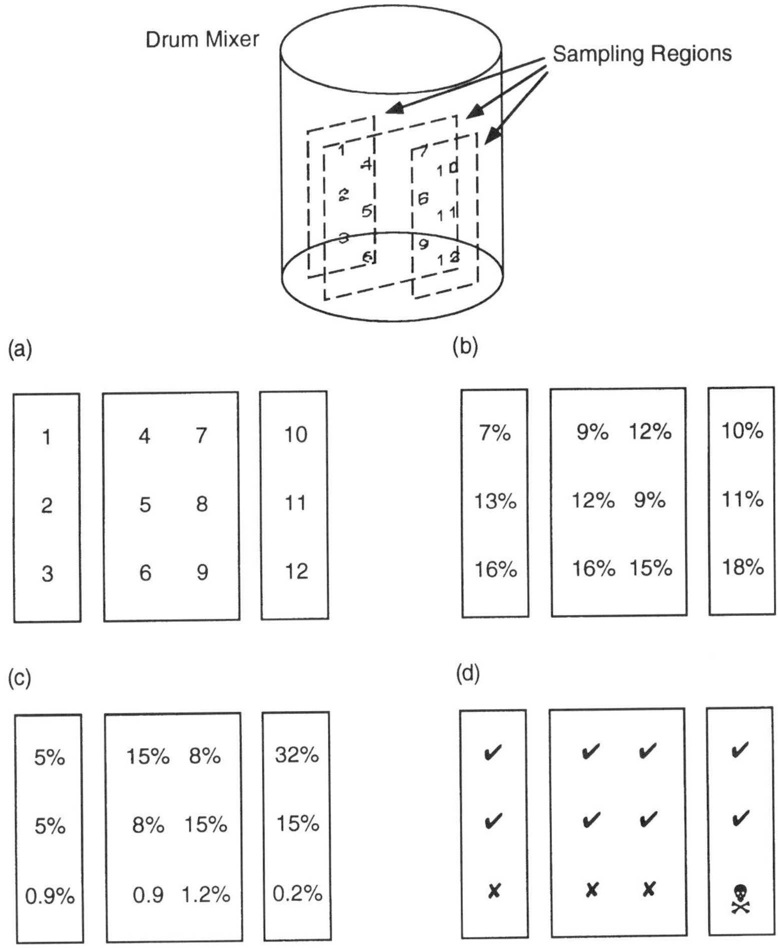

Figure 1.27 Systematic sampling of a powder mixture can locate bias (segregation present in the mixer). When monitoring the performance of a mixer the operator would like to have the data from the exploration of the mixer displayed in an action–verdict map. (a) Sample location map within the drum mixer. (b) Experimentally determined powder richness map (percentage of dispersed ingredients). (c) Probability of occurrence map for stated powder richness. (d) Action–verdict map showing samples of questionable richness highlighted.

when sampling systems for chemical analysis etc. the rule is to take as little as possible from the materials; however, in the case of powder mixtures, the graininess of the sample (the size of the quantum of the dispersed material) is relatively large with respect to the sampling volume. The variation of the sample richness for a specified expected level of richness can vary considerably within the limits set by statistical fluctuations in the sample constitution because of the inherent variation of a chaotically distributed substance. The first to realize that powder technologists were faced with a particular difficult situation when designing sample size for the assessment of the structure of mixtures was Buslik, who tried to have scientists accept what he called a **universal homogeneity of mixing index**, which also is known as a **Buslik's index**. The index is defined as the negative logarithm of the sample weight required to obtain a standard deviation of 1% in the measured richness of the samples withdrawn fron the mixer [78, 79]. That is, he recommended increasing the sample size until the stochastic variation in the samples arising from legal fluctuations is such that the standard deviation of the data is 1%, expressed in normalized units. Although Buslik did a service to the industrial community by stressing the need to take sample sizing to account when assessing the range of fluctuations being

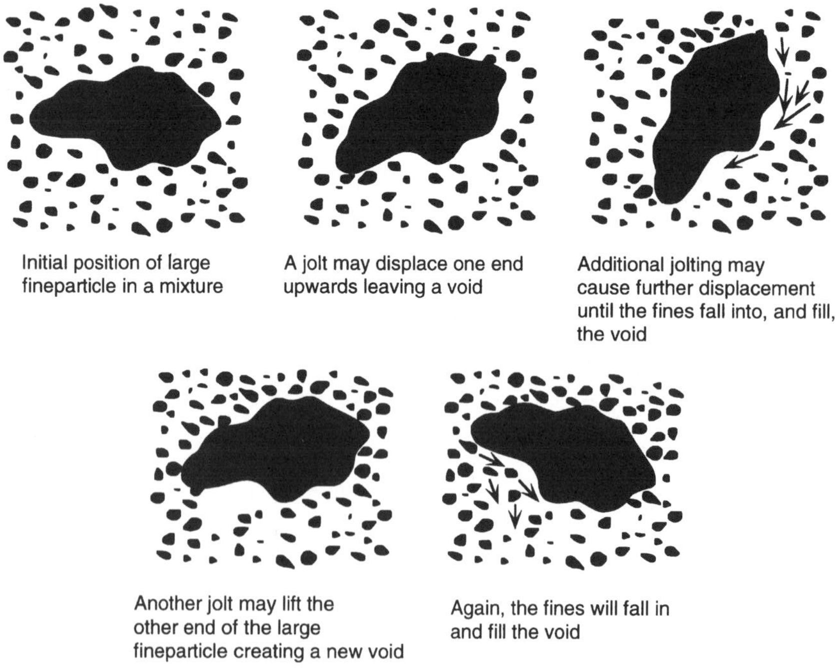

Figure 1.28 Vibration of a powder mixture can cause segregation with the coarser fineparticles migrating to the top of the mixer by 'wedge walking' [11, 12]. In 'wedge walking' the flow of adjacent fineparticles into local voids created by vibration lifts the larger fineparticles to the top of a standing mixture.

encountered, his index is somewhat *ad hoc* and not necessarily useful for modern industry [80].

To gain an appreciation of how sample richness can vary legally for a chaotically dispersed ingredient in a powder mixture we can model the variations in sample sizes using simulated mixtures. Thus in Figure 1.29(a) the technique for simulating the richness of a cube of powder taken as a sample from an overall mixture is indicated. Consider the case where the sample of powder was going to be a cube of side length 3*d*, where *d* is the size of the black fineparticles dispersed in the mixture. This is a relatively tiny sample, but the

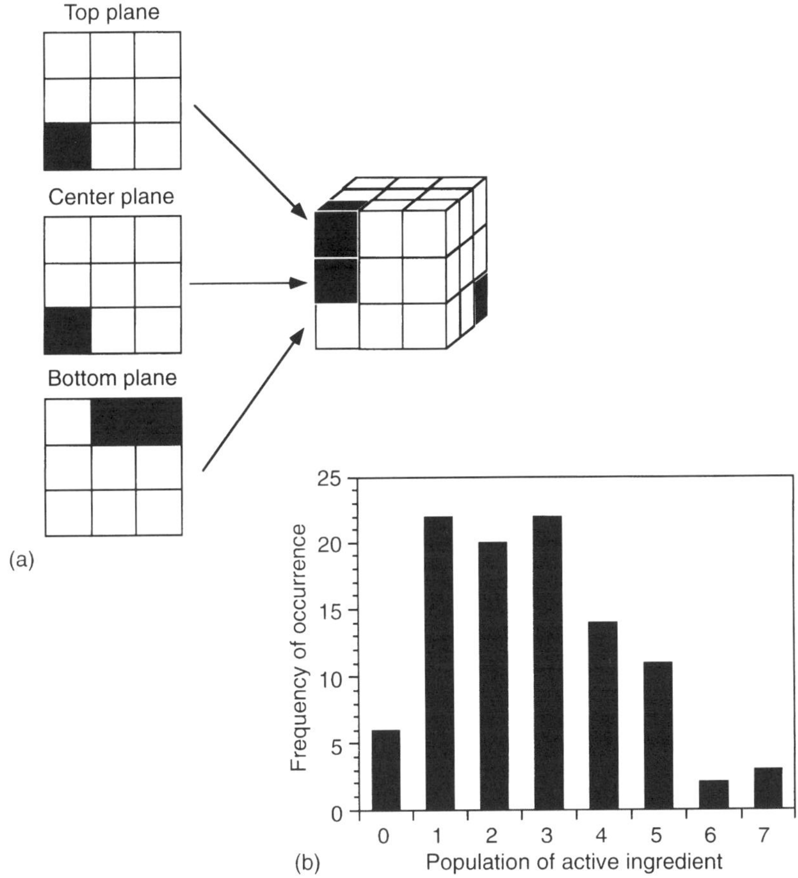

Figure 1.29 Simulated variations in mixture richness demonstrate that the fluctuations of samples taken from a chaotically dispersed monosized population of Fineparticles can be described by the 'bell curve' (data for simulated cubes of side length 3*d* when expected richness is 0.10 volume fraction). (a) The variation in the richness of sample cubes can be simulated using random number tables. (b) The bell curve fits the variation in sample richness histogram.

explanation of the principles of the simulation technique are easier to visualize with small rather than large simulated samples. After the principles of the procedure are understood, one can proceed to look at much larger samples. The sample of size $3d$ would consist of three slices each of size d with each slice containing nine unit pixoids. (A pixoid is the three-dimensional equivalent of a pixel. It is a small cube with each face the same size as the pixel area.) The volume of the sample is $81d^3$ and the volume of an individual pixoid is d^3. Therefore the sample size is 81 units when the unit is the volume of the dispersed fineparticles. To discuss the simulation procedure, we will consider the straightforward case of an expected mixture richness of 0.10 (10% by volume) so that we can simulate the volume percent in the actual sample by turning a specific digit into a pixoid in the three slices of the simulated cube. Thus for the cube of Figure 1.29(a) the simulated richness of the mixture in the $3d$ cube is 14.8%. We can now proceed to simulate the legal variations in cubes of this size by simulating 100 samples. The histogram of the richness values of the simulated cubes for the 100 cubes is shown in Figure 1.29(b). It can be seen that, within the limits of the small amount of data, the bell curve fits this type of variation. The bell curve is the popular term for a probability distribution known technically as the **Gaussian probability function**. It is also known as the **normal probability distribution**. For reasons that I have set out elsewhere, I prefer to use the term Gaussian probability distribution [41]. It can be shown that the legal variation in sample mixture richness is describable by the Gaussian probability distribution function. When working with this type of data it is convenient to work with a **cumulative less than distribution** for the variation in mixture richness. In Figure 1.30(a) the data are displayed in this manner on ordinary arithmetic scale graph paper. Special graph paper is available which is mathematically stretched in a 'probability of occurrence' scale so that, when the data presented in the format of Figure 1.30(a) are plotted on this graph paper, a straight line relationship is obtained. The basic structure of this graph paper, known variously as **Gaussian probability paper** or **arithmetic probability paper**, is shown in Figure 1.30(b). When using the Gaussian distribution it is usual to express the variable being investigated (in this case the richness of the mixture in the simulated sample cube) by means of the average value and the standard deviation of the data. It can be shown that, for a variable describable by the Gaussian probability distribution, 68.3% of all legal variations are located within the mean value plus or minus 34 (within plus or minus one standard deviation). Also 95.5% of all values lie within plus or minus two standard deviations and 99.7% of all possible values lie within plus or minus three standard deviations. These limits are shown on the graph paper of Figure 1.30(b). It is normal practice in decision making, based upon multiple sampling of a set of data that is describable by a Gaussian probability distribution, that a measured sample variation will be accepted as a legal variation (one which has a reasonable possibility of occurring by chance) if it lies within the mean plus or minus 2σ, where σ denotes the standard deviation. This translates into a probability of 1 in 20 of occurring by chance. Since this book is not about applied statistics, we will leave an explanation of this decision-making process to the textbooks on statistics [81]. Newcomers of this type of graph paper and statistical reasoning may find the presentations made in two other books of some assistance [9,41].

In Figure 1.30(c) the data from the histogram of Figure 1.29(b) are shown plotted on Gaussian probability paper. It can be seen that, for such a small sample, the legal variations in a measured richness in an expected 10% mixture can range from 5% to 25%. The legal variations are even more impressive if the variations are expressed as a percentage of the expected richness. Thus within the two standard deviation range, the samples can vary from 20% to 250% of the nominal richness. The data presented in this way are shown in Figure 1.30(d). It is obvious that to be able to decrease the legal variations in sample richness from a given expected richness one must take much larger samples than the $(3d)^3$ one of Figure 1.29. POMM, our expert system referred to earlier in this chapter, can simulate the variation in the expected richness with sample size as illustrated by

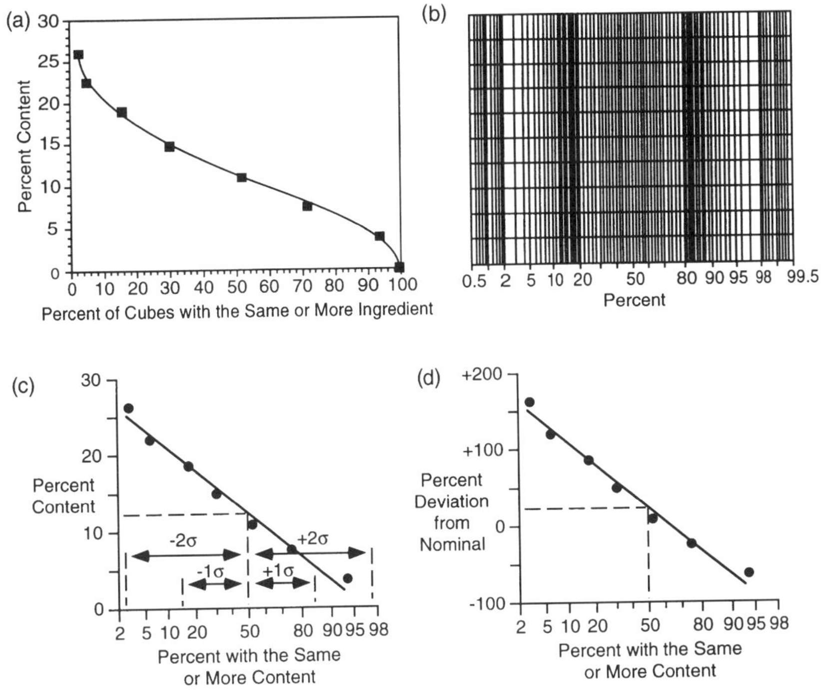

Figure 1.30 'Mixture variations' data distributions are describable by the Gaussian probability distribution. (a) 'Cumulative less than' distribution for the data points of Fig. 1.29 plotted on arithmetic paper. (b) Gaussian probability graph paper. (c) Simulated data of Figure 1.29 generates a straight line on Gaussian probability paper when expressed as a 'cumulative less than' distribution. (d) The data of part (c) with mixture richness expressed as a percentage of expected richness.

the displays of Figure 1.31. It was suggested earlier in this chapter that one should use a function defined as the **sampling efficiency factor**, which is the volume of the sample expressed as a fraction of the volume of the largest fineparticle present in the mixture. One of the problems with such an efficiency factor is the number of zeros that one must handle to describe the sample sizes one must use to reduce the possible legal variations to acceptable limits from an operational point of view. This is probably the reason why Buslik chose to

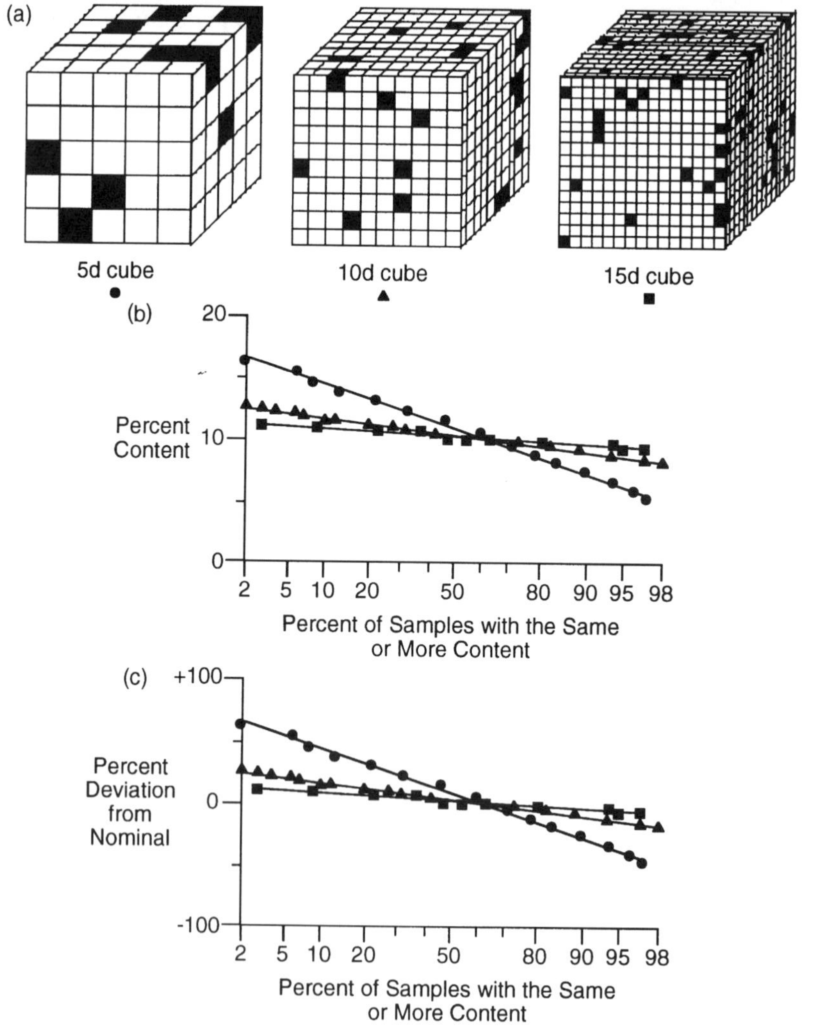

Figure 1.31 Permitted variations in measured mixture richness decline rapidly as the inspected sample increases. (a) Simulated cubes of various sizes. (b) Gaussian distribution of the variations in sample richness. (c) Gaussian distribution of the variaitons in sample richness expressed as a percentage of expected richness.

describe his index in terms of the reciprocal logarithm of the sample size required to reduce legal variations to a standard deviation, such that the mean plus or minus the standard deviation only permitted a range of 1% of the expected richness. Perhaps a simple way of dealing with this problem is to define a new term, the rationalized sampling efficiency factor, as the power of the number of units of ten required to multiply the raw sampling efficiency ratio to bring it to a 'one digit plus fraction' number. Thus the rationalized sampling efficiency factors of the three cubes shown in Figure 1.31 would be 2, 3 and 4 respectively. POMM will print out on demand the standard deviation of legal variations for a **rationalized sampling efficiency factor**.

As discussed earlier, the problem with many studies in powder mixing is that they use weight fractions and weights of samples, although the real variable in space is the volume of the sample when the mixture contains ingredients of different density. Thus to bring Buslik's index into line with this aspect of powder mixing technology, we define a **modified Buslik index**, which is the reciprocal of the volume of the sample required to reduce legal variations of chaotically organized mixtures to a standard deviation about the mean such that the permitted range of average plus or minus one standard deviation is 1% of the expected richness.

The phrase 'rationalized sampling efficiency factor' is rather long-winded, and one could refer to such an index as a **RASAEF index**. This is a pronounceable acronym. When spoken, the acronym gives the idea that when you have a large RASAEF index you are safe from random fluctuations!

When a technologist, as distinct from a research worker, explores the structure of a mixture in a container such as that of Figure 1.27, the display of the measured richness of the mix shown in part (b) of that diagram is not always a meaningful set of data. The technologist is not really interested in the variation of the sample richness with space, but whether or not the measured richness indicates that his or her mixer is performing as well as can be expected. To aid the working technologist in a search for answers to this type of question, POMM is programmed to convert the measured richness map of Figure 1.27(b) into the 'probability of occurrence map' shown in Figure 1.27(c). Thus for an expected richness of 10% of a black ingredient in a white matrix, POMM can show that the measured richness of part (b) has the probability of occurrence shown in the map of (c). A human operator surveying such data would probably surmise that the mixer is showing some segregation of the black powder in the bottom part of the container. Before POMM can endorse this situation it needs to be programmed with acceptance/rejection criteria for transforming the probability of occurrence map into an 'action–verdict' map of the type shown in (d). Thus if one goes on the usual statistician type of assumption that a probability of occurrence of less that 2.5% for low and high values of the richness of the mix consitutes danger, then the probability of occurrence map of (c) turns into a matrix of points shown in (d), showing the verdict that there is segregation in the bottom of the mixer. The segregation existence can be signaled to the operator either by a flashing line of symbols, as shown in the diagram, or a simple asterisk and exclamation mark. When POMM is working in real time, the color screen enables the different locations to be color coded. The map of Figure 1.27(d) would probably be an array of green lights with some yellows and a red. If an operator were to be presented with a control action–verdict map at a certain stage of the process of mixing of the type shown in Figure 1.27(d) he or she should probably proceed to mix the powder for a further length of time and, if the pattern persists, alter the process in such a way as to avoid this type of segregation.

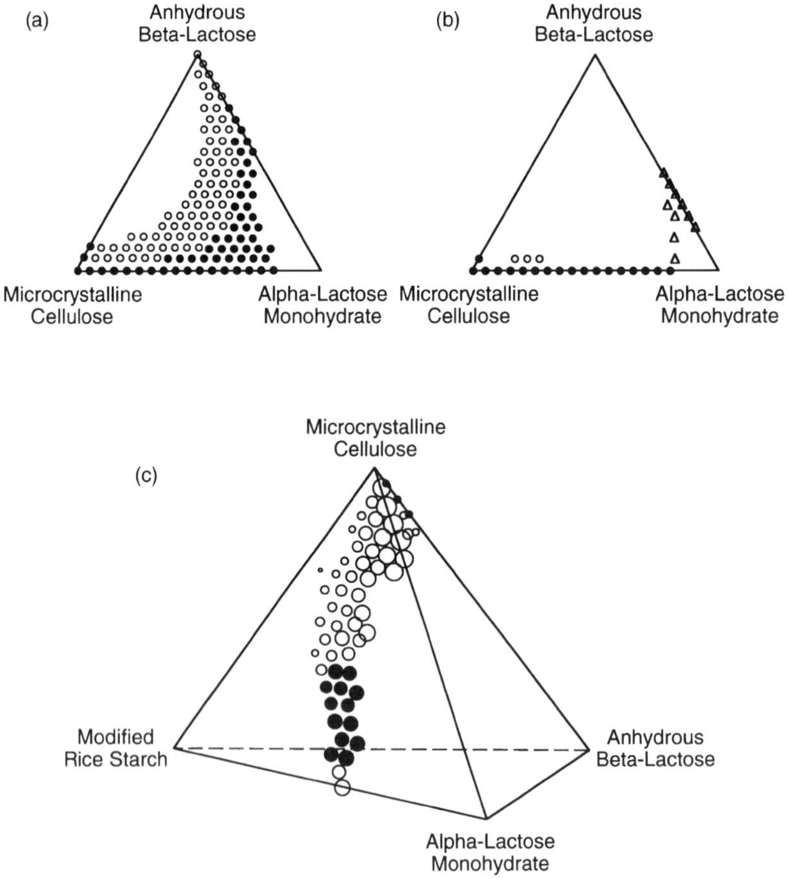

Figure 1.32 Modern computer graphics make it possible to summarize data on mixture structure and physical properties in multidimensional space. (a) Plot of crushing strength, disintegration time and friability of a placebo tablet. (b) Similar plot to (a) for a different formulation. (c) Similar plot to (a) but for a quaternary mixture percentage of the drug dissolved after 10 min. The size of the circles indicates their position in the depth of the graph (perspective). (From de Boer *et al.* [80]. Used by permission of J.H. de Boer.)

Research underway at Laurentian University is aimed at using a fiber optic–pneumatic lance probe to explore the *in situ* situation in a mixer with the automatic generation of the action–verdict map or any other display of Figure 1.28 to the operator. (The conceptual design of such a probe is discussed in Chapter 4.) However, such a relatively sophistic version of POMM lies in the future. Within the current 'state of the art' the generation of the action–verdict map of Figure 1.27(d) involves considerable labor. Recently, Bolhuis and co-workers have developed three and four dimensional data space displays to summarize the relationships between observed properties and mixture constitution (Figure 1.32) [81, 82].

NOTES

1. Beddow, J.K. (1980) *Particulate Science and Technology*, Chemical Publishing Co. Inc., New York.
2. Sommer, K. (1979) Statistics of mixedness with unequal particle sizes. *Journal of Powder and Bulk Technology*, **3**(4), 10–14.
3. Lacey, P.M.C. (1943) The mixing of solid particles. *Transactions of the Institution of Chemical Engineers (London)*, **21**, 53–59.
4. Dalla Valle, J.M. (1948) *Micromeritics*, Pitman, New York.
5. Orr, C. Jr (1966) *Particulate Technology*, Macmillan, New York.
6. Cooke, M.H., Stephens, D.J. and Bridgwater, J. (1976) Powder mixing–a literature survey. *Powder Technology*, **15**, 1–20.
7. Toor, H.L. (1987) Book review of reference 8, *American Scientist*, **75** (Nov–Dec), 594.
8. Kaye, B.H. (1993) *Chaos and Complexity: Discovering the Surprising Patterns of Science and Technology*, VCH Publishers, Weinheim, Germany.
9. McCauley, J.L. (1993) *Chaos, Dynamics and Fractals, an Algorithmic Approach to Deterministic Chaos*, Cambridge Nonlinear Science Series 2, Cambridge University Press, Cambridge.
10. Barker, G.C. and Grimson, M. (1990) The physics of muesli. *New Scientist* (26th May), 37–40.
11. Barker, G.C. (1994) Computer simulations of granular materials, in *Granular Matter, an Interdisciplinary Approach* (ed. A. Mehta), Springer-Verlag, New York, pp. 35–83. In this communication segregation in a powder mixture is simulated on a computer.
12. Rosato, A., Standburg, K.J., Prinz, F. and Swendson, R.H. (1987) Why Brazil nuts are on top: size segregation of particulate matter by shaking. *Physical Review Letters*, **58**, 1038–40.
13. Fayed, M.E. and Otten, L. (eds) (1995) *Handbook of Powder Science and Technology*, 2nd edn, Chapman & Hall, London.
14. Weidenbaum, S.S. (1973) Solid-solid mixing, in *Chemical Engineers Handbook*, 5th edn, (ed. J.H. Perry), McGraw-Hill Co., New York, pp. 21–30.
15. Williams, J.C. (1986) Mixing of particulate solids, in *Mixing, Theory and Practice*, Vol. 3, (eds V.W. Uhl and J.B. Gray), Academic Press, in Chapter 16.
16. Gordon, J.E. (1976) *The New Science of Strong Materials, or Why You Don't Fall*

Through the Floor, 2nd edn, Penguin Books, Harmondsworth. In the United States this book (1984) is available as a Princeton University Paperback.
17. Gordon, J.E. (1978) *Structures, or Why Things Don't Fall Down*, Penguin Books, Harmondsworth.
18. Hersey, J.A. (1981) Determination of interparticulate forces in ordered powder mixes. *Journal of Pharmacy and Pharmacology*, **33**, 485.
19. Hersey, J.A. (1975) Ordered mixing: a new concept in powder mixing practice. *Powder Technology*, **11**, 41–44.
20. Gleick, J. (1987) *Chaos: Making a New Science*, Viking Penguin Incorporated, New York.
21. MacDonald A.M. (ed.) (1967) *Chambers Etymological English Dictionary*, W. & R. Chambers Ltd., Edinburgh.
22. Ottino, J.M., Leong, C.W., Rising, H. and Swanson, P.D. (1988) Morphological structures produced by mixing in chaotic flow. *Nature*, **333**, (6172), 419–25.
23. Ottino, J.M. (1989) The mixing of fluids. *Scientific American* (Jan), 56–67.
24. Fan, L.T. and Chen, Yi-Ming (1990) Recent developments in solid mixing. *Powder Technology*, **61**, 255–87.
25. Hyman, A. (1976) *Computing, A Dictionary of Terms, Concepts and Ideas*, Arrow Books, London.
26. Kaye, B.H. (1991) Using an expert system to monitor mixer performance. *Powder and Bulk Engineering*, (Jan), 36–40.
27. Slagle, R.J. (1974) *Artificial Intelligence: 'The Heuristic Programming Approach'*, McGraw-Hill.
28. Schofield, C. (1967) A 5-channel photoelectric sampler and its use as an inline method of measuring mixture quality. Report LR 56 (CE), Warren Spring Laboratory, Stevenage.
29. Valentin, F.H.H. (1967) Mixing of powders and pastes: basic concepts. *Chemical Engineer (London)*, **45**, CE99–CE106.
30. Valentin, F.H.H. (1965) Mixing of powders and particulate materials. *Chemical Processing and Engineering*, (Apr), 181.
31. Allen, T. (1975) *Particle size measurement*, 2nd edn, Chapman & Hall, London, p.12.
32. Kaye, B.H. (1984) *Direct Characterization of Fineparticles*, J. Wiley & Sons, New York.
33. Kaye, B.H. Characterizing Powders, Mists and Fineparticles Systems. In preparation. (To be published by VCH, Weinheim.)
34. Kaye, B.H. and Sparrow, D.B. (1964) Role of surface diffusion as a mixing mechanism in a barrel mixer. Part I and Part II. *Industrial Chemist*, **40**, 200–205.
35. Harnby, N., Edwards, M.R. and Nienow, A.W. (1992) *Mixing in the Process Industries*, 2nd edn, Butterworths, London.
36. Described, for example, in the trade literature of Patterson-Kelley Co., Division of HARSCO Corp., P.O. Box 458, East Stroudsburg, PA 18301, USA.
37. Trade literature, GEMCO, 301 Smalley Ave., Middlesex, NJ 08846, USA.
38. The AeroKaye™ mixer is available from Amherst Process Instruments, Mountain Farms Technology Park, Hadley, MA 01035-9547, USA.
39. In North America enquiries regarding Luwa mixers should be addressed to LCI Corporation (formerly Luwa), Process Division, 4433 Chesapeake Drive, Charlotte, NC 28297, USA.
40. For a discussion of Brownian motion and its relationship with randomwalk theory, see Kaye [41].
41. Kaye, B.H. (1989) *A Randomwalk Through Fractal Dimensions*, VCH Publishers, Weinheim, Germany.

42. Pinmills are manufactured by several manufacturers, for example Alpine–Division of Hosokawa Micron Systems, Hosokawa Mikropul Environmental Systems, 20 Chatham Road, Summit, NJ 07901, USA.
43. Centridisc Systems of Day Mixing, 4932 Beech Street, Cincinnati, OH 45212, USA.
44. National Engineering Co., P.O. Box 6369, Chicago, IL 60680, USA.
45. Metropolis, K. and Ulam, S. (1949) The Monte Carlo method. *Journal of the American Statistical Association*, **44**, 335–41.
46. McCracken, (1968) The Monte Carlo method, in *Readings from Scientific American on Mathematical Thinking in Behavioural Sciences*, Freeman, San Francisco.
47. For a more detailed discussion of this type of clustering in a chaotic dispersion, see McCauley [9].
48. Stauffer, D. (1985) *Introduction to Percolation Theory*, Taylor and Francis, London.
49. Bridgwater, J. and Ingram, N.D. (1971) Rate of spontaneous inter-particulate percolation. *Transactions of the Institutions of Chemical Engineers*, **49**, 163.
50. Bridgwater, J., Sharpe, N.W. and Stocker, D.C. (1969) Particle mixing by pescolation. *Transactions of the Instiution of Chemical Engineers*, **47**, T114.
51. Unfortunately I did not keep a reference to the news story in *New Scientist.* My apologies to the authors of the study and the interested reader.
52. Hersey, J.A., Thiel, W.J. and Yeung, C.C. (1979) Partially ordered randomized powder mixtures. *Powder Technology*, **24**, 251.
53. Hersey, J.A. (1974) Powder mixing by frictional pressure: specific example of the use of ordered mixing. *Journal of Pharmaceutical Sciences*, **63**(12), 1960.
54. Hersey, J.A. and Cook, P.C. (1974) Homogeneity of pharmaceutical dispersed systems. *Journal of Pharmacy and Pharmacology*, **26**, 126–33.
55. Hersey, J.A. (1974) Application of a proposed universal homogeneity and mixing index to tableting operations. *Powder Technology*, **10**, 97–98.
56. Hersey, J.A., Cook, P.C., Smyth, M., Bishop, E.A. and Clarke, E.A. (1974) Homogeneity of multicomponent powder mixtures. *Journal of Pharmaceutical Sciences*, **63**, 408–411.
57. Hersey, J.A. (1972) Avoiding powder mixing problems. *Australian Journal of Pharaceutical Sciences*, NSI, 76–78.
58. Hersey, J.A. (1970) Sampling and assessment of powder mixtures for cosmetics and pharmaceuticals. *Journal of the Society of Cosmetic Chemists*, **31**, 259–69.
59. Hersey, J.A. (1967) The assessment of homogenmity in powder mixtures. *Journal of Pharmacy and Pharmacology*, **19**, 168S–176S.
60. Orr, N.A. (1974) Quality control and pharmaceutics of content uniformity of medicines containing potent drugs with special reference to tablets, in *Progress in the Quality Control of Medicines* (eds P.B. Deasy and R.F. Timony), Elsevier/North Holland Biomedical Press, Chapter 9.
61. Orr, N.A. and Shotton, E. (1973) The mixing of cohesive powders. *The Chemical Engineer*, (Jan), 12–18.
62. Orr, N.A., Hill, E.A. and Sallam, E.A. Mixing of small amounts of cohesive powders in both solid and semi-solid systems, in *Fluid Mixing*, The Institution of Chemical Engineers (Great Britain), Symposium Series No. 64.
63. Staniforth, J.N. (1985) Ordered mixing of spontaneous granulation. *Powder Technology*, **45**, 73–77.
64. Staniforth, J.N. (1982) Determination and handling of total mixes in pharmaceutical systems. *Powder Technology* **33**, 147–59.
65. Staniforth, J.N. and Rees, J.E. (1981) Powder mixing by triboelectrification. *Powder Technology*, **30**, 255.

66. Staniforth, J.N. and Rees, J.E. (1982) Effect of vibration time, frequency and acceleration on drug content uniformity. *Journal of Pharmacy and Pharmacology*, **34**, 700–706.
67. Staniforth, J.N. and Rees, J.E. (1982) Electrostatic charge interactions in ordered powder mixes. *Journal of Pharmacy and Pharmacology*, **34**, 69.
68. Staniforth, J.N., Rees, J.E., Lai, F.K. and Hersey, J.A. (1982) Interparticle forces in binary and ternary ordered powder mixes. *Journal of Pharmacy and Pharmacology*, **34**, 485.
69. Kaye, B.H. (1992) Microencapsulation: the creation of synthetic fineparticles with specified properties. *KONA*, **10**, 65–82. (*KONA* is an international powder technology journal published once a year by the Hosokawa Company, 10 Chatham Road, Summit, 07901, USA.)
70. Nara Machinery Co. Ltd., 5–7, 2-Chome, Jonau-Jima, Ohta-Ku, Tokyo, Japan 143.
71. Yokoyama, T., Urayama, K., Naito, M. and Kato, M. (1987). The Angmill Mechanofusion system and its applications. *KONA*, **5**, 59–68.
72. Gratton-Liimatainen, J.L. (1995) Characterizing the structure of cohesive powder mixtures by optical methods. MSc thesis, Laurentian University, Sudbury, Ontario.
73. Koishi, M., Ishizaka, T. and Nakakjima, T. (1984) Preparation and surface properties of encapsulated powder pharmaceuticals. *Applied Biochemistry and Biotechnology*, **10**, 259–62.
74. Furnas, C.C. (1931) Grading aggregates. *Industrial Engineering Chemistry*, **23**, 1052.
75. Gray, W.A. (1968) *The Packing of Solid Particles*, Chapman & Hall, London.
76. Mandelbrot, B.B. (1983) *The Fractal Geometry of Nature*, W.H. Freeman & Company.
77. Pape, H., Riepe, L. and Schopper, J.R. (1984) The role of fractal quantity as specific surfaces and tortuosities for physical properties of porous media. *Particle Characterization*, **1**, 66–73.
78. Buslik, D. (1973). A proposed universal homogeneity and mixing index. *Powder Technology*, **7**, 111–16.
79. Buslik, D. (1950) Mixing and sampling with special reference to multi-sized granular material. *ASTM Bulletin*, **66**, (Apr).
80. de Boer, J.H., Bolhuis, G.K. and Doornbus, D.A. (1991) Comparative evaluation of multi-criteria decision making and combined contour plots in optimization of directly compressed tablets. *European Journal of Pharmacy and Biopharmacy* **37**(3) 159–65.
81. Moroney, M.J. (1953) *Facts from Figures*, 2nd edn, Pelican Books, Harmondsworth.
82. For a description of how to use triangular graph paper of the type used by Bolhuis and colleagues, see Chapter 13 of McCauley [9].

2

Powder and powder mixture characterization technology

2.1 SAMPLING A POWDER MIXTURE

A primary task of the powder technologist concerned with mixing operations is the sufficient characterization of the size distribution function of the ingredients of a powder mixture. To obtain a representative sample of the powder to be characterized, there are several sampling procedures available. An efficient device available for the sampling of a powder supply is the **spinning riffler** shown in Figure 2.1(a). The powder to be sampled is fed as a steady stream to a rotating basket of containers. It has been shown that, provided the sampling time divided by the time of one revolution of the basket is a large number, each sample chamber contains a representative subsample of the powder flow [1]. This powder sampling system is excellent for free flowing powders and for powders which do not contain significant percentage of fines, which can become fugitive from the sampling devices due to air currents as the device rotates and the powder flows into the containers. If a powder containing fines is to be sampled, the free-fall tumbling mixer described in Chapter 1 is an excellent device for prehomogenization of the powder to be characterized. A sample container mounted with a ladle in the lid of the container will contain a representative sample when the tumbling ceases.

Studies of the efficiency of the free-fall tumbling mixer establish that a short period of tumbling creates such chaotic conditions inside the container that any subsample caught in a sampling device placed in the mixing chamber contains a representative sample of the powder to be examined [2]. Unfortunately the free-fall tumbling device is not available commercially, but one can anticipate the commercial version being available in the not too distant future.

The power of the system to act as a mixer/sampler is illustrated by the data of Figure 2.2 [3]. A crushed calcium carbonate powder nominally 15 μm was sampled after tumbling a container of the powder for 10 min. The sample was characterized by the Aerosizer®, an instrument to be described later in the text. The measured size distribution and that of the subsequent sample taken after a further 10 min are shown in Figure 2.2(a). In Figure 2.2(b) the size distributions of nominally 6 μm and 15 μm powders as measured by the Aerosizer® are shown, along with the size distribution of a mixture prepared from these two components in the proportion 25% 6 μm powder to 75% 15 μm powder. In Figure 2.3(c) the mathematically calculated size distribution of the mixture based on the known size distributions of the two ingredients is indistinguishable from that of the mixture as obtained from the Aerosizer® after the mixture had been tumbled for 20 min in the mixer/sampler. Since the powders were not free flowing, the

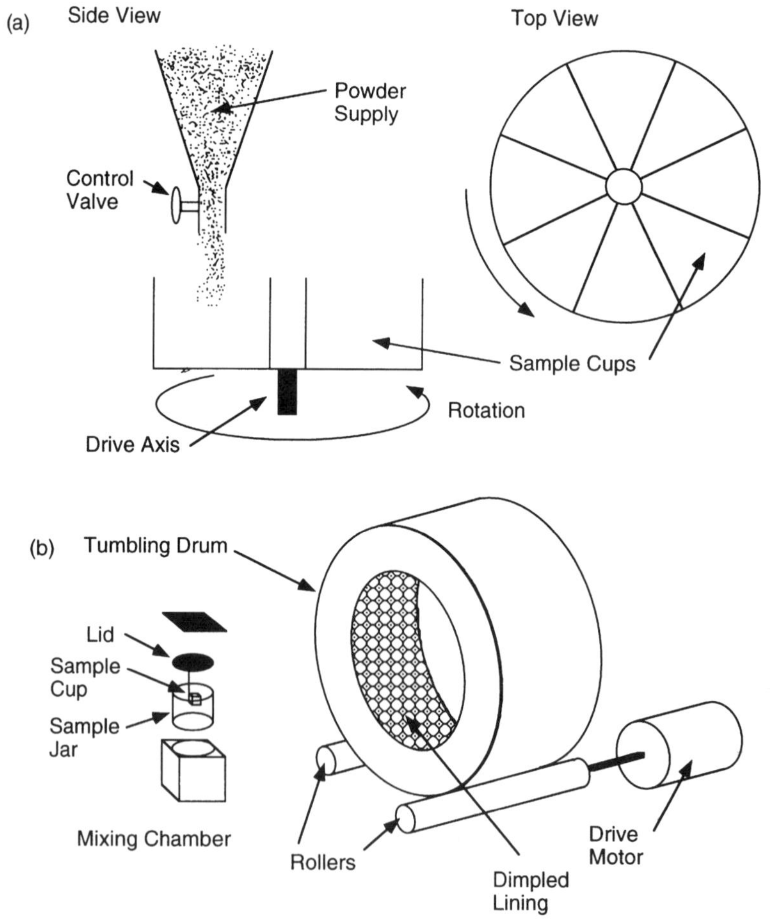

Figure 2.1 Obtaining a representative sample of a powder ingredient is an essential step in any powder characterization study. (a) A spinning riffler is a process sampling device for selecting a representative sample of a free flowing powder. (b) The AeroKaye® mixer/sampler can be used to obtain representative samples of a mixture.

ability to mix these two powders so that a representative sample matched exactly the predicted structure of the mixture is a good indication of the power of the system to homogenize a powder which had segregated during previous handling [3]. (See also discussion on powder mixing monitoring in Chapter 5.)

Although the spinning riffler is a useful device for obtaining a representative sample of a single powder ingredient, it may not be suitable for taking samples

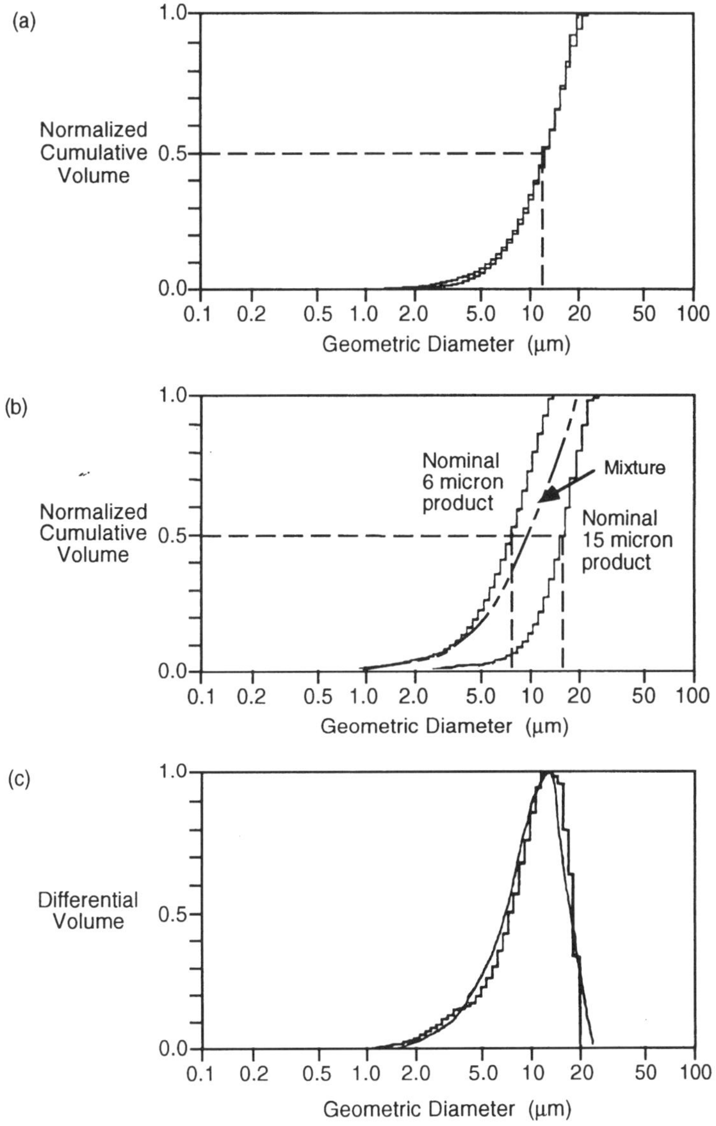

Figure 2.2 The AeroKaye® mixer can function as an efficient sampling device [3].

of powder mixtures because of segregation occurring during the flow of the powder into the sampling cups. The sampling cups, although containing representative samples, may have segregated sizes within the cup due to the fact that the flow characteristics and the bulk supply of powder may not be homogeneous to start with. For this reason, even the subsamples must be carefully homogenized by shaking or some other procedure before proceeding to take a subsample. In specific situations, sometimes it is necessary to use a cascade of spinning rifflers of diminishing cup volume in order to be able to obtain a small subsample which is representative of a large bulk supply [4].

Sometimes it is not economic or feasible to carry out a cascade of spinning riffler subdivisions and it is necessary to obtain samples from within a large bulk supply of powder. In such situations it is sometimes possible to use a pneumatic lance. In Figure 2.3, sketches of a classical pneumatic lance and a new type of pneumatic lance for taking mixture samples are shown. When the new lance is thrust into the powder mixture, a gentle flow of air out of the sampling nozzle lubricates the movement of the probe through the powder. When a specific location is reached, the air flow is reversed to pull a sample up into the bottom part of the lance. The powder sample builds up against a porous placed across the tube, as shown in the plate diagram. With the reverse flow of air still

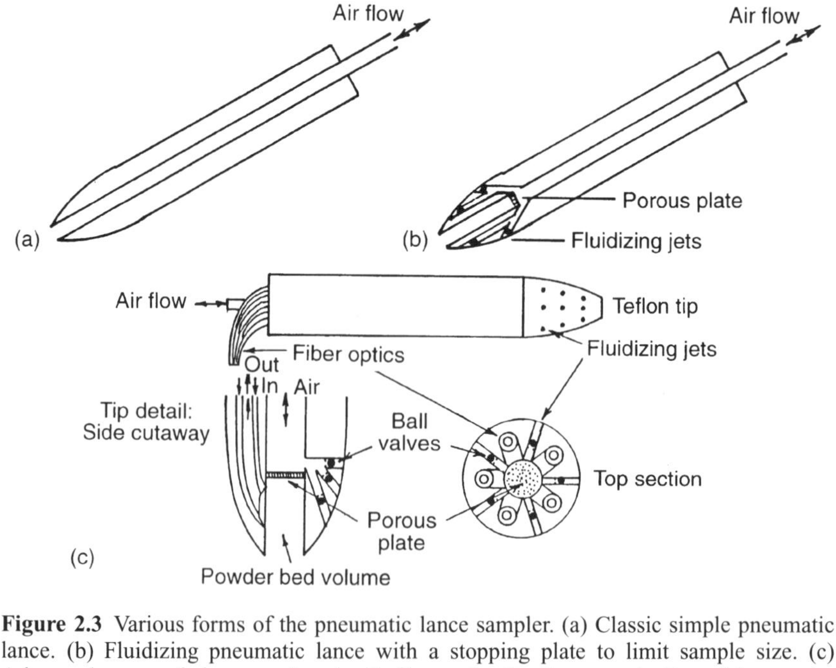

Figure 2.3 Various forms of the pneumatic lance sampler. (a) Classic simple pneumatic lance. (b) Fluidizing pneumatic lance with a stopping plate to limit sample size. (c) Advanced pneumatic lance equipped with fibre optics to allow examination of the powder sample within the probe.

operating, the lance is then withdrawn and the sample emptied before reversing the air flow for a second entry in the bulk powder supply.

Many workers have used the thief sampler shown in Figure 1.1(b) to obtain supplies of powder from inside a bulk supply of an ingredient. Sometimes it is difficult to push the thief sampler into the powder and hence the reason for the suggestion to use lubricating air with the pneumatic lance as it enters the powder. Secondly, as mentioned earlier, it is not a good idea to have too many close fitting, moving surfaces when one is handling a powder. I have known instances where people attempted to use the thief sampler with a relatively abrasive powder, only to find that powder entered the gap between the moving surfaces, and rotation of the inner tube caused binding of the two walls by the abrasive biting into the surfaces. As a result of this powder binding, a rather expensive thief sampler was never used again. The pneumatic lance obviously suffers from the problem that if one sucks too hard when trying to take a sample, one might exaggerate the fines content of the powder by preferential aspiration of the fines into the sampling device. Traditional pneumatic lances do not contain the porous plate. When equipped with the porous plate the problem of fines bias is minimized. It is recommended that the reversal of the air flow from lubricating airflow to sample intake airflow of the powered sample takes place slowly, with airflow into the sampling device being increased by stages, to minimize any airflow-induced segregation. Efficiency studies of this pneumatic lance have yet to be published. However, this device appears to minimize the disturbing effect of the act of sampling when looking into a supply of a single size distribution ingredient or a powder mixture. The use of the modified pneumatic lance with fiber optic inspection devices will be discussed in more detail in Chapter 5 [5].

2.2 TECHNIQUES FOR CHARACTERIZING THE GRAIN SIZES OF A POWDER

After obtaining a representative sample of the size required for the characterization of the ingredient, it is often necessary to process that sample prior to its insertion in a characterization device. For example, a small amount of powder may have to be dispersed in oil on a glass slide prior to inspection of the dispersed fineparticles by automated image analysis. Again, it may be necessary to disperse the powder in a liquid for use in one of several devices, such as the X-ray sedimentometer, or a stream method of characterization, such as the Coulter counter, or one of the many photo-zone counters [1]. When dispersing the subsample for inspection it is necessary to respect the **operational integrity** of the fineparticles. This subject was touched on earlier in Chapter 1 when discussing the dispersion severity with which the mixture should be treated during inspection. Technologists not sufficiently experienced in size characterization can often use an inadvertently severe method of dispersion which completely changes the operational size of the powder ingredient. Thus it is not always realized that, in an ultrasonic dispersion of powder in liquid, the actual dispersive forces in the cavitating zone of the ultrasonic bath can be extremely severe, with the breakdown of agglomerates which normally would cohere together throughout the process to which the mixture is to be subjected. Again, one sometimes reads instructions such as

> Gently disperse the fineparticles in a drop of oil on a microscope slide using a rod to shear out the suspension drop.

The forces in a shear zone underneath a glass rod moving over a microscope slide can be very severe, with subsequent degradation of the operational sizes of the fineparticles to be characterized.

When setting out to characterize an ingredient by a new method of size characterization, the technologist should vary the sample preparation techniques to explore the effect on the system of such techniques. No universal rules can be given for the best method for dispersing of a powder prior to characterization; one can only urge caution and the experimental study of the effect of the preparatory technique adapted to disperse the fineparticle [1].

It is useful to divide powder characterization techniques into two groups, 'direct methods' and 'indirect methods'. Direct methods of size characterization measurements are made on individual fineparticles in a relatively direct manner. Thus sieving would be classified as a direct method of characterization because the physical dimensions of the individual grains determine the probability of passage through a set of apertures in a sieve. On the other hand, interpreting the diffraction pattern generated by passing laser light through a random array of the fineparticles in suspension is a relatively **indirect method of size characterization** since it is necessary to make assumptions about refractive index, shape and orientation when interpreting the diffraction pattern. Whole books have been written on different methods of size characterization of powders, and in this chapter we will seek only to highlight the different types of methods that can be used by the technologists concerned with powder mixing studies and to signpost some dangers they may run into when selecting their method of characterizing their ingredients. Normally the powder mixing technologist requires a knowledge of the size range of grains in a powder, because the wider the range of sizes in a given powder the more likely that one will run into segregation problems. (As a rule of thumb, size range in ingredients of greater than 3 : 1 will cause percolation segregation.) Furthermore, the finer the ingredients, the more likely they are to give problems in the theological aspects of feeding the powder into a mixer and to any subsequent process. When discussing powder mixing problems in the course of workshops, it has become apparent to me that, very often, technologists are surprisingly passive in their acceptance of raw materials provided to them. Apparently they often fail to appreciate that a word with their supplier is all that is needed to change the characteristics of the powder provided, with subsequent easing of powder mixing problems. For example, a powder supplier who is using an airswept micronizer to produce a powder can sometimes adjust the settings on the mill to reduce the amount of fines in the powder being supplied to the customer. Furthermore, the supplier may even be able to create premium markets for other customers by separating a powder into coarse and fine, and selling the fines to another user. The technologist should also realize that, very often, the information generated in a size distribution study is limited. It may not be sufficient for him or her to control their powder. For example, I was once asked to look into the problems being experienced in a mixing situation, because powder obtained from one vendor was proving to be a satisfactory ingredient, whereas the other was causing hangups in the mixer and problems in the die compression when making tablets. It turned out that both powders were meeting the pharmaceutical user's specifications which stated that 'the powder be such that no more than a certain amount be retained by a sieve of a given aperture'. It was then discovered that one vendor was providing a spray-dried powder and the other was providing a powder which had been crystallized and dried. The appearance of the two different powders is shown in Figure 2.4 [6]. To those experienced in handling such powders, it would be obvious that the rheological properties of these two powders would be different, even though they were meeting the same commercial purchase specification.

In another situation, a purchaser of an ingredient in a powder metallurgical mix being used to make alloy compacts discovered that his powder was no longer functioning properly in the fabrication process. Very low strength compacts were being generated in a new powder supply in the manufacturer's process. When powders from satisfactory and unsatisfactory products were examined under the microscope, it was obvious that the shape of the powder grains had changed. Further investigation revealed that the purchaser of the powder was gradually increasing its order for the supply of powder to the point where the vendor could no longer supply that amount of powder by ballmilling the raw materials. The supplier had therefore switched to a higher capacity micronizing device, which changed the shape distribution but not the average size of the powder being produced. These two case histories demonstrate how important it is to take a quick look through a microscope at any powder being supplied, so that one can gain some appreciation of the fineness and shape of the powder grains.

2.3 QUANTITATIVE DESCRIPTION OF THE SHAPE OF POWDER GRAINS

Quantitative description of the shape of the grains is a very difficult task, and only a brief outline of the possible strategies that can be used to specify the shape of a grain will be given here. Classical methods of characterizing the shape of the two-dimensional projected profiles involve geometric measurements of the width and breadth of the profiles. Thus one widely used geometrical factor is the **aspect ratio**, which is the maximum length of the profile divided by the width of the profile at right angles to that measurement. It should be noted that in occupational health and hygiene, a fineparticle with an aspect ratio of greater than 3 is considered to be a fiber. Thus many of the profiles of the crystalline drug powder of Figure 2.4(b) would be considered to be fibers. However, whether one regards the crystalline grains as being equivalent to fibers or not (which they may well be if inhaled), it is fairly obvious that the powder such as that of Figure 2.4(b) will, in general, be more difficult to pour than that of Figure 2.4(a). Compacts made from the crystalline powder generally become more open structured than those made from spray-dried powder. The spray-dried product of Figure 2.4(a) will probably flow relatively easily, although the more grains there are of highly irregular structure the poorer will be the rheological properties of the powdered material. The properties of suspensions of the two powders of

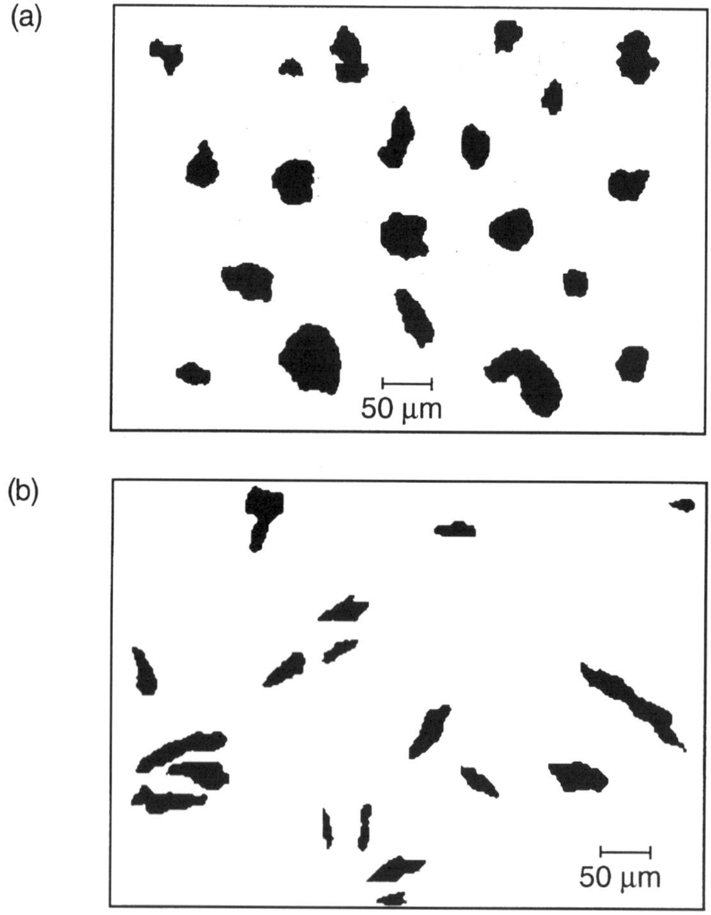

Figure 2.4 (a) Spray-dried and (b) crystallized drug powders, which satisfy two different regulating agencies' size specifications, behaved very differently in industrial powder processing because of their different shapes [6].

Figure 2.4 in liquids will be highly dependent on the shape factors and the surface characteristics of the grains. The shape of the grains in a powder is a particularly important property of the powder if it is eventually incorporated into a slurry, since it is a very important contribution to the viscosity and non-Newtonian behaviour of the slurry (Chapter 9).

In recent years, two different methods for characterizing the shape of powder grains, such as that in the powder of Figure 2.4(a) have been developed [7–13]. In one method, Fourier analysis of what is known as the geometric signature waveforms of the profile has been developed. The other method uses the theorems of fractal geometry to characterize the profiles. The basic principles of the Fourier analysis of signature waveforms of two-dimensional fineparticle profiles are shown in Figure 2.5(a) [7, 8, 12]. The data for this figure were originally generated by A. Flook. The first step in this technique for generating a shape description of the profile is to locate the centroid of the profile when it is treated as if it were a thin laminar structure. (This is often the most difficult part of the automated characterization procedure.) A reference direction is then chosen and the value of the vector $\boldsymbol{R}$ measured at a series of angles θ with respect to this direction. The resultant values of $\boldsymbol{R}$, normalized with respect to the largest value of $\boldsymbol{R}$, are then plotted to generate the **geometric signature waveform** shown in Figure 2.5(a). This signature waveform is then treated as if it were one cycle of a long complex wave so that it can be subjected to **Fourier analysis** (a mathematical technique used to break a complex wave into contributory simpler waveforms [10]). Fourier analysis breaks the geometric signature waveform into its contributory harmonics. Flook and other workers showed that the basic shape of the profile is contained in the first five harmonics of the Fourier analysis. Thus in Figure 2.5(b) the profile constituted from the first five harmonics, derived from the analysis of the signature waveform of the original profile, has been synthesized. It can be seen that these first five harmonics contribute the basic shape of the profile. As one adds higher and higher harmonics into the synthesized profile the higher harmonics only contribute to the texture of the fineparticle. Thus up to the first ten harmonics, the basic shape of the five harmonics is modified by a couple of bumps added to the profile, as shown in Figure 2.5(b). When the first 25 harmonics are used to synthesize the profile, the resultant additions regenerate most of the rugged structure of the original profile, as shown in Figure 2.5(b). This technique for measuring the shape of the profile has been used to study the effect of shape on the flow and packing properties of coarse granular material. Application of the Fourier analysis geometric signature waveform method to very rugged profiles is limited by the fact that the value of the vector $\boldsymbol{R}$ becomes difficult to interpret, since it often crosses convolutions on the profile. Thus if one were to attempt to construct the signature waveform for the carbonblack profile of Figure 2.5(c), vectors crossing the convolutions of the profile would be indeterminate in magnitude. It should also be noted that it is difficult to relate any of the various harmonics present in the signature waveform to the formation dynamics of the

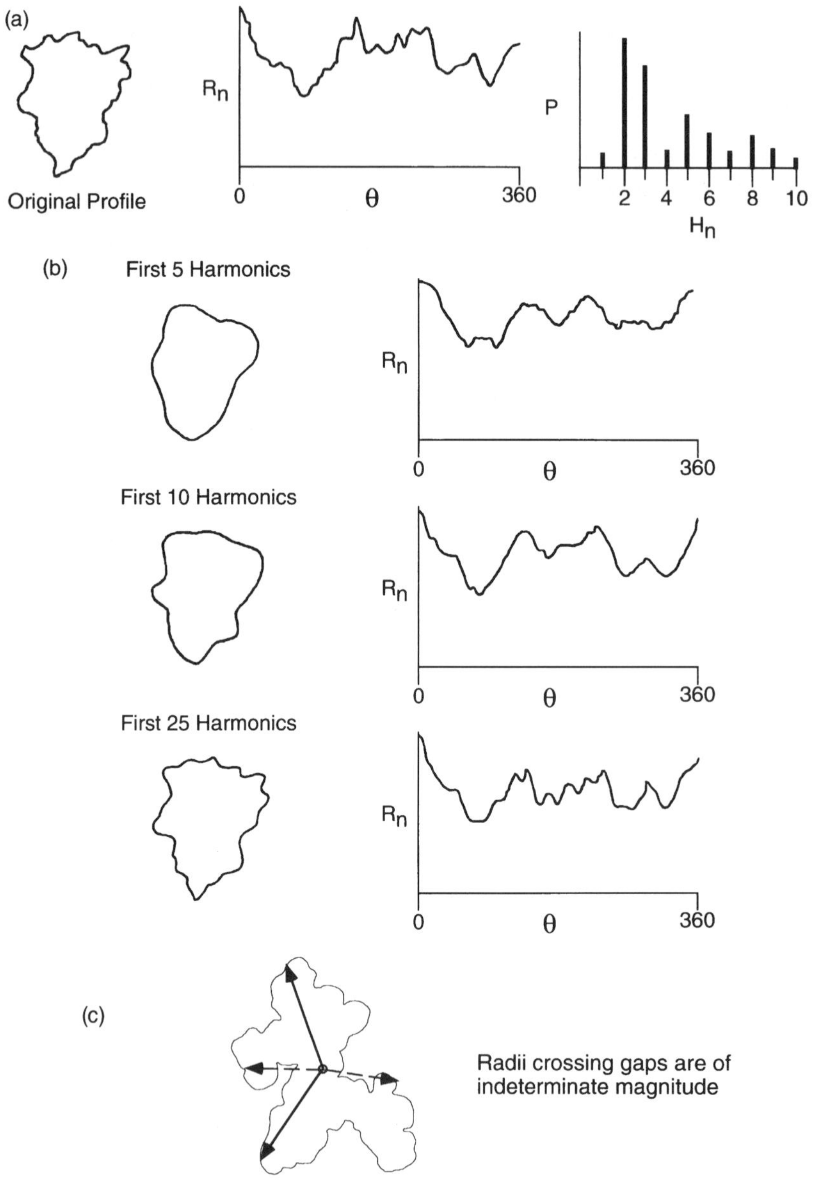

Figure 2.5 Shape characterization by Fourier analysis of geometric signature waveforms can be useful with some fineparticle profiles. (a) Fourier analysis of a profile. (b) Profiles reconstructed from the stated number of harmonics [7, 8]: θ = angle at which the radius vector is generated with respect to the reference vector; R_n = normalized magnitude of the radius vector at angle θ; H_n = harmonic number from Fourier analysis; P = relative strength of the stated harmonic. (c) Carbonblack profile; dashed lines indicate that vectors crossing the gaps are of indeterminate magnitude.

fineparticles. In other words, the relation between the contributory harmonics and the formation dynamics of the profile is not anticipated to be a useful field of study.

Until very recently the direct generation of the two-dimensional Fourier transform of fineparticle profiles, such as those shown in Figure 2.5, was too expensive and only of academic interest. However, the continuing increase in the sophistication of many computers and their falling price makes it likely that the direct two-dimensional Fourier transform of profiles will become an important technique for measuring the shape of fineparticles. Thus in Figure 2.6, the two-dimensional Fourier transform of the three Flook profiles and the carbonblack profile of Figure 2.5(c) are shown. It can be seen that as sharp edges develop in the profiles, energy levels increase in the outer region of the Fourier transform map, that is in the higher harmonics of the two-dimensional Fourier transform.

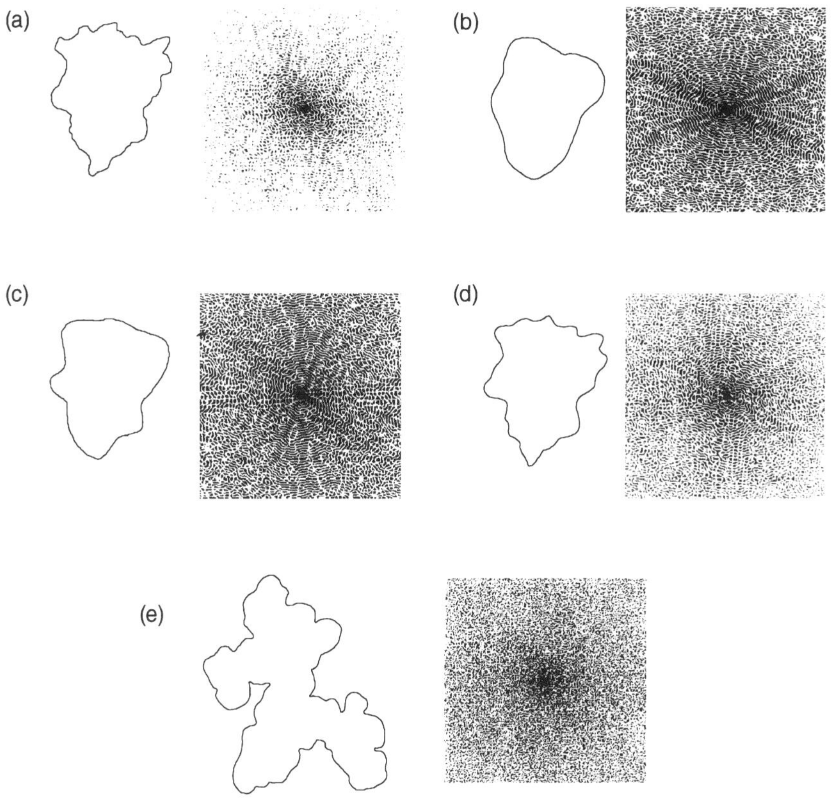

Figure 2.6 The falling cost and increasing power of personal computers and the sophistication of commercially available processing algorithms are making it possible to consider using two-dimensional Fourier transforms as shape characterization procedures [9].

The two-dimensional Fourier transform is essentially the same as the Fraunhoffer diffraction pattern of a profile [10]. Therefore the two-dimensional Fourier transform can be used to look at the type of errors that might be encountered if the shape of a fineparticle changes when its size is being estimated using commercially available diffractometers, which generally assume that the fineparticles being examined in the instrument are homogeneous transparent spheres. In general the presence of sharp edges on the diffracting profiles will tend to throw energy further out in the diffraction pattern of the fineparticle. This 'far out' energy will be interpreted by the logic of most commercial machines as coming from finer particles than those actually present in the array of profiles. In other words, pseudofines will be reported as being present in the powder, when in fact the diffraction patterns are being generated by sharp edges present in the profiles. This is a problem if one is interested in the effect of shape on the flow of the powders into and around mixing equipment. The reader is also advised that some commercially available, laser-diffraction-pattern-based size analysis equipment takes short cuts in the logic of processing the data by assuming that a certain size distribution function is likely to be found in powders generated by a given production process. This curve fitting can produce an inaccurate result for the distribution of sizes in the powder. The reader will also appreciate that all of the work that is currently going on with regard to the shape of fineparticles usually deals with two-dimensional profiles. Obviously information on the three-dimensional structure of the system is lost by the image projection process. However, in the absence of low-cost three-dimensional techniques for characterizing shape there are still a great deal of useful investigation that can be carried out relating two-dimensional information to powder properties. Techniques for recovering the basic shape information from the 'noisy signal' signature waveform are delayed until section 5.6 because it is convenient to discuss the mathematical procedures for auto- and cross-correlation data processes for one- and two-dimensional space in one comprehensive presentation.

The other major innovation in shape characterization brought about by the availability of low-cost computer-aided image analysis is the use of fractal dimensions to characteize the structure of rugged fineparticle profiles. A full discussion of the concepts of fractal geometry is beyond the scope of this book, and all that can be presented here is a brief outline of the basic concepts, as applied to the determination of the shape of fineparticle profiles [11–13]. The basic concept of fractal geometry is that if one is looking at a profile, such as that of carbonblack in Figure 2.5(c), or of a rock fragment, then as one looks at it with higher and higher resolution inspection systems, more and more detail is revealed in the surface texture and structure, with the consequent conclusion that the surface area of such a profile is infinite at infinite resolution. This uncomfortable paradox can be resolved by realizing that what one is really saying about such highly magnified images is that a quantity such as surface area cannot be determined, and that all one can do is give operational estimates of the surface area or boundary length at a given magnitude of inspection. Mandelbrot, in a pioneering book entitled *Fractals: Form, Chance and Dimension*, suggested that, when faced by the paradox of an indeterminate ruggedness of a surface area, one can describe the ruggedness by the rate at which the magnitude of the surface estimate or boundary length increases as a function of the resolution parameter of inspection [11]. In other words, Mandelbrot suggests that we switch attention from trying to determine the absolute surface area or boundary length of a powder and try to determine the ability of the boundary to fill space. Mandelbrot suggested a basic innovation in the dimensional description of a system, which can be understood for fineparticle boundaries by considereing the systems of lines shown in Figure 2.7 [12]. **Topology** is the mathematical study of relationships in geometric systems that remain the same when the space supporting them is distorted [13]. Thus if one looks at all of the lines of Figure 2.7, if they were drawn on rubber graph paper, then as the two-dimensional space supporting them is distorted they all basically remain lines, and the mathematician will state that all of the lines of Figure 2.7 are **topologically equivalent** and are of **topological dimension** one. Mandelbrot has suggested that we describe the ruggedness of a system by adding a fractal addendum to the topological dimension of a system to describe its space filling ability. Thus if we look at the lines of Figure 2.5 the fractal addendum needed to describe the structure of the lines is shown as a quantity labeled the **fractal dimension** of the various lines. Mandelbrot showed that if we plot a graph of the estimates of the perimeter of a profile against a measure of the inspection resolution at which the perimeter is estimated, then the positive magnitude of the slope of the dataline is the fractal addendum used to estimate the fractal dimension of the boundary. He illustrated this fact by considering a famous problem, 'How long is the coastline of Great Britain?' He imagined that a series of investigators, from giants with large strides down to small-scale giants with much smaller step size, all investigated the coastline of Great Britain by counting the number of strides that they had to take by walking around the coastline. Their estimate of the perimeter of the coastline would be larger the smaller the size of step they took because the smaller giants would have to go around the peninsulas and bays, whereas the larger giants would step across them. One can simulate the sets of estimates made by such a group of giants in the way illustrated in Figure 2.7(a) using a pair of compasses. One then takes the estimates of the coastline and the step size (both normalized with respect to the maximum projected length of the island) and plots the data on log–log scales, as illustrated in Figure 2.7(b).

The plot of perimeter estimates versus resolution parameter on log–log scales used in the estimation of the fractal dimension of a boundary is known as a Richardson plot. This name honors Lewis Fry Richardson who was one of the pioneers of studies concerned with the problems of tackling the characterization of rugged boundaries. Richardson was concerned with actual coastlines and political boundaries, but his work had tremendous implications for all rugged boundaries.

It can be seen that the datapoints manifest a straight line relationship. Mandelbrot has shown that the slope of such a line describes the ruggedness of the boundary. More recent high-resolution studies have shown that the coastline dataline slope changes at high-resolution inspection and that the graph of

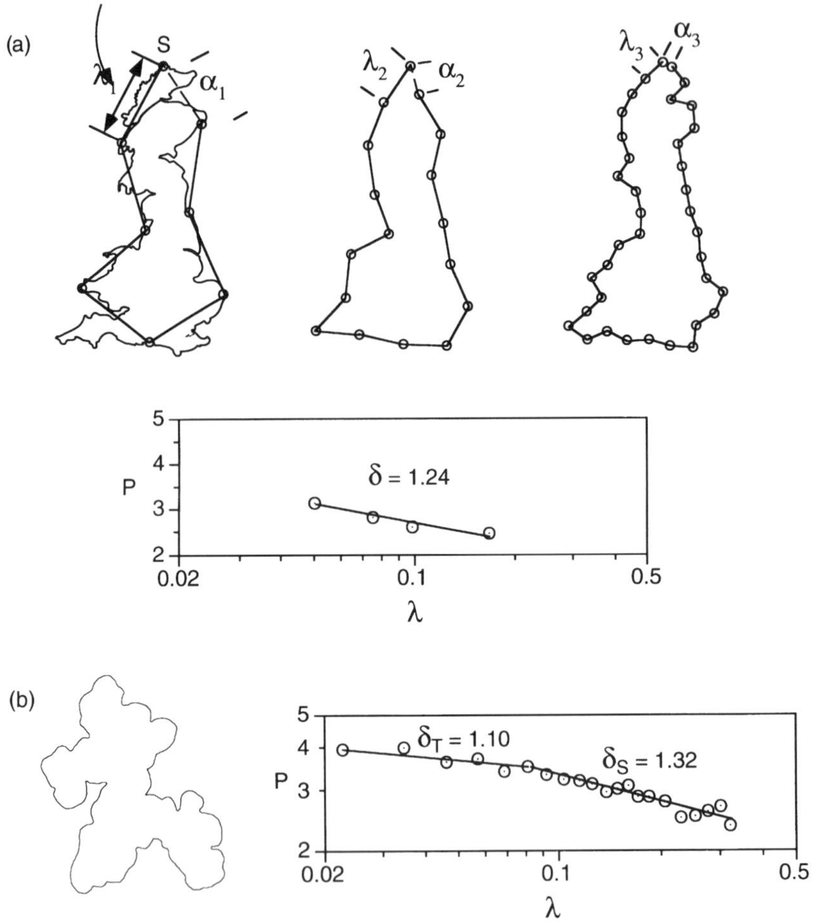

Figure 2.7 The fractal addendum needed to estimate the fractal dimension of a rugged boundary can be deducted from a graph of perimeter estimates versus resolution parameters – a graphical display called a Richardson plot. The 'yardstick' or structured walk method was used by Mandelbrot to explore the rugged structure of the coastline of Great Britain at a series of resolutions [12]. (a) Polygons formed around the perimeter of Great Britain by walks using various yardsticks. (b) Richardson plot of the data from the exploration of the coast of Great Britain. (c) Different aspects of a carbonblack profile structure can be described by different fractal dimensions. λ is the resolution of inspection used to construct the polygon, P = perimeter estimate generated at resolution λ. δ = boundary fractal dimension derived from the slope of the dataline.

Figure 2.7(b) is characteristic of the promontories and bays which dominate the data at coarse resolution (large exploration steps) of the coastline, whereas higher-resolution studies are more descriptive of short-range ruggedness of the actual coastline itself [7]. Some of the earlier workers in the field were somewhat disturbed to find that a natural fractal such as a coastline had more than one ruggedness fractal dimension at different resolution inspections. However, it is now realized that the coastline ruggedness is produced by a variety of interacting forces, and that in the formation of the features of exploration of the coastline at coarse resolution certain forces may dominate, whereas their interactions change when looking at the coastal features at high-resolution scales of scrutiny. Thus the modern approach to the fractal characterization is that the range over which a straight line relationship is manifest on a graph such as that in the diagram is information on the dynamic forces interacting to produce the structure. This can be appreciated from the studies that were carried out on the carbonblack profile of Figure 2.5(c). The data from this experiment are summarized in Figure 2.7(c). In the latter figure the same 'stride around' system to that used to explore the coastline of Great Britain has been carried out on the carbonblack profile using a set of compasses. Two clear datalines are obvious on the Richardson plot. The line at the coarse resolution can be linked to the structural features of the carbonblack, whereas the dataline at high resolution can be linked to the texture of the agglomerate formed by the unit capture spheres forming the soot agglomerate packing together [1]. For this reason the two different fractal dimensions deduced from different regions of the Richardson plots are described as structural and textural fractal dimensions. Thus the magnitude of both of these fractal dimensions can be linked to the underlying dynamics producing the carbonblack [7]. The structural dimensions probably describe the flow and packing properties of a powder, whereas the textural fractal is related to intergrain friction and chemical reactivity.

An alternative method for evaluating the fractal dimension of a rugged fineparticle boundary, known as the **equispaced method**, is illustrated in Figure 2.8. In this technique, the boundary to be characterized is first digitized, as illustrated in Figure 2.8(a). The polygon which is to be the estimate of the boundary perimeter at a given inspection resolution is created by stepping along the digitized point for a specified number of steps and then drawing a line between the first and the last step. Thus in Figure 2.8(a) the polygons for a five-step estimate and a 15-step estimate are shown. These perimeter estimates, and the resolution parameters (the number of steps times the space in between the dots) are then normalized with respect to the maximum Feret's diameter of the profile, as shown in the figure. When these perimeter estimates are plotted against the step size, the set of datapoints, shown in Figure 2.8(b), generates two straight line relationships giving the structural and textural fractal of carbonblack boundary. The utility of this method depends upon the fact that, although we have physically drawn the lines in Figure 2.8(b) to show the construction of the polygon used to estimate the perimeter when this technique is linked to a computer-based system, the length of the chords between the points stepped out along the boundary can be calculated using the coordinates of the point and Pythagoras's theorem. All of the various step size estimates of the perimeter can be carried out within the computer, and the graph plotted along with the slopes of the lines used to estimate the fractal dimensions of the profile. Thus once the profile boundary is digitized the operator of the system is free to look for more profiles to inspect as the computer manipulates the digitized data to estimate the fractal dimension(s) of a profile. This is the logic used in a commercially available system for characterizing the fractal dimensions of fineparticle profiles [4]. (For a full discussion of this method see references 7 and 12.)

The fact that there is no simple relation between the size of complicated fineparticles measured by image analysis and their size characterized by other methods is illustrated in Figure 2.9. These profiles are of fineparticles that were

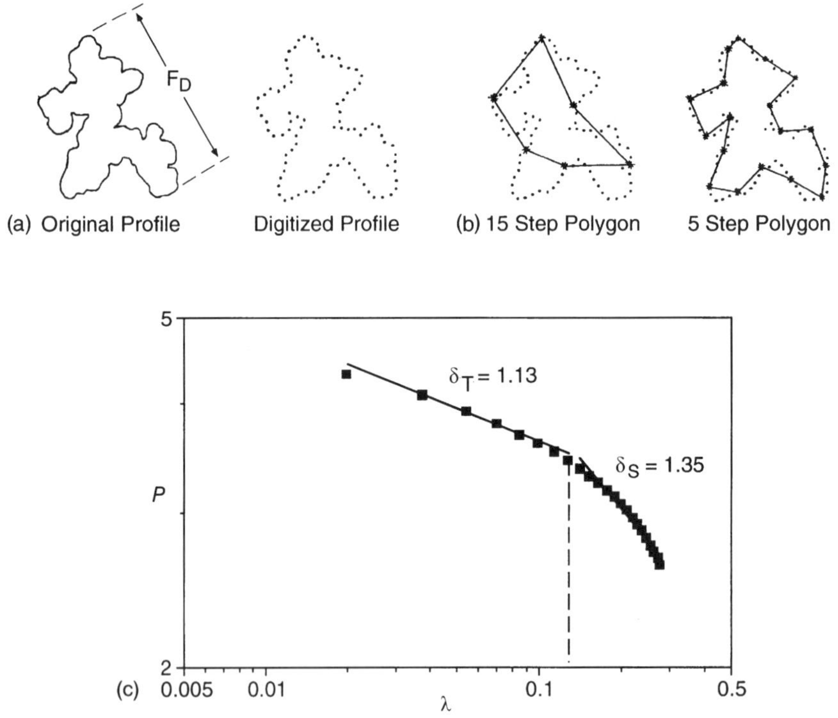

Figure 2.8 An alternative technique to the method of Figure 2.7 is known as the equispaced technique. A series of equally spaced points are derived from the profile and polygons are constructed using chords generated using specific numbers of points to define the chords. (a) Equispaced points defining the profile. F_D = Feret's diameter. (b) Polygons constructed with five-and 15-step chords. (c) Richardson plot of the data generated from the method of (a). λ = resolution of inspection used to construct the polygon. P = perimeter estimate generated at resolution λ. δ is the boundary fractal dimension derived from the slope of the dataline.

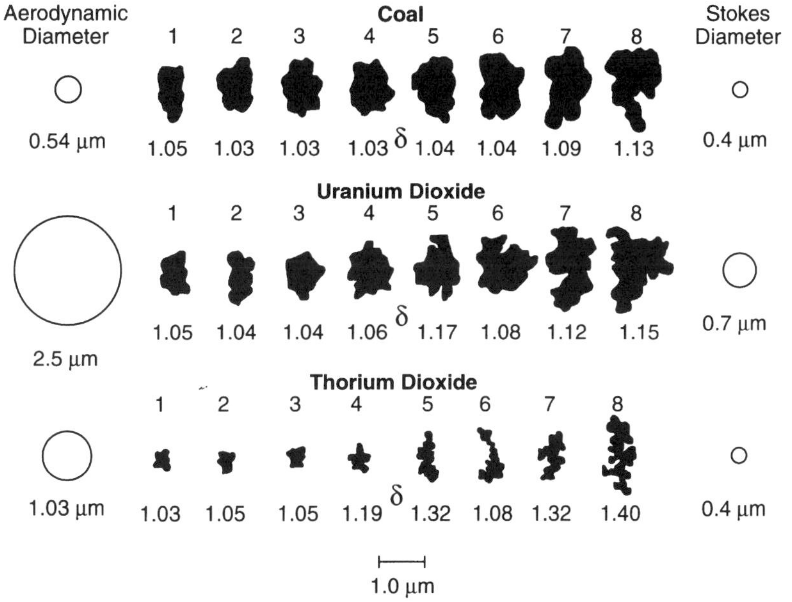

Figure 2.9 Within groups of isoaerodynamic fineparticles, increasing physical size appears to be accompanied by increasing fractal structure.

fractionated by various workers in the Stober spiral centrifuge [1]. The **aerodynamic diameter** is a concept used in many powder size distribution studies when the density of fineparticles settling in air or fluid is unknown. The measured settling speed is interpreted in terms of the size of the sphere of unit density which would have the same settling speed. The **Stokes diameter** is the size of the sphere of the same material which would have the same settling speed in the same fluid. Thus data from X-ray sedimentometer, air classification and elutriation methods are all presented in terms of Stokes diameters. The coal fineparticles differ in size from their Stokes diameter not only because of the shape, but also because they are probably porous containing blind internal pores. (**Pores** are classified as **open** or **blind** depending on whether or not they have access to the surface of the fineparticle.) It can be seen that for powders such as coal, small dense fineparticles have physical dimensions close to that of their Stokes diameter, but that more rugged and convoluted profiles have much larger physical dimensions than their Stokes diameters. This is very important when considering properties such as packing and flow. The coal powder was produced by a shattering process. The uranium dioxide powder of the diagram was created by precipitation and ballmilling. The thoria (thorium dioxide powder) was prepared by a precipitation process. Interrelating the light diffraction properties, physical behavior and physical dimensions as measured by light diffraction studies with stream counters and sedimentation counters is obviously a very difficult task for anything other than simple spheres. In most cases, the powder technologists undertaking powder mixing studies will probably have to choose a size characterization procedure and stay with that procedure to establish empirical relationships between flow characteristic and mixing behavior through direct studies [15].

Typical of the kind of empirical study which can be used to look at properties relevant to mixer performance and the flow of powders are the studies carried out by Butters and Wheatley. They studied the flow properties of a substance known as mass PVC (polyvinyl chloride powders). They used the Malvern diffractometer (one of the several commercially available size characterization instruments based upon the study of optical diffraction patterns) to characterize their PVC powders [15, 16]. They reduced their size distribution data to a mean particle size denoted by *ms* and what they called 'the fines of the powder', which was defined as the amount of materials less than 61 μm. This latter quantity they denoted by *f*. They also measured the time for a given quantity of powder to flow out of a funnel under fixed conditions (*ft*). An ASTM standard for this test has been set up for the plastics industry [17]. The initials **ASTM** stand for American Society for Testing Materials which is the standards setting body for many of the industrial operations carried out by companies in the United States of America. Butters and Wheatley discovered empirically that the funnel flow time could be related to the mean particle size and the fines, as determined by the Malvern size analysis equipment, by

$$ft = 54.5 - 0.2335\, ms + 0.95 f$$

in other words they found that the funnel time decreased with increased particle size and increased as the fines content of the powder increased. In their scientific report on their investigation they present the graphs shown in Figure 2.10 and comment

> The agreement between calculated and measured funnel flow data is good and gives confidence in the value of size analysis data from the Malvern ST 1800. To date well over 1,000 suspension PVC (a different type of PVC) and mass PVC samples have been analyzed. There is now considerable confidence in the data given by the Malvern ST 1800 in the analysis of these materials [18].

One of the things that the alert reader may have noticed is that to obtain their empirical relationship Butters and Wheately convert the size distribution data generated by an expensive instrument costing the order of $60 000 to a two-parameter equation. They could have measured their mean particle size much more simply and inexpensively with a device known as the permeameter and their fines content by an empirically controlled sieve analysis. One of the major problems with devices such as the diffractometer is the small size of the sample actually used in the characterization study, so that the sophistication (and thus cost) of the sampling equipment has to be high, whereas the permeameter uses a sample of 100 or more grams of powder, as does the sieve. This method makes the sampling easier and less expensive. Of course the instrumentation is far less glamorous than the diffractometer equipment, but I suspect a cost–benefit analysis would show that the permeameter–sieve combination would win hands down when both sampling and analytical studies are taken into account.

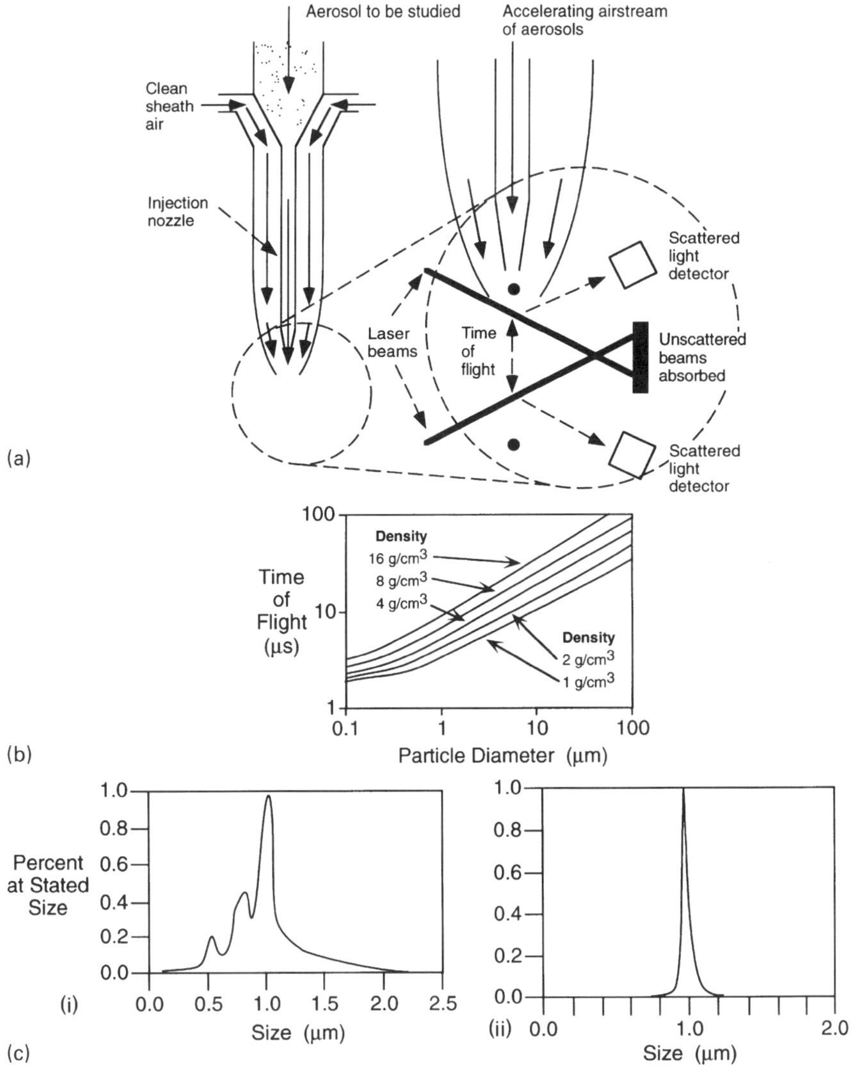

Figure 2.10 In the Aerosizer®, the aerodynamic diameters of fineparticles are measured by recording the time of flight of aerosolized fineparticles as they cross two laser beams. (Reprinted by permission of Amherst Process Instrument [19].)

2.4 FINGERPRINTING POWDER MIXTURES USING AN AEROSOL SPECTROMETER

So far in this chapter our concern has been the characterizing of the size distribution of ingredients in a powder mixture. Recently the structure of mixture has been studied using an instrument known as the **aerosol spectrometer**. Currently two different companies manufacture this type of instrument, Amherst Instruments and TSI Incorporated [19, 20]. In this discussion we will use data generated using the **Aerosizer®** manufactured by Amherst Process Instruments Inc. to describe and discuss techniques for characterizing mixture structure. For reasons which will become apparent in the following discussion the procedure is called fingerprinting the mixture sample.

The basic system of the Aerosizer® is illustrated in Figure 2.10(a). An aerosol of the powder grains to be characterized is created by the dispersion systems shown in the diagram. This aerosol is sucked into an inspection zone operating at a partial vacuum. As the air leaves the nozzle at near sonic velocity the fineparticles in the stream are accelerated across this inspection zone. It should be noted that, as the aerosol stream emerges into the inspection zone, it is surrounded by a stream of clean air that confines the aerosol stream to the measurement zone. The use of a stream of clean air to focus an aerosol stream to be characterized is a widely used technique in aerosol science known as **hydrodynamic focusing**. The term is somewhat confusing, since it was originally developed with instruments employing liquid streams to examine a series of fineparticles. Over the years the term has been extended to clean gas sheaths that serve the same function to improve the efficiency of a size characterization equipment. In the aerosol spectrometer the smaller the fineparticle, the faster the acceleration through the measurement zone. The individual fineparticles are characterized by the time they take to travel across two laser light beams monitoring the inspection zone (Figure 2.10(b)). As they pass through the laser beams, they scatter light which is detected and converted into electrical signals by the two photomultipliers. A computer correlation procedure establishes which peaks in the second laser inspection system constitute the matching signal to the initiation peak as the fineparticle crosses the first beam. This cross-correlation editorial process enables the machine to operate at very high fineparticle flow densities. The equipment can measure fineparticles at a rate of 10 000^{-1} s. The instrument is calibrated using standard fineparticles, as shown by the data of Figure 2.10(b), in which the time-of-flight between the two laser beams is related to the size and density of the standard fineparticles. The Aerosizer® measures the geometric diameter of the aerosolized fineparticles if the density of the fineparticles is known, however, if the density is not known, the data can be interpreted as aerodynamic diameters. The geometric diameters are essentially Stoke's diameters.

As an example of the resolution of the instrument, a sample was prepared from a mixture of three different standard latex samples, 0.5 μm, 0.75 μm and

1.0 μm latex were mixed and the resulting size distribution obtained from the Aerosizer® is shown in Figure 2.10(c)(i). For comparison the 1.0 μm sample was run alone and the results are shown in Figure 2.0(c)(ii). A useful feature of the instrument is that the system used to generate the aerosol for inspection can have variable shear rate dispersion force so that one can study the force needed to disperse a given material into an aerosol. In Figure 2.11 some typical data generated for a difficult cohesive powder are shown. The instrument allows the information on size to be printed out either in differential or cumulative form and by number or volume. The powder data of Figure 2.11 are taken from a study of the size distribution of paint pigments. In the differential display of the data by number, the fines dominate the chart, whereas if the data are presented by volume, there appears to be a small amount of agglomerated powder which may be dispersible by higher shear dispersion study. The particular sample of titanium dioxide used in this experiment had been stood on a shelf for several

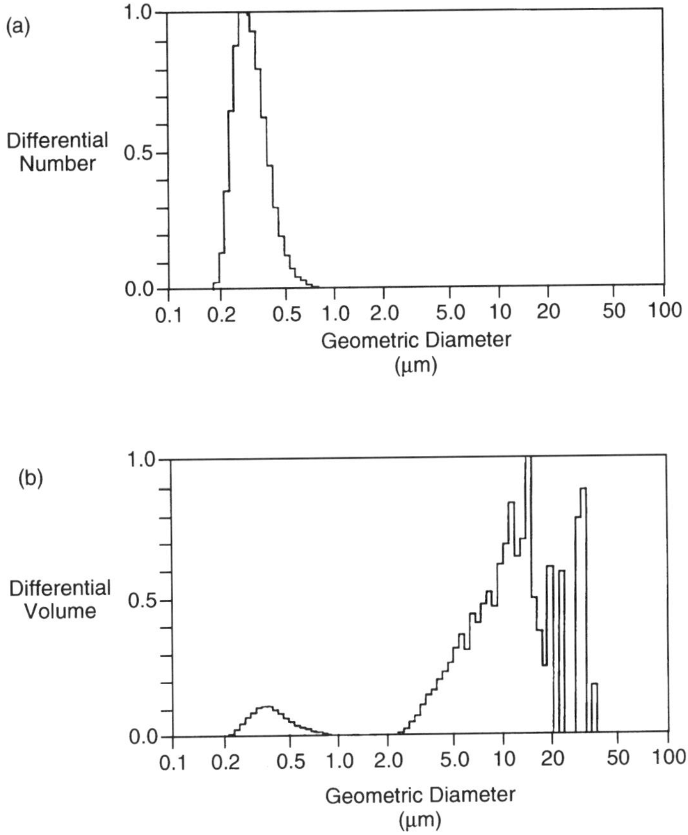

Figure 2.11 Typical data generated by the Aerosizer® in a study of cohesive paint pigment.

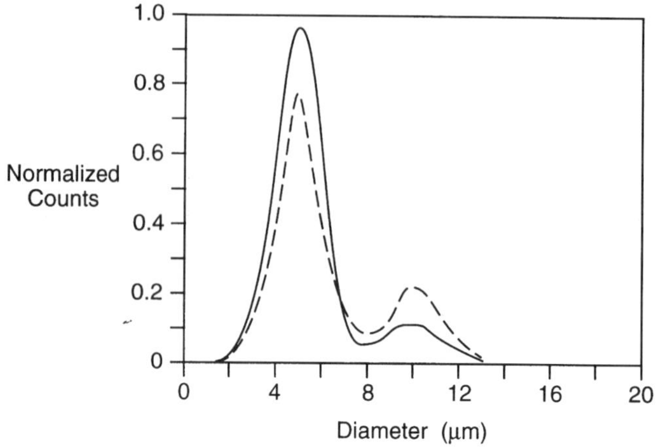

Figure 2.12 Data from the Aerosizer® size analysis of a mixture of two different-size spheres, illustrating the concept of a mixture fingerprint.

years and it may well have agglomerated over that period. It should be noticed that pigments such as titanium dioxide are notoriously difficult to disperse into a dry aerosol form, and one needs to study the measured distribution of the pigment at different shear rates before one can evaluate the physical significance of data such as that displayed in the Figure 2.11.

To understand what is meant by fingerprinting a powder mixture, consider the data of Figure 2.12. A mixture of 5 μm silica fineparticles and new 10 μm silica spheres was passed through the Aerosizer®. In this number count size distribution the two ingredients of the powder mixture can be clearly seen. The relative proportions of the two components can be comprehended from the data display (note if the data were transformed into a weight distribution the relative size of the two humps would change drastically). If a sample had been taken from the mixture which was biased and had more of the coarse spheres than anticipated, the measured curve would have looked like the dotted dataline of Figure 2.12. The two curves of Figure 2.12 are for a mixture in which the ingredients are of the same density, so that the two curves would be actual size and distributions. If, however, the densities of two or more ingredient mixtures were involved in a mixing process, the curve could be regarded as a fingerprint characteristic of a sample taken from the mixer. The monitoring system could be calibrated with known mixture variations, and interfaced with an expert system which could examine the curves of Figure 2.12 for mixing variations between samples taken from a mixer. One needs to make a decision as to when a mixing process is satisfactorily completed. The generation of graphs such as those of Figures 2.11 and 2.12 requires a few minutes and uses samples of a fraction of a gram. Therefore the feedback of information as to the performance of a mixture can be extremely fast and highly resolved. Information data as to variations between samples taken from various parts of a mixer can be generated very rapidly. Work is underway to investigate the quality of this monitoring technique with mixtures of cocoa, flour and sugar.

The fingerprinting data of Figure 2.12, makes use of the differential size distribution display data. Figure 2.13 shows data on a mixing experiment involving two cohesive powder mixes in the AeroKaye™ sampling device in

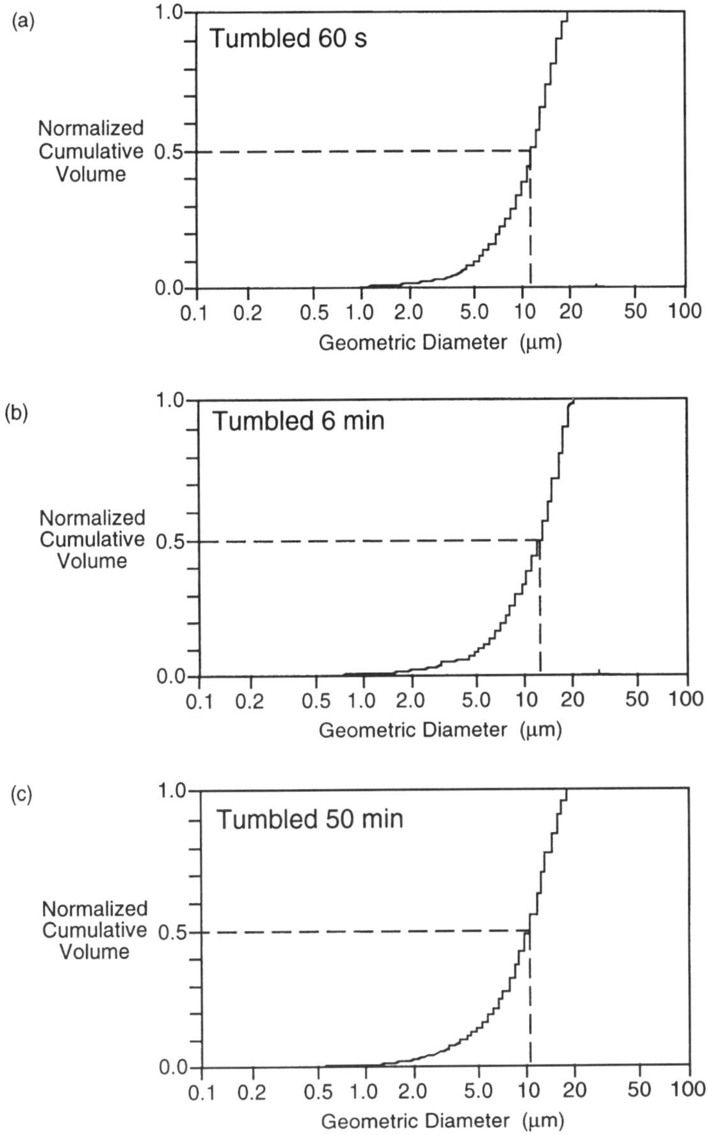

Figure 2.13 The progress of a mixing process can be followed by fingerprinting samples taken from the mixture.

which the cumulative size distribution data version of the information is generated by the Aerosizer®.

2.5 CHARACTERIZING A POWDER MIXTURE BY ITS PERMEABILITY

One of the problems faced in the design of efficient studies of fineparticle size of a powder is that scientists are trained to attempt to gain absolute knowledge and control over their systems, whereas very often in industrial science, quality control and relative knowledge are the order of the day. If this latter fact is realized before a design study is carried out, one can often achieve adequate control of a system at much lower cost than that incurred when one moves the sophisticated research equipment from the research laboratory into the processing work area.

For many years, the permeametry equipment was the workhorse of many industrial laboratories concerned with systems such as cement, food powders, explosives and many other industrial processes. The problem in using such equipment in the modern laboratory stems from the fact that the early permeameters were designed in the late 1930s and have never really been automated and dressed up to perform their job quickly [21].

The basic concept employed in permeability methods for characterizing the fineness of powders is that one assembles a plug of the powder and then measures the resistance to air flow of the powder plug. The finer the powder, the more the air resistance of the plug, i.e. the pneumatic resistance of the powder plug increases with fineness. The basic concepts of the method can be appreciated from the pioneer equipment constructed and described by Lee and Nurse in 1939, shown in Figure 2.14(a). The pressure drop across a plug of cement was compared to the pressure drop across a standard air resistance. This equipment was modified and became the operating principles of the **Fisher subsieve sizer**. This instrument was very widely used in industry, and in fact many people use the **Fisher number** or average diameter derived from the equipment without realizing that they are quoting a permeability-derived measure of fineness. The cement industry adopted the use of a simplified air permeability equipment designed by **Blaine**. This equipment was so widely used that cement powders were often described by their Blaine fineness number. The Blaine fineness tester is shown in outline in Figure 2.14(b). The Blaine fineness tester used the pneumatic circuit equivalent to the electrical circuit which measures the time required for a voltage on a capacitance to decay through a resistance. The powder to be tested was assembled as a plug in a metal cylinder, and the time taken for a pressure head to decay when the air was allowed to flow through the plug was measured. The basic data from permeability methods can be interpreted using an equation known as the **Kozney–Carmen equation**.

The Kozney–Carmen equation has a term in it related to the **voidage** (the open pores in the powder) and empirical adjustment (fudge) factors derived from a study of such systems as glass beads. There have been many studies aimed at improving the interpretation of the measured permeability data to the fineness of the powder with more sophisticated equations, but a lot of this effort was misplaced, since the air resistance is probably more related to the structure of the

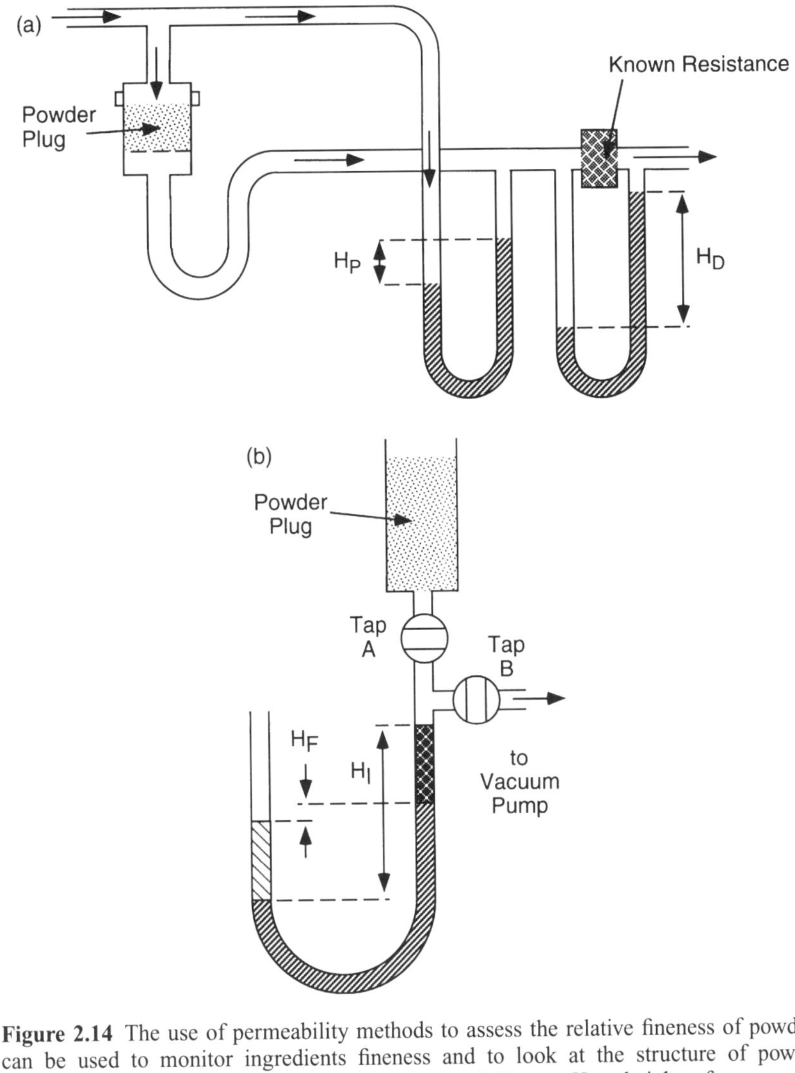

Figure 2.14 The use of permeability methods to assess the relative fineness of powders can be used to monitor ingredients fineness and to look at the structure of powder mixtures. (a) Permeameter developed by Lee and Nurse: H_P = height of manometer column due to the pressure drop across the powder plug; H_D = height of the manometer column due to the pressure drop across the known resistance. (b) The Blaine fineness tester: H_I = height of the manometer column at the start of the test; H_F = height of the manometer column at the end of the test.(Reprinted from Kaye [21], with kind permission from Elsevier Science SA, Lausanne, Switzerland.

voidage paths (percolation paths) in the powder plug than to the exact surface area of the powder. In my opinion, one should treat the permeability method as a useful empirical system which can be used to monitor the fineness of powders. It should never be treated as a primary method of measuring the fineness of a system. It can be calibrated using samples of known fineness and, provided the basic structure of the powder systems being used does not change, the empirical data can be used to monitor quality of ingredients. Thus if one is buying a powder from a supplier who produces it using a ballmill, the permeability equipment will be a useful monitor. If, however, one attempted to compare powders that had been generated by micronizing with those that had been developed by spray drying, the data from a permeameter could be confusing.

In the past, one of the major problems associated with using permeameters to monitor powder fineness was the difficulty of producing a plug of powder in a standard manner. Recent studies have resulted in the development of a permeameter in which the plug of powder is assembled by means of hydrostatic pressure operating on a rubber membrane surrounding the powder plug. The advantages of this type of permeameter make it possible to consider continuing to use the permeameter for many quality control situations [22, 23].

The failure to appreciate the fact that the permeameter does not measure surface area but the structure of the air pathways through the powder mixture has obscured the fact that the permeability study of powder mixtures can actually generate information on the changes in the internal pore structure of the mixture as the ingredients of the mixture intermingle. To illustrate this fact, we will consider some data generated in the late 1960s on the structure of powder mixtures used in rescue alert for flares. In these particular problems there was a need to develop a technique for measuring the quality of the mixture of two ingredients as a quality control procedure for producing the mixture used to drive the flares. The modified Blaine tester shown in Figure 2.14(b) was actually designed to study this problem and the powder plug could be assembled at a remote location and brought for testing at the permeameter. Three different **propellant mixtures** were investigated. The first mixture was a combination of aluminum powder and molybdenum oxide powders. Unfortunately, at the time the study was carried out, the size distribution data of the ingredients were not available, so that we can only report partial data for the investigation. In Figure 2.15(a) the data for earlier mixtures of aluminum and molybdenum powders as measured with the Blaine permeameter are shown. If the surface area of the powder measured by permeability was an additive function, then the dataline linking the measured surface area of the mixture to the mixture richness would be a straight line of the type that was actually generated for the mixture of aluminum powder and molybdenum oxide powder. The datapoints show that in fact this curve could be used for this mixture to monitor the progress of mixture, since samples taken out of the mixture could be placed in the permeameter with relatively fast generation of information on the mixture richness of the sample. When we come to look at the modified pneumatic lance described in Chapter 5, we will discuss how, for simple interacting mixtures of this type, it may be possible to measure the mixture richness *in situ* by measuring the pneumatic resistance of the powder sucked into the sampling chamber of the pneumatic lance. In Figure 2.15(b) the permeability-estimated surface areas of a mixture of powdered aluminum and vanadium peroxide are shown. When interpreting the physical significance of this graph it should be remembered that an increase in airflow resistance offered by the plug of mixture is interpreted as an increase in fineness of the powder. It appears that the addition of a small amount of aluminum powder to the vanadium peroxide greatly increases the resistance to air flow offered by the plug of powder mixture. This is probably because the

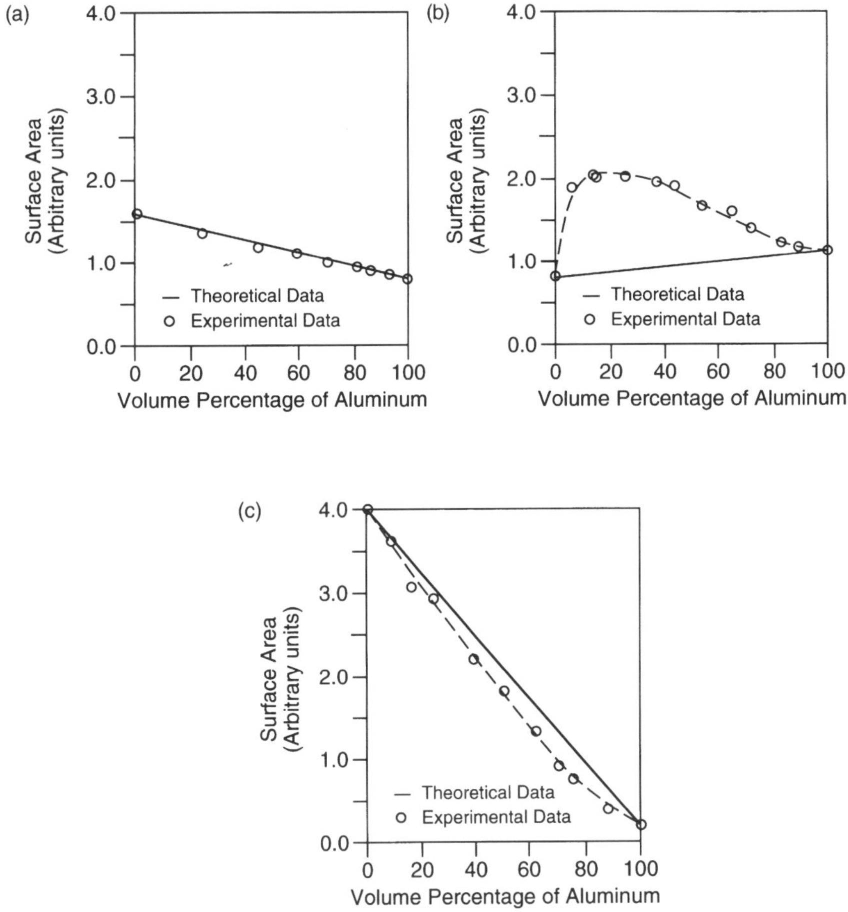

Figure 2.15 The change in permeability of a powder mixture with the richness of the ingredients is a complex problem which can be fingerprinted rapidly, and at low cost, using the Blaine fineness tester. (Reprinted from Kaye [21], with kind permission from Elsevier Science SA, Lausanne, Switzerland.

aluminum powder was finer than the vanadium peroxide and was able to occupy the interstitial spaces between the larger grains of vanadium peroxide. This reduced the available flow paths for the air moving through the powder plug. As the richness of the aluminum powder in the mixture increased, this effect tended to dissipate and the curve of Figure 2.15(b) can actually be regarded as a method for exploring the interpacking of the two ingredients in the mixture. Furthermore, since one is interested in combustion rates of such mixtures, it is probable that the shape of the curve would be indicative of burning rates and density of compacts. If a similar curve had been generated by components of metal alloys or ceramic powders, the data would indicate interpacking of the ingredients. The opposite behavior was detected in the case of a mixture of aluminum and copper oxide powders, as demonstrated by the data of Figure 2.15(c). Here the measured surface area appeared to decrease when the aluminum powder was added to the copper oxide powder. A possible explanation of this phenomenon is that the two powders were relatively similar in size but that the copper oxide was slightly coarser. This would cause a dilation of the mixture, opening up air paths for the flow through the powder compact when the copper oxide was added to the mixture. Remember what is reported as a decrease in surface area is physically a decrease in pneumatic resistance, which means that the pathways have opened up to some extent. If one were hoping to use the permeability of the powder mixture to explore the performance of the mixer there would be some difficulty in using the curve of Figure 2.15(b) since there would be a non-unique value for the mixture richness. However, if one were to be looking for an expected value of, say 50% aluminum in the mixture, one could probably use the technique provided the percentage in any given sample did not drop below 15% by volume. (This would be a bad mixture anyway and would probably be visibly low in aluminum.) There is tremendous scope for exploring the structure of mixtures using permeability measurements and we can expect to see data in this area in future scientific publications.

NOTES

1. Kaye, B.H. (1981) *Direct Characterization of Fineparticles*, J. Wiley & Sons, New York; and Allen, T. (1990) *Particle Size Analysis*, 4th edn, Chapman & Hall, London.
2. The free-fall tumbling mixer is available from Amherst Process Instruments, Mountain Farms Technology Park, Hadley, MA 01035, USA.
3. Kaye, B.H. (in preparation) Efficient powder sampling by means of powder mixing and snatch samples.
4. The Gilson Co. Inc., Box 677, Worthington, OH 43085, USA, supplies a range of powder sampling devices.
5. Kaye, B.H. (1991) Optical methods for measuring the performance of powder mixing equipment, in *Proceedings of the 1991 Powder and Bulk Solids Conference/Exhibition*, Rosemont, Illinois. 6–9 May. Published by Cahners Exhibition Group, PO Box 5060, Des Plaines, IL 60017-5060, USA

6. Kaye, B.H., Leblanc, J.E., Moxam, D. and Zubac, D. (1983) The effect of vibration on the rheology of powders, in *Proceedings of the 1983 Powder and Bulk Solids Conference/Exhibition*, May, Rosemont, IL, Cahners Exposition Group (now Reed Exhibition Companies), Chicago.
7. Kaye, B.H. (1993) *Chaos and Complexity: Discovering the Surprising Patterns of Science and Technology*, VCH Publishers, Weinheim, Germany.
8. Flook, A.G. (1981–82) Fourier analysis of particle shape, in *Particle Size Analysis 1981–1982* (eds N.G. Stanley Wood, T. Allen), Wiley Heyden Limited, London. (From *Proceedings of the Fourth Particle Size Analysis Conference*, Loughborough University of Technology, 21–24 September 1981.)
9. Kaye, B.H. (1993) Applied fractal geometry and the fineparticle specialist, part 1: rugged boudaries and rough surfaces. *Particle and Particle Systems Characterization*, **10**, 99–110.
10. See discussion of Fraunhoffer diffraction patterns in Kaye, B.H. (1994) *A Randomwalk Through Fractal Dimensions*, 2nd edn, VCH Publishers, Weinheim, Germany.
11. Mandelbrot, B.B. (1977) *Fractals: Form, Chance and Dimension*, Freeman Press, San Francisco.
12. Kaye, B.H. (1994) *A Randomwalk Through Fractal Dimensions*, 2nd edn, VCH Publishers, Weinheim, Germany.
13. Stewart, I. (1975) *Concepts of Modern Mathematics*, Pelican Books, Middlesex, UK. (A readable introduction to geometric set theory and topology for those unfamiliar with these subjects.)
14. Galai Instruments, Inc., 577 Main Street, Islip, NY 11751, USA.
15. For a recent review of particle size analysis methods see Fayed, M. and Otten, L. (1995) *Handbook of Powder Science and Technology*, 2nd edn, Chapman & Hall, London.
16. Malvern Instruments Inc., 10 Southville Road, Southborough, MA. 01772, USA.
17. ASTM Funnel flow test, *ASTM Standards* Part 35, ANSI/ASTM D1895–69, 1975.
18. Butters and Wheatley.
19. Amherst Process Instruments, Mountain Farms Technology Park, Hadley, MA 01035, USA.
20. TSI Incorporated, Industrial Test Instruments Group, PO Box 64394, 500 Cardigan Road, St Paul MN 55164-6877, USA.
21. Kaye, B.H. (1967) Permeability techniques for characterizing fine powders. *Powder Technology*, **1**(1), 11–22.
22. Kaye, B.H. and Legault, P.E. (1978) Realtime permeametry for the monitoring of fineparticle systems, in *International Symposium on In-Stream Measurements of Particulate Solid Properties*, Bergen, Norway, 22–23 August, Vol.1.
23. Hoffman, A.D. (1989) A Soft-wall permeameter for online characterization of grinding circuits. *M.Sc. Thesis*, Laurentian University, Sudbury, Ontario, Canada, P3E 2C6.

3
Powder rheology

3.1 A NEW ANGLE ON POWDER FLOW CHARACTERIZATION

It ought to be a self-evident truth that the flow characteristics of a powder ingredient entering a mixer should be an important parameter of study for the powder mixing technologist. However, many technologists use the powders they are given and only investigate the flow properties of the powder when they run into trouble. It is useful when looking at powder flow to distinguish between **low-pressure rheology** and **high-pressure rheology**. High-pressure rheology is the study of the powder flow when the moving powder at the exit from the container is subjected to considerable pressure from the powder above it. The term 'considerable pressure' is intentionally vague, since it is difficult to give an exact delineation between high-pressure rheology and low-pressure rheology. In general, however, the powder mixing technologist, at the mixing stage and subsequent processing, is generally involved in low-pressure rheology in which the pressure being sustained by the powder under study is relatively low, and factors such as cohesiveness, environmental vibration and the roles of additive are much more important than in the case of high-pressure rheology. In general terms, one can state that high-pressure rheology is the study of powder flow where powder is being processed in storage bins and where several tons of powder are being handled. Usually the powder technologist is not interested in systems where applied pressure to any powder process is of the order of many thousands of newtons per square meter. (These systems are used in, for example, tableting and in ceramic fabrication, but such topics are not covered in this book.) Engineers who design large bin holders for industrial powder processors have made a speciality study of the flow properties of powders using a device known as the **Jenike shear cell** [1]. There have been several books and review articles written on data generated by the Jenike shear cell and similar instruments. It would be redundant to review these further here [2]. The study of low-pressure rheology has received surprisingly little study and there are virtually no data on the effect on low pressure flow systems from changes in humidity, electrostatics and vibration and temperature changes. In this chapter we will look at some basic data describing the properties of powders flowing under low pressure and at how this behavior is modified by vibration and/or the use of what are known as 'flowagents' and the relevance of such facts to powder mixing.

Flowagents are additives, present usually in small percentages by volume, placed in powder systems to change the flow characteristics of a powder [3–5]. Different types of flowagents are available, and the pharmaceutical industry uses the term **glidants** to describe such additives. A useful device for studying the important characteristics of powder flow under low-pressure systems is shown in Figure 3.1(a). This is the apparatus used by Kaye *et al.* to study the effect of vibration, flowagents and aeration on the low-pressure rheology of powders [5]. When the powder is allowed to flow out of a container such as that of Figure 3.1,

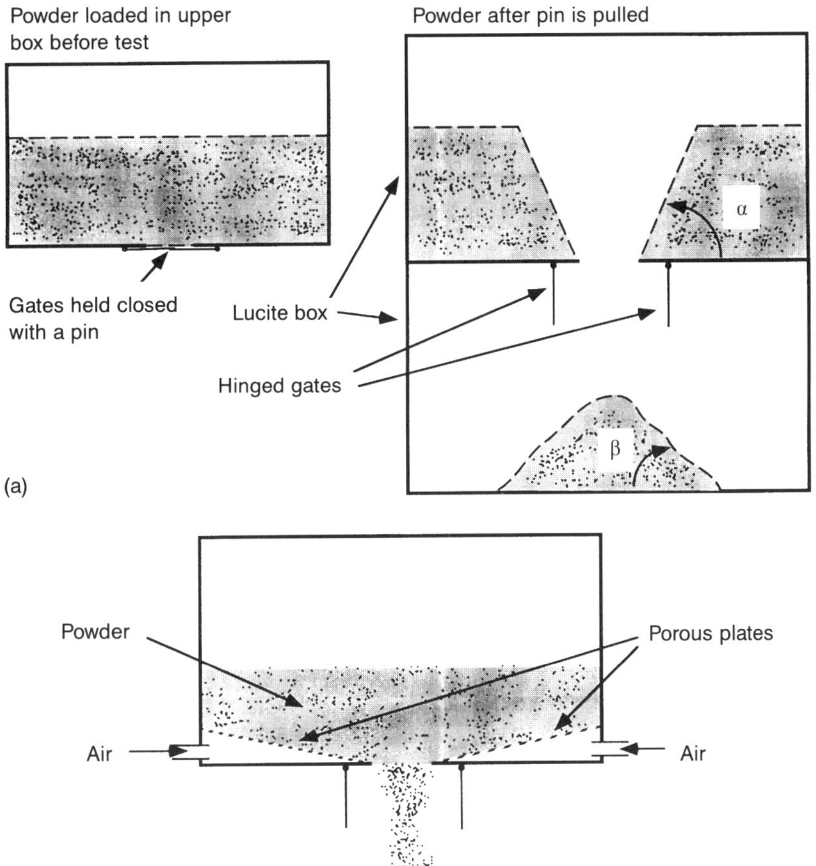

Figure 3.1 Apparatus used by Kaye *et al.* to study the effect of vibration, flowagents and aeration on the low-pressure rheology of powders [5]. α = angle of drain; β = angle of repose.

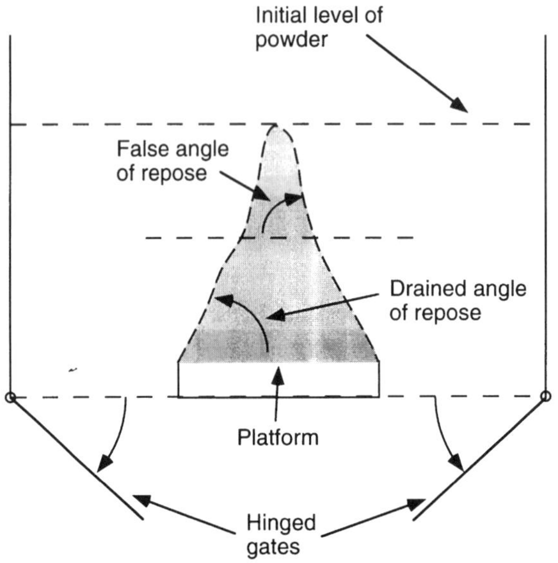

Figure 3.2 The drained angle of repose can differ from the poured angle of repose.

the angle β of the cone of powder building up under the exit orifice is known as the **angle of repose**. Sometimes the angle of repose is measured in a different way, using equipment such as that of Figure 3.2. At the start of the experiment the platform is buried in a container of powder. When the trapdoors are opened the powder drains away, leaving a cone of powder on the cylinder platform, as shown in the figure. Particularly with cohesive powders, this cone may not be uniform in structure and may have a peak. The angle of such a peak is sometimes known as the **false angle of repose**, as shown in the diagram. If an untreated powder is poured into a heap, then if the measured angle of repose is less than 30°, the powder is considered to flow easily. When the angle of repose is between 30° and 50°, the powder is considered to flow with difficulty. If the value of the angle of repose is greater than 50°, the powder is considered to be a cohesive powder which needs assistance to make it flow into and out of any piece of equipment.

Another measurement which is used by experimentalists to describe the bahavior of the powder is the angle α of Figure 3.1(a) which is described by different workers as the **angle of sliding** or simply the **angle of drain**. The angle of drain is influenced by the treatment that the powder has sustained before a pouring experiment is carried out. Vibratory consolidation of the powder and the effect of moisture can increase the angle of drain to the point that the powder leaving the storage bin will rat-hole, a phenomenon briefly mentioned in Chapter 1. Obviously when a powder rat-holes the angle of drain has reached 90°. Other terms used for flowagents are **flow conditioner**, **lubricant**, or **anti-caking**

agent. As will be discussed later, the use of the term lubricant to describe the action of some flowagents, such as finely divided silicas, is misleading. The term will not be used further in this text. Industrialists, in an attempt to provide themselves with tests for controlling the properties of raw materials, have developed *ad hoc* flowability tests. Thus the ASTM (American Society For Testing Materials) has a standard for measuring the flow of many powders, particularly those such as the PVC powders used in the plastics industry. The test method entitled 'Apparent Density, Bulk Factor and Pourability Of Plastic Materials'. (ASTM D1895-69 (reapproved 1975)) [6]. In this standard the dimensions of a funnel to be used to measure the pourability of a powder are strictly specified, along with information on the surface finish to be given to the funnel (the friction between the flow powder and the funnel walls will affect the time taken for a fixed amount of powder to exit from the funnel). The technologist who wishes to follow this standard is told:

> Take a sample of the plastic material weighing in grams 100 times its specific gravity. Work this sample on a paper until there is no tendency for the material to pack or cake. Close the small end of the funnel with the hand or with a suitable flat strip and pour the sample of powder lightly into the funnel avoiding any tendency to pack it. Then quickly open the bottom of the funnel and start a stop watch or timer at the same instance. Allowing the material to run from the funnel as freely as it will, stop the watch or the timer at the instant the last of it leaves the funnel. The report shall state the time in seconds required for the funnel to discharge to the nearest 0.2 second or if so found that the material would not run through the funnel [6].

These instructions to the technologist illustrate the importance of the powder pretreatment and also the avoidance of delay between filling the funnel and opening the orifice at the base of the funnel. I was once asked a problem that a company was experiencing with this particular test. They had modified it slightly by putting an automatic gate on the bottom of the funnel and were operating the equipment in a laboratory at the top of a large building. The top floors of a large building vibrate far more than the basement (hence the location of vibration sensitive equipment in rock bottom basements). The problem arose because the technician, after filling the funnel with powder, was allowing a slight delay of varying duration to occur before operation of the equipment. This was sufficient to allow slight vibratory compaction of the powder in the funnel, with subsequent variation of flow time for repeat tests on any particular powder. The type of funnel testing described briefly above is widely used by industries using large quantities of powders [7, 8].

A variation in the same approach to the *ad hoc* characterization of the flow properties of a powder is embodied in the equipment illustrated in Figure 3.3(b). These funnels are described in some experiments carried out by Taubmann. [9]. In the English version of the Taubmann's paper these sets of funnels are described as **sand glasses**. A better term in English is flow **characterization funnels**. (The English translation of the German term used by Taubmann may have been referring to the fact that the operation of the measuring glass is similar to that of a sand timer or hourglass of the type discussed in Chapter 1.) The set of flow measuring funnels have different-sized orifices. If

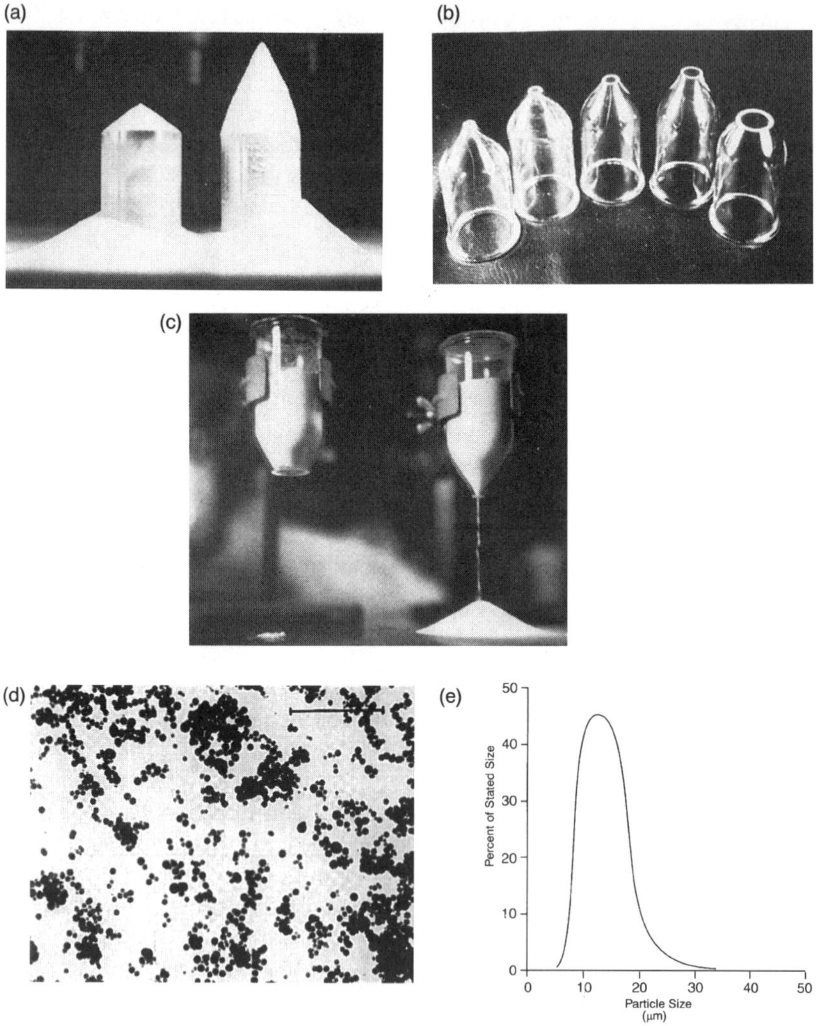

Figure 3.3 Taubmann has reported measurements on the flow characteristics of powder, under low-pressure conditions, with and without silica flowagent added to the powder [9]. (a) Drained angle of repose for two powders with different flow properties. (b) Funnels used to assess the flow properties of a powder. (c) Typical performance of a cohesive and a free flowing powder. (d) Electronmicrograph of Aerosil silica flowagent. (e) Size distribution of Aerosil flowagent. (Used with permission from Taubmann [9].)

the powder will flow out of number 1 funnel, with the smallest orifice, then the powder is described as having excellent flow behavior. Powder flowing from a number 1 funnel is shown in Figure 3.3(c). It should be noted that the poured angle of repose is much smaller than the bin bottom angle of the funnel, showing that the powder is leaving the test funnel under mass flow conditions, as demonstrated by the fact that the upper surface of the powder in this figure is a horizontal surface. If the powder will just flow out of number 5, the rating of the powder is the number 5, and if it will not flow out of 5 it is given the number 6, indicating that it does not flow without other methods of assistance. In Figure 3.3(d) the drained angles of repose for two powders reported by Taubmann are shown. He describes the technique for measuring these angles of repose in the following manner.

> The powder to be characterized is poured through a metal screen on to a cylinder of diameter 50 millimeters and 8 millimeters high, the distance between the powder flow device and the cylinder should be slightly higher than the poured cone which forms on the cylinder [9].

Thus his measurements are what we have called the poured angle of repose. In the next section we will discuss how vibration can alter the angle of drained repose.

3.2 USING FLOWAGENTS: A FAUSTIAN BARGAIN?

Faust is a character in German literature who makes a deal with the Devil so that he can have all he wants in life for a certain period, but at the end of that period he must give the Devil his soul. From this character the term Faustian bargain has evolved to mean any deal which at first looks good, but which later has a very high price tag attached to it. We use the term Faustian bargain here to describe the use of flowagents in powder systems because, all too often, a temporary improvement of flow is achieved by using the flowagent but then as one might say, there is the devil to pay later on in the process. Thus in the pharmaceutical industry, one group of technologists might add a silica flowagent to a powder being poured into a mixer to obtain better flow characteristics into and out of the mixer and to enhance the speed of mixing. But the presence of such a flowagent will considerably increase friction and wear in a tableting machine to which the powder mixture is fed. This may be an acceptable price to pay for better mixing and flow, but the mixing technologist should be aware of the ultimate price to be paid for his or her innovation. On the other hand, if the powder technologist uses the stearate type of flowagent, the doctored powder will flow better, but the presence of that type of flowagent in a pharmaceutical mix may interfere with the bioavailability of the drug in a subsequent tablet. Not all uses of flowagents are Faustian bargains, and the use of a stearate flowagent, if a powder is going to be subsequently used in making

Table 3.1 The effect of various amounts of the flowagent Aerosil on the poured angle of repose of a heap of powder in the work carried out by Taubmann [9]

	Slope angle (deg.) for Aerosil content (%) of									
Material	0.0	0.01	0.05	0.1	0.25	0.5	1.0	5.0	10.0	Optimum (%)
Cornstarch	49.6	42.9	39.7	33.9	32.4	31.4	37.8	46.5	46.5	0.5
Lactose	49.6	39.2	37.3	38.0	38.7	38.7	39.5	42.9	43.6	0.05
$MgCO_3$	42.3	40.0	39.7	40.0	41.0	41.6	42.3	45.0	46.8	0.05
ZnO	45.0	41.0	38.0	37.8	36.7	37.8	37.8	41.0	41.6	0.25
TiO_2	48.2	43.9	42.9	42.6	41.9	41.6	41.0	45.0	46.5	1.0
MgO	50.5	48.3	47.2	49.6	47.8	47.8	48.3	49.1	49.5	0.05

a paint or plastic, may be an advantage because the presence of the stearate may not only promote flow, but may enhance the dispersibility of the colored powder mixture into an oily or waxy substance. It is because of this latter possibility that the question mark has been used in the title of this section. The dramatic title is intended to alert the technologist to the need to consider the long-term consequences of adding a flowagent to a powder system at the mixing stage of any subsequent processing or product properties.

Taubmann's experiments were discussed briefly in section 3.1. He has discussed the effect on various powders of a flowagent which is a commercial form of colloidal silica. It is manufactured by Degussa and known by the trade name Aerosil [10]. His experimental data are summarized in Table 3.1.

From the data of Table 3.1 it can be seen that the incorporation of Aerosil into the powder system systematically reduces the poured angle of repose up to a certain level of Aerosil. After that point, increasing the concentration of Aerosil actually can increase the poured angle of repose. In some cases, high concentrations of Aerosil generated poured angle of repose values higher than that of the untreated powder. Taubmann gives the usual explanation for the operation of the silica as a flowagent (one which is also given in the trade literature of Cab-O-Sil®, a silica flowagent manufactured by the Cabot Corporation [11]), that the silica flowagent acts as a lubricant to make the grains of the **host powder** (a term used for the powder which is doctored with the flowagent) move more easily past each other in the flowing powder. If this were the correct description of the major mechanism of the silica flowagent, it is hard to understand why the poured angle of repose starts to increase as the percentage of flowagent in the overall mixture increases above a certain value. Surely if the additive were a lubricant powder the measured angle of repose would continue to decline until it reached the value typical of the flowagent, which would be lower than that of the undoctored powder. Again, if the silica flowagent is a lubricant, it is hard to explain why the addition of silica flowagent to pharmaceutical powders increases wear and friction in a tableting die and piston assembly. It can be demonstrated that in fact the silica flowagent promotes flow by preventing interpacking of powder grains, through an increase in friction between the grains of the host powder [12]. In fact using silica to promote flow is truly a Faustian bargain because, if at any stage after the powder has been treated with a silica flowagent, the powder is allowed to rest in a container subjected to vibration, the powder may never leave the container easily in any subsequent operation, because vibratory compaction of the system has taken place, with the high-friction flowagent binding the grains together. Before moving to a discussion of evidence which supports the claim that silica flowagents promote flow by preventing packing, it is necessary to discuss what happens to the structure of a loosely assembled powder bed which is subject to vibration.

3.3 SETTLING DOWN IN A VIBRATED BED

If one takes a powder such as artificial creamer used to whiten coffee and shakes it in a bottle, then immediately after shaking the powder appears to behave in the bottle like a fluid. Similar behavior may be observed in a powder poured into a container. This liquid type behavior of the freshly poured or shaken powder arises from the fact that the freshly agitated powder bed contains a considerable amount of air trapped between the powder grains by the action of shaking the bottle. If one watches the behavior of such a freshly poured aerated system, it collapses fairly rapidly to a lower level. Subsequent tapping on the outside of the container will result in further reduction of the volume of the powder bed. It is for this reason that when technologists assess the bulkiness of a powder, they carry out a **tap density test** in which the powder is placed in a cylinder and tapped a specific number of times (with specified vigor) and the volume of the powder bed measured after such a controlled amount of vibration. This volume is used to calculate what is known as the **tapped bulk density** of a powder, a term often shortened to **bulk density** of the powder. It is the weight of powder divided by the volume of the tapped powder bed. Some quantitative data on the behavior of a freshly poured bed of aluminum powder subjected to vibratory compaction is summarized in Figure 3.4 and Table 3.2 [5]. When the powder is tapped ten times, the powder settles to a voidage of approximately 0.52. During the consolidation after the first ten taps, one can almost see air bubbling out of the solid. Therefore, one can describe the initial collapse of the powder bed as taking place by the elimination of **bubble voids**, which are gross islands of air trapped within the powder. After approximately 20 taps, the vibrating collapse of the powder proceeds at a different rate, with the datapoints defining a logarithmic collapse of the powder at a rate described by the slope.

The main changes occurring in the packed powder bed under the effect of vibration in the second region of the collapse curve are that the grains of the powder are starting to pack into regular structures, which form islands of regularity, with the remaining gross voids located along and at the junctions between these islands. As the vibration continues up to point B of the collapse

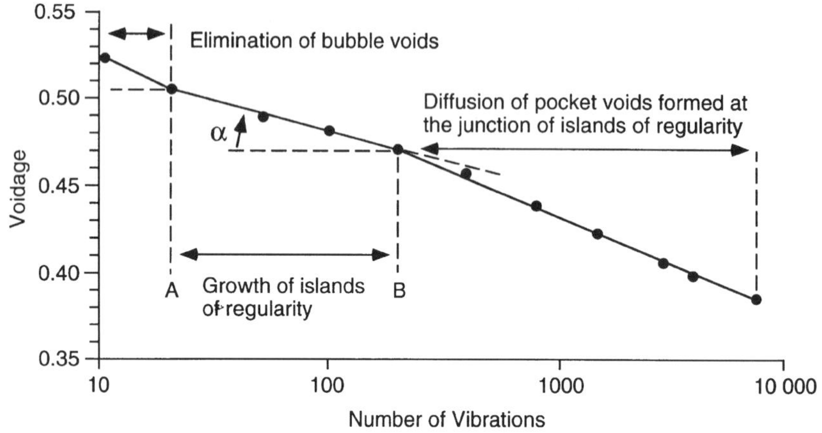

Figure 3.4 Studies of the collapse of a bed of freshly poured aluminium powder exhibits three regions of voidage reduction. The postulated mechanisms dominant in each region are indicated in the diagram [5].

curve, the islands of regularity are growing in size and the sustainable strength (a concept discussed later in this section) of the powder bed is increasing. The appearance of the structure of these different types of collapse phenomena, as demonstrated in a two-dimensional powder bed are shown in Figure 3.5. These

Table 3.2 Collapse rates of a bed of aluminum powder with and without flowagents

	Voidage of aluminum powder		
Number of taps	Untreated	With Cab-O-Sil	With magnesium sterate
5	0.536	0.520	0.550
10	0.521	0.505	0.535
20	0.503	0.488	–
25	–	–	0.509
50	0.488	0.475	0.497
100	0.480	0.468	0.486
200	0.470	0.460	0.476
400	0.457	0.449	0.464
800	0.439	0.439	0.453
1 500	0.423	0.429	0.444
3 000	0.406	0.418	0.437
4 000	0.399	–	–
5 000	–	0.409	0.435
7 500	0.387	–	–
8 000	–	–	0.431
10 000	0.381	0.398	0.430

pictures are taken from some interesting model work carried out by Nowick and Mader on the crystalline changes that occur when a piece of metal is heated and annealed [13].

The very slow collapse that takes place after point B of Figure 3.4 appears to be dependent upon the strength of the vibration given to the powder bed. This is because the powder bed seems to have to be dilated by the act of vibration before there is any movement in the powder bed. The slow decrease in the voidage in the third region of consolidation arises from the migration out of the powder bed of interstitial voids such as those visible in Figure 3.5(c).

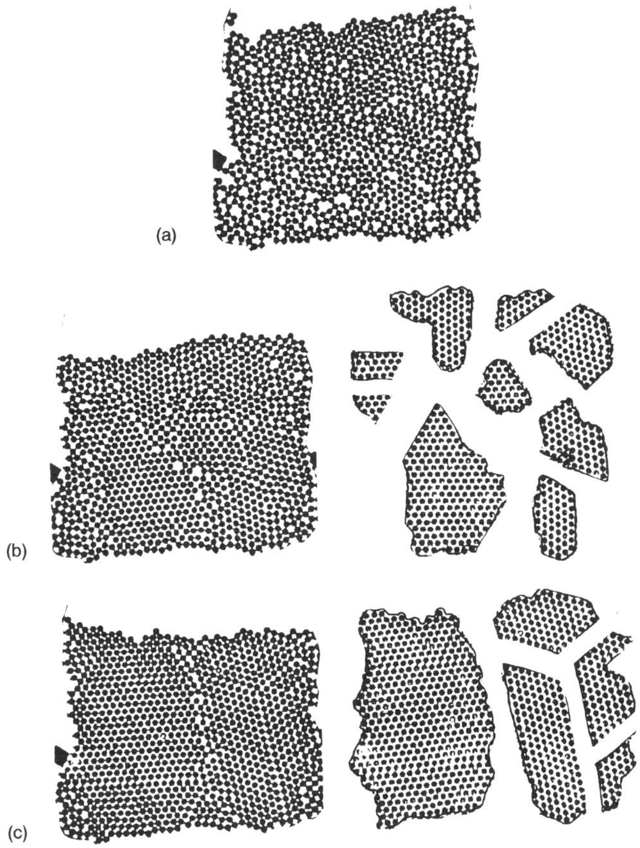

Figure 3.5 Vibration of a powder bed results in the growth of islands of regularity. The very slow changes in voidage in the last stages of Figure 3.4 are due to the migration of small voids [13]. (Used by permission.) (a) After elimination of gross bubble voids, at the second stage of consolidation in Figure 3.4, the voids to be eliminated are widely spaced through the system. (b) As the second stage of consolidation proceeds, islands of regularity are formed. (c) The third stage of consolidation, corresponding to the third region of Figure 3.4, is the slow migration of small voids out of the bed.

An empirical technique used to evaluate the changing physical properties of a powder bed subjected to vibrational consolidation, called evaluating the sustainable strength of the powder bed (ESSOP for short), has been developed by Kaye and co-workers [5, 14–16]. In this technique a small piece of paper, about the size of a postage stamp, is coated on both sides with the powder being studied. This piece of paper is then hung from the top piece of an Instron tester [17], equipped with a low-level strain gage. The coated paper is suspended in a small cylinder placed on the lower base of the Instron tester. The powder is then poured gently into the cylinder. When the Instron tester is operated, the coated paper is pulled upwards and the force needed to remove the paper strip is recorded. Since this force is dependent on many factors, including the size of the coated paper specimen and the ambient conditions, the force measured is referred to as the sustainable strength of the powder bed around the coated strip. Any series of investigations of the changing properties of a vibrated powder bed should be carried out under specified ambient conditions. As shown by the data on the vibrated powder beds reported here, the ESSOP test has proved to be a useful *ad hoc* test for studying the holistic behavior of powder beds [18, 19]. A coated needle, if it is more convenient, may be substituted for coated paper in the ESSOP test.

In Figure 3.6 some measurements on the changes of the sustainable strength of a powder bed of dry nickel ore tailings powder reported by Kaye and Akhter are shown [14, 15]. This bed is subjected to vibratory consolidation and its strength determined by measuring the force required to withdraw a needle coated with the

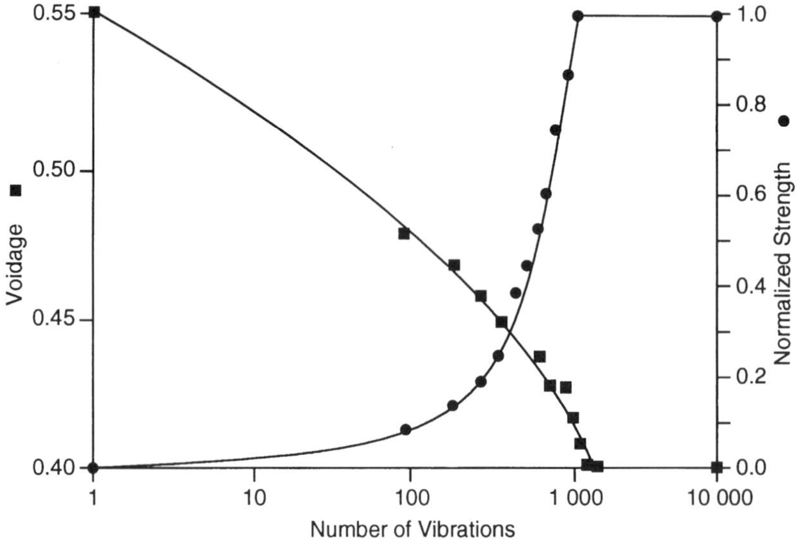

Figure 3.6 Data reported by Kaye and Akhter show how measuring the sustainable strength of a powder bed can demonstrate the changes in physical structure of the collapsing powder bed [14,15].

powder, buried in the powder bed. It can be seen for the first 100 taps there is a very dramatic drop in the voidage of the powder with very little change in the sustainable strength of the powder bed. (Note that the measurements of the sustainable strength of Figure 3.6(b) have been expressed in normalized units with respect to the maximum measured sustainable strength of the powder bed.) Then between 100 and 1000 taps, which is similar to the region of (a) to (b) of Figure 3.4, there is a very rapid rise in the sustainable strength of the powder bed until 1000 taps have been administered. From that point on, even a tenfold increase of the vibration of the bed does not change significantly the sustainable strength of the powder bed. The porosity strength curve versus vibrations of Figure 3.6(b) would appear to be a very useful source of information on the behavior of a powder when it is being stored or processed. The technique has been used by Biing Lee for PVC and other plastic powders [18].

The shape of the consolidation–vibration behavior of a powder does, however, depend upon the shape of the grains being consolidated. Thus Figure 3.7(b) shows the changes in the sustainable strength of the vibrated powder beds of two pharmaceutical powders essentially the same size when one powder was produced by spray drying and the other by crystallization. It can be seen that the crystalline powder did not change drastically with vibration, whereas the spray-dried powder, although it flowed easily, also consolidated relatively easily. It is rather interesting to note that Kaye *et al.* report that, although the spray-dried powder was initially free flowing after vibratory compaction in the equipment of Figure 3.6(a), the powder would not move out of the orifice. The crystalline powder, although it did not flow as readily as the spray-dried powder (even after being consolidated with 2000 taps) would still flow out of the equipment. They attribute the difference in behavior to the shape of the powder grains, with the comment that the crystalline powder did not appear to consolidate under vibratory compaction. If one looks at the withdrawal curves for the spray-dried powder, it can be seen that up to 200 taps the powder behaved in a manner analogous to a viscous fluid, resisting the movement of the buried needle, whereas after 1000 taps the powder bed was snapping like a brittle solid. On the other hand, the withdrawal curves for the crystalline version of the powder show no such changes in behavior from viscous fluid to brittle solid.

Having explored the behavior of powder beds which have not been treated with flowagents, we can now consider the interpretation of comparable data obtained for treated powders. Thus in Figure 3.8(a), (c) the vibratory collapse of a powder treated with silica flowagent and magnesium stearate are compared to collapse curves for the untreated aluminum powder. It can be seen that the initial voidage of the powder with silica flowagent is lower than that of the untreated powder, but after approximately 2000 taps the untreated powder actually consolidates further than the treated powder. The action of the silica on the powder is probably a complicated combination of desiccant action (absorption of moisture), discharge of electrostatic phenomena and increase in the friction between the grains to prevent the interpacking. Thus in the initial stages of consolidation, the fact that the silica agent takes the moisture and electrostatic forces out of the interplay of causes, causes the silica-treated material to pack better than the untreated, but note that it does not show any third region of consolidation. This

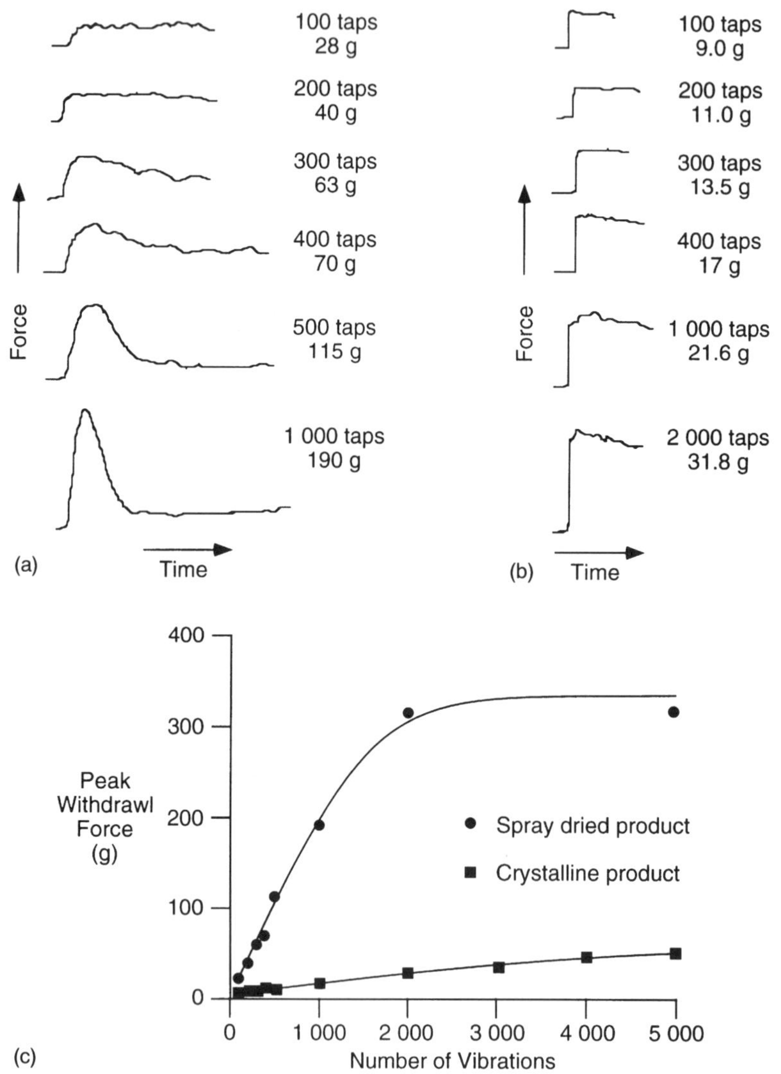

Figure 3.7 The measured 'sustainable strength' of two different pharmaceutical powders in the experimental system of Figure 3.6(a) demonstrates the different structures of the beds of the two differently shaped powder grains after some vibratory treatment. (a) Instron withdrawal curves for spray-dried pharmaceutical powder after vibratory consolidation. (b) Instron withdrawal curves for crystalline pharmaceutical powder after vibratory consolidation. (c) Sustainable strength versus taps for the two powders. (Note: curves in (b) are enlarged to five times the scale of (a).)

means that the migration of the voids under the continued vibration of the bed is suppressed by the presence of the silica agent. My attention to this type of phenomenon was originally stimulated by an inquest into a situation where a silica-treated powder had been left in a mixer for five hours because of a

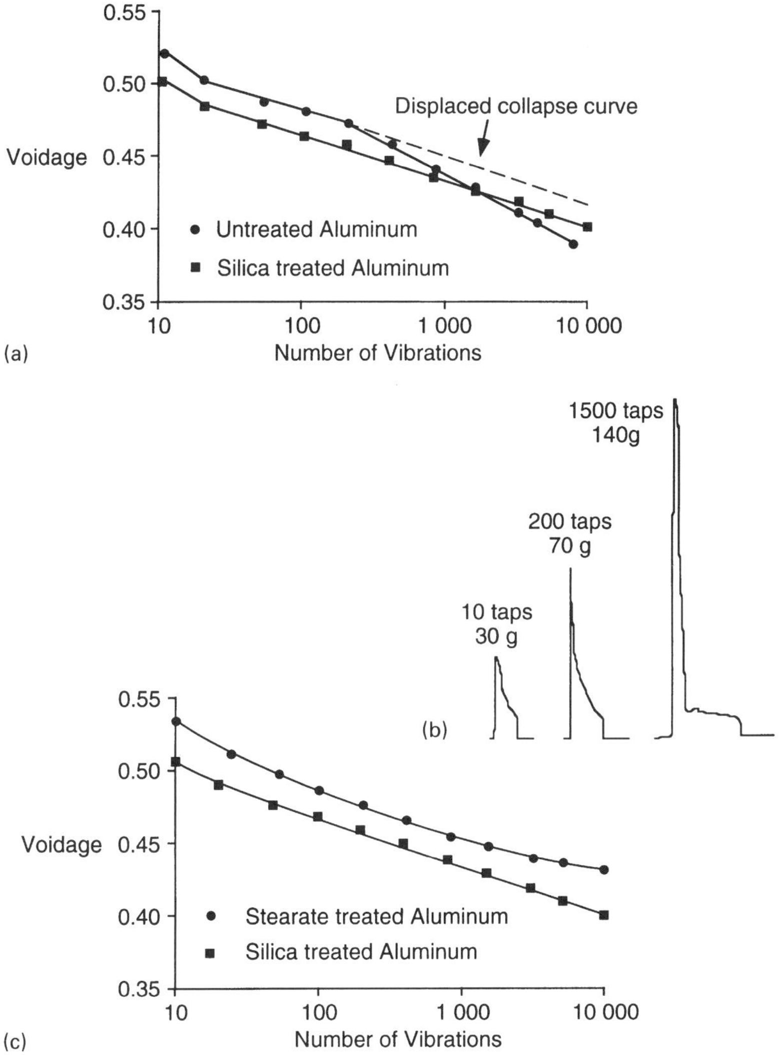

Figure 3.8 Vibrated beds of powder to which flowagents have been added manifest different packing properties to those manifest by the untreated powder. (a) Comparison of collapse rate curves for aluminum powder with and without silica flowagent. (b) Sustainable strength curves for silica flowagent treated aluminum powder. (c) Comparison of the collapse rates for aluminum powder treated with silica flowagents and magnesium stearates.

malfunction in another part of the plant. Although normally the powder flowed out of the mixer, after the five-hour stand the powder was so interlocked that it had to be chipped out of the top of the mixer. The graph of Figure 3.8(a) shows this difference in behavior of the treated and untreated powder. The powder appears to be prevented from reaching its ultimate porosity by the silica powder, again an unexpected phenomenon if needed the silica was acting as an intergrain lubricant. The data for the vibratory collapse curves of Figure 3.8 and the original graph of Figure 3.4 are summarized in Table 3.2. In Figure 3.8(b) the sustainable strength of the vibrated powders containing silica are shown, and again it can be seen that the vibratory compaction of the powder makes the powder bed snap like a brittle solid.

A surprising feature of the collapse of the powder bed when the aluminum powder is treated with magnesium stearate is that the voidage of the vibrated stearated powder is always a lot higher than that of the silica-treated powder. The ultimate voidage at 10 000 taps is considerably more than that of either the untreated powder or the silica-treated powder. What appears to happen with the stearate-treated powder is that its wax structure seems to promote flow by agglomerating the fines in the powder to the larger grains. Thus the stearate treatment seems to dedust the powder and make the powder bed bulkier. This is further demonstrated by the fact that if one were to take a stearate-treated powder (several different chemical stearates can be used as flowagents (glidants)) and tumble the powder in a ballmill the powder will gradually form relatively large balls. Such behavior is often observed in **fillers** (material added to bulk out a plastic material or to increase its strength when handled in an industrial process) after they have been treated with stearates. This is sometimes called **spontaneous pelletization** or **spontaneous balling** in different industries. Again, when one is working with the pigment which is going to be dispersed in a paint in a high-shear environment, this spontaneous pelletization is not a penalty, and can be an advantage. However, if one treats a pharmaceutical powder with a stearate, the more the powder is allowed to tumble around the larger agglomerates formed by spontaneous pelletization, the more difficult it is for these grains to disperse when used in a medication. An extensive study of the role of flow agents in the food industry has been made by Peleg and colleagues [3, 4]. In Figure 3.9 some of the excellent electron micrographs that Peleg has published are shown. Peleg has also looked into the effect of shape of the individual grains on the flow of various food powders; important information when measuring such powders.

Although flowagents have been widely used and continue to be used in industry, one should always consider the fact that aeration of the storage–mixer system may be more advantageous than the use of flowagents. This is demonstrated by the data summarized in Tables 3.3–3.7. These data were reported by Kaye *et al.* using the equipment shown in Figure 3.1(b). They measured the angle of drain from the storage bin for different powders treated with silica flowagents or treated by aerating the bottom of the storage bin (fluidization). The

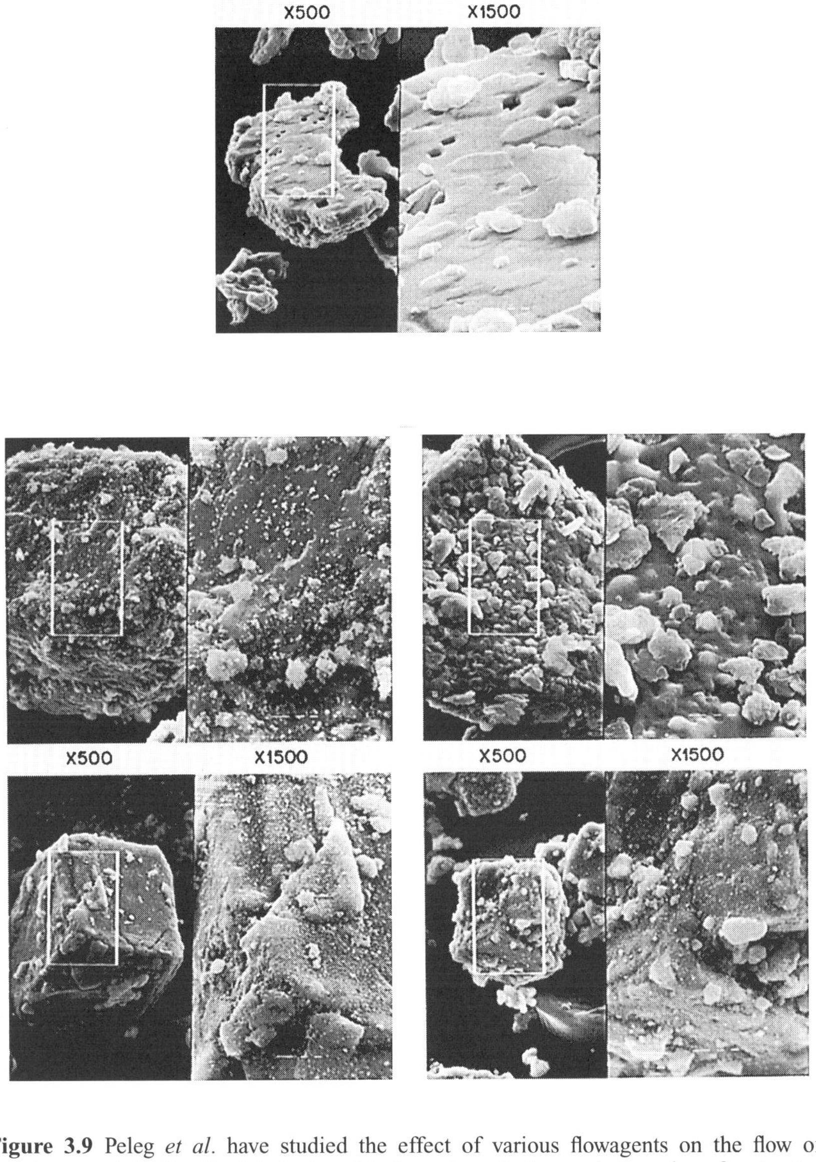

Figure 3.9 Peleg *et al.* have studied the effect of various flowagents on the flow of powders used in the food industry. (a) Scanning electron micrographs of untreated powdered salt. (Note that the fineparticle surface is active and attracts fines.) (b) Scanning electron micrographs of powdered salt conditioned with 2% sodium silico-aluminate (top left), calcium stearate (top right), silicon oxide (bottom left), and tricalcium phosphate bottom right). (Note the adherence of the conditioner to the salt fineparticle surface.) (Reprinted from *Powder Technolgy*, Vol. 35, No. 1, May/June, A.M. Hollenbach, M. Peleg and R. Futner, Interparticle surface affinity and the bulk properties of continued powders, 51–62, 1983, with kind permission from Elsevier Science SA, Lausanne, Switzerland.)

Table 3.3 Angle of drain α: measurement for spray-dried pharmaceutical powder with and without silica flowagent and with and without fluidization

% Cab-O-Sil® by weight	Number of taps	α after fluidization[a]
0.00	100	Free flowing $\alpha = 0$
0.00	200	Free flowing $\alpha = 0$
0.00	300	Free flowing $\alpha = 0$
0.00	400	Free flowing $\alpha = 0$
0.00	500	Free flowing $\alpha = 0$
0.00	1000	Some caking on sides
0.00	2500	Some caking on sides
0.32	300	Free flowing
0.32	400	Free flowing
0.32	500	Free flowing
0.32	1000	Free flowing
0.32	2500	Free flowing

[a] Powder did not flow without fluidization.

surprising feature of these data is the very high drain angles and the very low angles of repose that were obtained when such powders were treated with Cab-O-Sil®. (Such low angles of repose were also found using other flowagents; data not shown.) This shows how any studies of the effect of flowagents on the flowability of a powder must take into account the history of the system, and that

Table 3.4 Angle of drain α: measurements for crystalline pharmaceutical powder with and without silica flowagent and with and without fluidization

% Cab-O-Sil® by weight	Number of taps	α after fluidization[a]
0.00	100	Free flowing $\alpha = 0$
0.00	200	Free flowing $\alpha = 0$
0.00	300	Free flowing $\alpha = 0$
0.00	400	Free flowing $\alpha = 0$
0.00	500	Free flowing $\alpha = 0$
0.00	1000	Some caking on sides
0.00	2500	Some caking on sides
0.12	100	Free flowing
0.12	200	Free flowing
0.12	300	Free flowing
0.12	400	Free flowing
0.12	500	Free flowing
0.12	1000	Free flowing
0.12	2500	Free flowing
0.12	5000	Free flowing

[a] Powder did not flow without fluidization.

Table 3.5 Angle of drain α and poured angle of repose β (deg.): measurement for spray-dried pharmaceutical powder with various concentrations of flowagent and vibration

% Cab-O-Sil® by weight	No taps		100 taps		200 taps	
	α	β	α	β	α	β
0.10	48	29	–[a]	–[a]	–[a]	–[a]
0.19	48	31	–[a]	–[a]	–[a]	–[a]
0.32	50	31	78	29	82	30
0.41	50	30	78	29	–[a]	–[a]
0.52	47	31	73	29	–[a]	–[a]
0.64	48	37	79	31	86	32
0.79	43	32	66	36	75	32
1.83	48	31	5	9	78	30
3.23	60	27	78	28	80	30

[a] Powder did not flow.

studies of freshly poured heaps can generate misleading data concerning the ultimate behavior of the system in equipment being used to process the material. It should be noted too, that in all cases the aeration from the bottom of the bin did far more for the flowability of the powder than any of the flowagents used in the study.

Table 3.6 Angle of drain α and poured angle of repose β: measurement for crystalline pharmaceutical powder with various concentrations of flowagent and vibration

% Cab-O-Sil® by weight	No taps		100 taps	
	α	β	α	β
0.00	–[a]	–[a]	–[a]	–[a]
0.02	–[a]	–[a]	–[a]	–[a]
0.04	–[a]	–[a]	–[a]	–[a]
0.07	–[a]	–[a]	–[a]	–[a]
0.10	77	30	–[a]	–[a]
0.12	79	27	–[a]	–[a]
0.21	76	21	–[a]	–[a]
0.32	71	13	–[a]	–[a]
0.43	79	14	–[b]	–[b]
0.76	80	20	–[b]	–[b]
0.99	87	21	–[b]	–[b]
2.81	76	18	–[b]	–[b]

[a] Powder did not flow.
[b] The vibratory consolidation was leaving the powder bed considerably compacted and some difficulty was encountered in removing the powder from the apparatus.

Table 3.7 Results from investigations with various powders in the fluidized box of Figure 3.1(b)

	Did powder flow				
	after N taps?				
Powder	$N = 0$	100	200	500	after fluidization?
Amax iron powder	Yes	No	No	No	Yes in all cases
Canadian coal	No	No	No	No	Yes in all cases
Bronze filter powder	Yes	Yes	No	No	Yes in all cases
Spherical silica dust	Yes	No	No	No	Yes in all cases

3.4 CHARACTERIZING THE FLOW BEHAVIOR OF A POWDER BY STUDYING AVALANCHING BEHAVIOR

In the late 1980s a new subject entitled the study of critically self-organized systems developed. This subject is a branch of catastrophe theory [20–25]. **Critically self-organized systems** are nonlinear interactive systems which undergo an abrupt change in behavior under the influence of smoothly evolving multiple causes. Perhaps the most popular image of a critically self-organized system is the camel carrying a load of straw. As the straws are added one at a time, one suddenly reaches the point at which a single straw breaks the camel's back. This is a catastrophe both from the point of view of the camel and the mathematicians who study such systems! One of the systems studied extensively by scientists active in critically self-organized systems study is the behavior of what was known as **pseudostatic sand heaps**. In such studies the behavior of a sand heap, to which one grain of sand is being added at a time, is studied. Eventually one reaches a stage where one grain of sand precipitates an avalanche of the sand heap. The successive sand grains constitute the evolving forces and the avalanching is the catastrophic change in behavior. Pseudostatic sand heaps were studied originally as potential models for explaining the behavior of other catastrophic systems such as earthquakes [20].

Soon after publication of the studies of pseudostatic sand heaps, Kaye and co-workers extended the technique to study the behavior of dynamically evolving powder heaps, in which powder was poured onto the top of the heap. A study of the flow of powders in such heaps is described as **holistic rheology** to distinguish it from ideal rheology. In holistic rheology, one takes into account not only the flow properties of the powder, but the effect on the flow properties of the environment, such as any vibration and/or conditions such as humidity and electrostatic charging [25, 26]. In Figure 3.10(a) an early version of equipment used by Kaye and co-workers to study the avalanching behavior of powders is shown. Data obtained using this equipment is summarized in Figure 3.10(b).

The percentage of avalanches greater than or equal to a given weight is prescribed as a cumulative occurrence graph. The slope of the portion of the line describing the major percentage of the avalanching behavior is known as the fractal dimension in data space of the avalanching system, and can be used to summarize the rheological behavior of the powder. The system of Figure 3.10 has been applied to the study of the affect of flowagents on the holistic rheology of a powder system. Thus in Figure 3.11 the dramatic change in the avalanching behavior of a pulverized rock powder, caused by the addition of a silica flowagent, is shown. Currently studies are underway to assess the affect on the rate of mixing of two powders resulting from the addition of flowagents to the mixing systems. In these studies a possible link between the change in the fractal

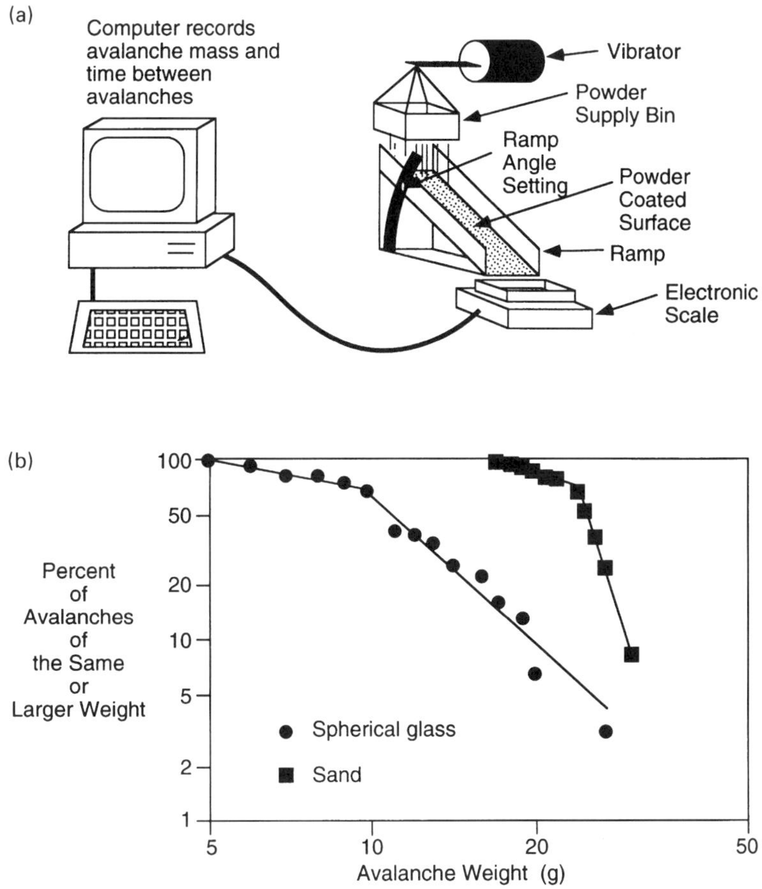

Figure 3.10 Equipment used to study the avalanching behavior of various powders and a graph of typical results. (a) The ramp and related equipment used to the study avalanching of powders. (b) Weight distribution of the avalanches for various powders.

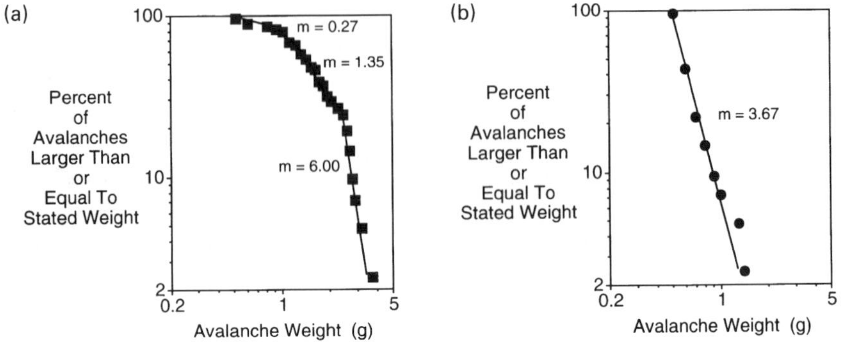

Figure 3.11 The avalanching behavior of a rock tailings powder, with and without flowagents, differs significantly. (a) Size distribution of avalanches for rock powder. (b) Size distribution with 0.5% silica flowagent by weight.

dimension in data space of the powder created by the addition of the flowagent to the increased rate of mixing is being explored.

In the study underway at Laurentian University we are also making use of avalanching data presented in a different format, using what is known as a **strange attractor plot** in data space [25, 27]. To understand the basic concepts underlying the use of strange attractors to summarize the observed behavior of critically self-organized systems consider the data shown in Figure 3.12. One of the systems which has been studied extensively in the study of critically self-organized systems is the behaviour of a dripping faucet (or tap). The system qualifies as a critically self-organized system because the drop grows gradually and changes shape under the interaction of the fluid supply, surface tension and gravity. The catastrophic behavior is manifest when the drop suddenly falls away from the faucet after growing smoothly. In Figure 3.12 the percentage of intervals between droplets greater than or equal to a stated time is presented in the classical form (an extensive discussion of why there are three regions to the data graphs is beyond the scope of this book, but the regions essentially delineate three different interactions of contributing factors affecting the stability of the drop–faucet system). The alternative way of presenting the data is to develop what is known as a **discrete time map** of the type shown in Figure 3.12(b). To generate this graph, T_{n+1} is plotted against T_n, these being adjacent members of the time series. As the system evolves, lines are drawn linking the successive datapoints on the graph. As can be seen from the graphs of Figure 3.12(b), these lines connecting successive datapoints in this discrete time map seem to cluster about a point in space. Early workers in the field of physically self-organized systems found this strange and in their imagination it seemed that the lines were being attracted to the center point. They therefore gave the name **strange attractor** to the pattern of lines in such a map. The name is somewhat misleading to students, since there is nothing strange about the behavior of the system, and there is certainly no physical attraction to the center point of the map. However, the term, although illogical, is now in common usage. In the case of the dripping faucet, if the drops were generated in such a way that there was exactly the same time interval between each drop, a discrete time map would be a single point located at what is the center of the datalines of the graph, around which they seem to cluster. Thus the centroid of the strange attractor is the time

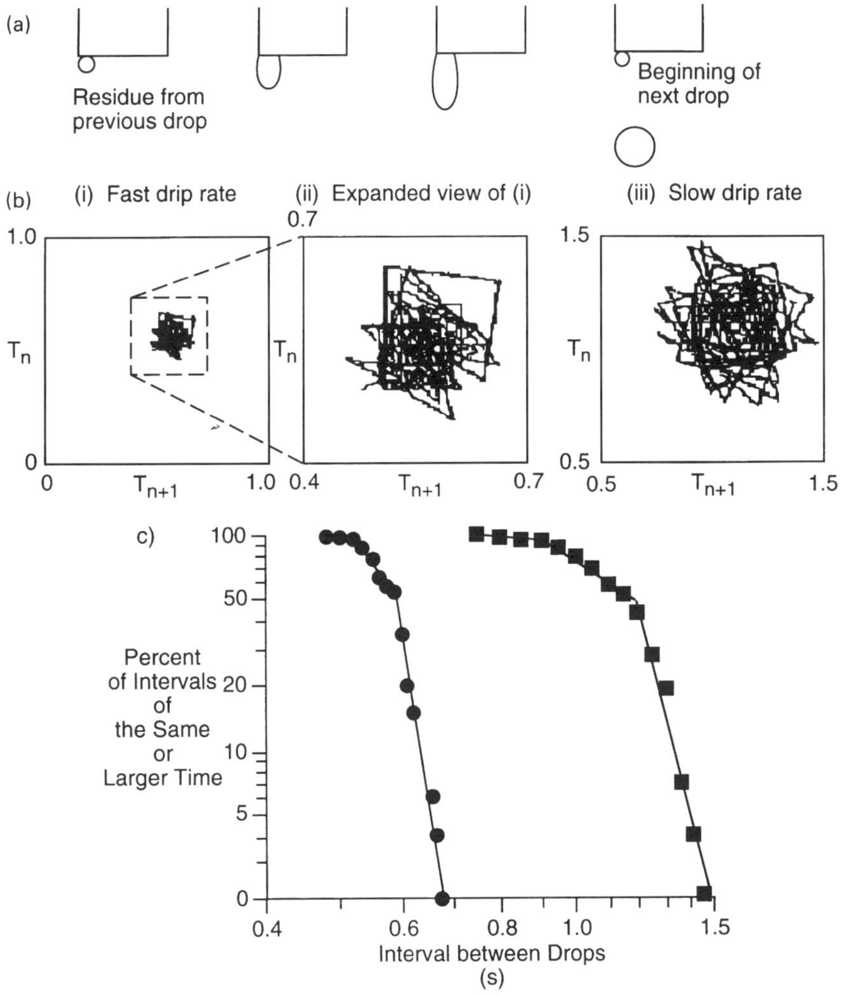

Figure 3.12 The stochastic variation in the time series generated by recording the behavior of a dripping faucet can be used to generate attractor diagrams and descriptive fractals in data space. (a) The progress of growth of a water drop in a faucet. (b) Attractor diagrams for a faucet dripping at two different rates. Note that the first and last graphs cover the same magnitude of time on their axes but that the center graph has been magnified to demonstrate the self-similarity between the two regimes. (c) Date of (b) presented as time distributions show that the dripping faucet is a scaling phenomenon.

period of the equivalent simple oscillator of a system such as the dripping of the faucet. The structure of the strange attractor is a measure of the variability of the behavior of the system. If one imagines all of the data points projected down onto the abscissa of the discrete time map, the centroid is equivalent to the average of the data and the scatter of the strange attractor is related to the standard deviation of the data. The advantage of the discrete time map is that it shows the scatter of the data, and drawing the lines between the points preserves the sequential data of the time intervals between the catastrophic events being studied. Figure 3.13 shows a set of data generated by studying the avalanching of the plastic powder down a ramp using the equipment shown in Figure 3.10. In this experiment the weight of the successive avalanches was recorded and the discrete data map represents the size of sequential avalanches. As can be seen from the data, there are a few large avalanches, but in the main the avalanches

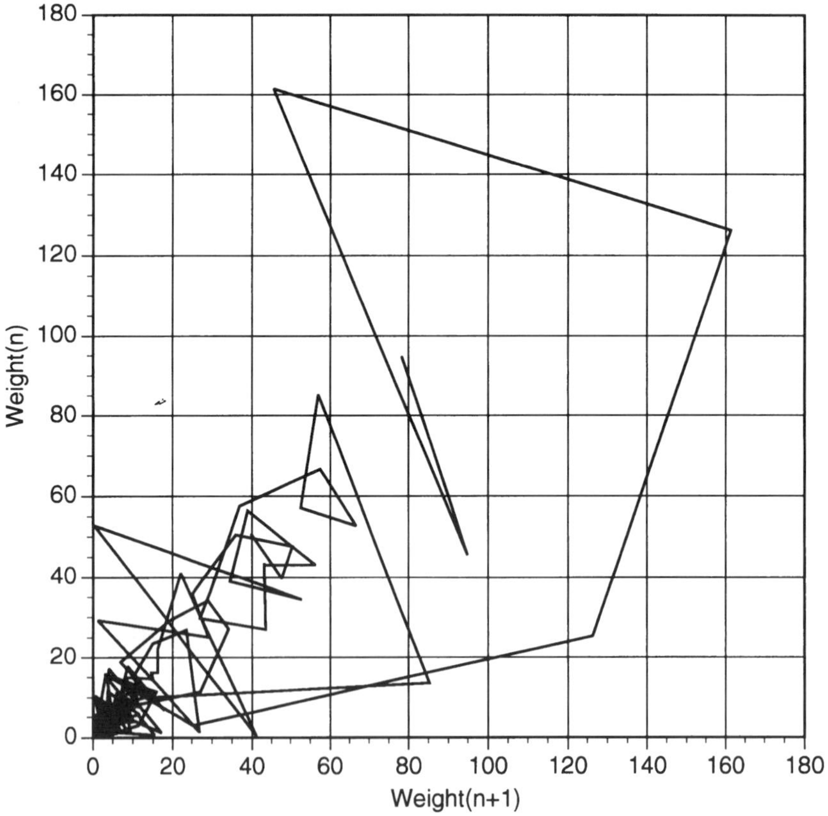

Figure 3.13 Discrete data map in two-dimensional phase space of the sequential weight of avalanches of a plastic powder, as studied using the equipment of Figure 3.10 [27]. (a) Attractor map of the data for a series of avalanches. (b) Size distribution of the powder used in (a).

cluster around the centroid of approximately 15 units. The effect of the flow-agents is demonstrated dramatically by the strange attractor diagrams, as illustrated by the data of Figure 3.14. In the sequence presented in part (a) of this diagram, the shift in this avalanching behavior caused by the addition of two different levels of silica flowagent. In holistic rheology, studies of the patterns of the strange attractor data have been described as **attractor fingerprints**. Although there is nothing fundamental about the patterns of data, because they

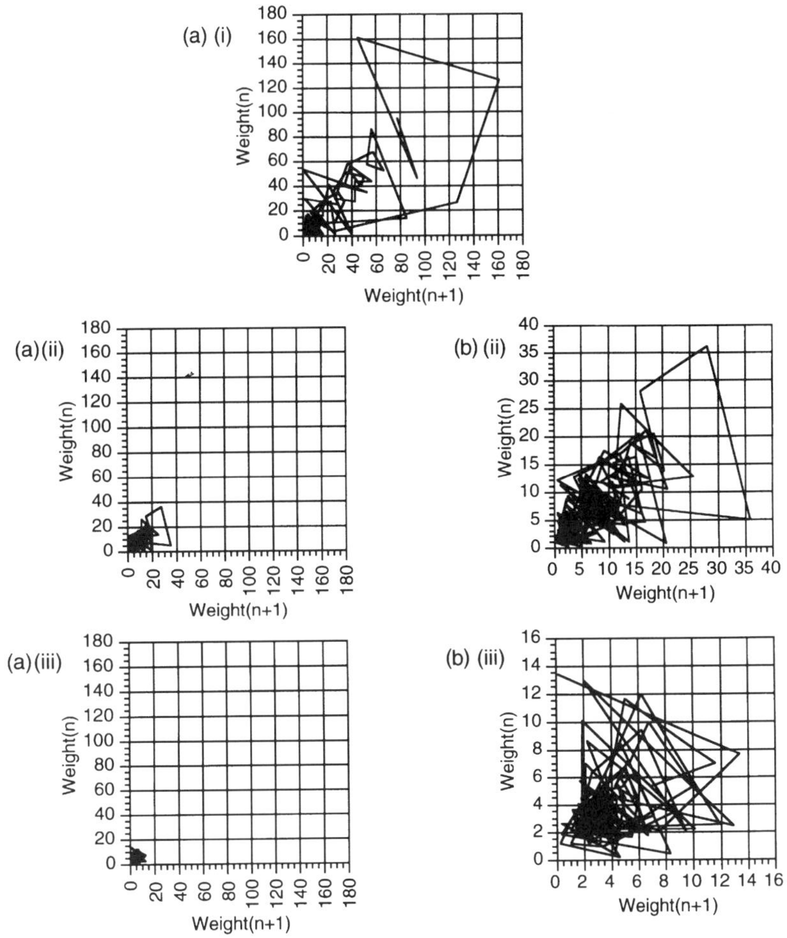

Figure 3.14 The changes in the avalanching behavior caused by the addition of flow-agents exhibit self-similarity [27]. (a) Changes in the structure of the strange attractor as flowagent is added to a powder are dramatic. (i) No flowagent. (ii) 0.1% by weight flowagent. (iii) 0.2% by weight flowagent. (b) Enlargement of the strange attractors for the powder with flowagent demonstrates that the changes in avalanching behavior are self-similar, confirming that we are studying a fractal system.

are dependent upon the extract configuration of the equipment, the feed rate and the powder etc., they appear to be a distinctive and useful record of behavior which helps to identify the significant parameters affecting the rheology of the system. An interesting feature of the avalanching data as generated in these studies is illustrated by the graphs of Figure 3.14(b). These strange attractors are parts (ii) and (iii) of the sequence in part (a) which have been enlarged so that the largest avalanche represents the unit 1 on the abscissa. It can be seen that, although the strange attractor shrinks and shifts in the sequence (a) when scaled for comparison in sequence (b) the graphs are obviously self-similar. This is an important property that identifies the system as being a fractal system from a behavioral point of view (see discussion of statistically self-similarity in the references given in notes 25 and 28).

In the general study of powder mixing, although it has been appreciated that the powder rheology is an important aspect of the system, the lack of relatively easy-to-operate, low-cost techniques for characterizing the flow properties of the powder have tended to deter workers from studying the rheology of the powders in their mixing process. Some of the developments of instruments for measuring the flow properties of a powder with relatively low-cost equipment, described in the foregoing paragraphs, can be expected to stimulate several studies on the effect of the flow properties of the powder on the mixing process and, in particular, to quantify the effect of flowagents on the mixing process [29].

NOTES

1. The first publication dealing with flow cells is Jenike, A. W. (1961) *Bulletin 108*, Utah Engineering Experimental station.
2. A good starting point for a study of the work carried out with the Jenike flow cell is Beddow, J. K. (1980) *Particulate Science and Technology*, Chemical Publishing Company, New York.
3. Peleg, M., Mannheim, C. H. and Passy, N. (1973) Flow properties of some food powders. *Journal of Food Science*, **38**, 959–964.
4. Peleg, M. and Mannheim, C. H. (1973) Effect of conditioners on the flow properties of powdered sucrose. *Powder Technology*, 45–50.
5. Kaye, B. H. (1993) Applied fractal geometry and the fineparticle specialist, part 1, rugged boundaries and rough surfaces. *Particle and Particle Systems Characterization*, **10**, 99–110.
6. ASTM D1895–69. (1980) Apparent density, bulk factor and pourability of plastic materials. Annual ASTM Standards Part 35, 605–610.
7. Matocha, S. R., Crooks, J. H., Zediak, C. S. and Tejehman, T. (1990–91) The flow funnel test determinators, in *Alumina Properties Characterized Light Metals*, The Minerals, Metal and Materials Society, pp. 179–85.
8. Staffa, K. H., John, J. and Clausseen, N. (1977) Flowability of powders under the influence of vibrations. *Powder Metallurgy International*, **9**, 20–23.
9. Taubmann, H. J. (1982) Beeinflussung der Fliesseigenschaften von Schüttgütern durch Beimischung von Fliesshilfsmitteln. (The influence of flow activators on the flow bahavior of powders.) *Aufbereitungs-Technik* (*Journal for Preparation and Processing*), **8** (Aug), 423–28. This paper has a column-by-column English translation alongside the German text.

10. Commercial details of this product available from Degussa Corp., 150 Springside Drive Suite, Akron, OH 44333, USA.
11. For a discussion of the structures of Cab-O-Sil® flowagents and an introduction to the commercial literature in flow agents produced by the Cabot Corporation see pages 97–98 of Kaye, B. H. (1993) *A Randomwalk Through Fractal Dimensions*, 2nd edn, VCH Publishers, Weinheim, Germany.
12. The fact that the use of silica flowagents increases the wear of tableting machines in the pharmaceutical industry is not, to my knowledge, documented in scientific publications, but I have been assured by several attendees at workshops on powder technology in the pharmaceutical knowledge that the wear effect of added silica flowagents is a well known fact amongst industrial scientists and technologists.
13. Nowick, A. S. and Mader, S. A. (1965) A hard sphere model to simulate alloy thin films, *IBM Journal,* Sep/Nov, 358–374.
14. Kaye, B. H. and Akhter, S. K. (1982) Low pressure rheology of dry powder systems in *Proceedings Powder Technology Conference*, May, Chicago, Illinois, International Powder Institute, 1350 East Touhy Avenue, PO Box 5060, Des Plaines, IL, USA.
15. Akhter, S. H. (1982) Fineparticle morphology and the rheology of suspensions in powders systems. M.Sc. Thesis, Laurentian University.
16. Kaye, B. H. (1977) Silica flowagents and the movements of powder, in *Proceedings PowTech Conference*, Chicago, Illinois, International Powder Institute, 1350 East Touhy Avenue, PO Box 5060, Des Plaines, IL, USA.
17. The Instron tester is marketed by Instron Corporation, 2500 Washington Street, Canton, MA 02021, USA.
18. Lee, Biing-Lin (1990) Thermo-mechanical test methods for polymeric powders. *Powder Technology,* **63**, 97–101.
19. Lee, Biing-Lin (1988) Low pressure rheology of granular powders using a drawing plate technique. *Polymer Engineering and Science*, Mid. April, **28**(7), 469–76.
20. Bak, P. and Chen, K. (1991) Self organized criticality. *Scientific American* (January), 46–53.
21. Bak, P., Tang, C. and Weisenfeld, K. (1988) Self organized criticality. *Physical Review,* **38**(1), 364–74.
22. Held, G. A., Solina, D. H., Keane, D. T., Haig, W. J., Horn, P. M. and Grinstein, G. (1990) Experimental study of critical mass fluctuations in an evolving sandpile. *Physical Review Letters*, **65**, 1120–23.
23. Nadis, S. (1990) Sandbox scholars. *Technology Review* (Feb–Mar) 21–22. This news story discusses and reviews Leo Kadanoff and Sydney Nagel's work on the avalanching behavior of sand.
24. Zimmer, C. (1991) Sandman. *Discover* (May), 58–59. A popular review of the work on sandpiles carried out by G. Held and colleagues.
25. Kaye, B. H. (1993) *Chaos and Complexity: Discovering the Surprising Patterns of Science and Technolgy*, VCH Publishers, Weinheim, Germany, Chapter 13.
26. Kaye, B. H. (1993) Fractal dimensions in data space; new dimensions for fineparticle systems. *Particle and Particle Systems Characterization*, **10**, 191–200.
27. Kaye, B. H., Gratton-Liimatainen, J. and Lloyd, P. J. The effect of flowagents on the rheology of a plastic powder. *Particle and Particle Systems Characterization*, **12**, 194–197.
28. Kaye, B. H. (1994) *A Randomwalk Through Fractal Dimensions*, 2nd edn, VCH Publishers, Weinheim, Germany.
29. Kaye, B. H., Gratton-Liimatainen, J. and Faddis, N. Studying the avalanching of a powder in a rotating disc. *Particle and Particle Systems Characterization*, **12**, 232–236.

4

Can ingredient modification expedite mixing strategies?

4.1 ALTERNATIVE INGREDIENT STRATEGIES FOR SOLVING POWDER MIXING PROBLEMS

Modification of the physical state of the powder may be all that is necessary to solve the powder mixing problem. When I moved from Great Britain to North America, I was introduced to a research development procedure which scientists in the United States called 'brainstorming'. In brainstorming a group of people are put into a room, provided with plenty of coffee and donuts, and then they are told to discuss freely whatever comes into their head concerning the problem in hand. Initially in a brainstorming session most people are guarded and circumspect in what they say, but then as the exchange of ideas escalates, they start to be much more willing to say things which they feel might be stupid but at least are worth discussing. Brainstorming tends to loosen up the intellect and be a source of many new ideas. When I conduct workshops on powder mixing I always try to have a brainstorming session near the end of the presentation to try and get people to discuss their problems and to look at their problems in a new way. One of the surprising things that I have found when brainstorming in powder mixing workshops is that, very often, it does not seem to occur to a busy technologist that a powder mixing problem can be solved by modifying the physical state of some of the ingredients to be assembled in a complex mixture. For example, in one discussion, a food mixing specialist was having difficulties trying to add a small amount of liquid flavoring to a food mixture. He was surprised when it was suggested that the problem could be avoided by preparing a premix of the flavor ingredient with one of the gelatin additives (used to turn the food mixture into a jelly), by encapsulating the flavor with the gelatin to create a free flowing ingredient with particle size similar to that of the majority of the other ingredients. In another situation, one of the powders being supplied to a process was difficult to add to the mixture because it did not flow freely. In discussions with the people who purchased the powder, it was discovered that the material was being generated in a device the operating parameters of which could be adjusted to produce a powder with far less fine material in the supply of powder. The new more granular powder, produced with the new production parameters on the machine, flowed easily. In fact the subsequent mixture was much easier to handle, since there was no fugitive dust from the mixture as there had been in the past. Obviously one cannot reveal details of these particular commercial processes, but these two examples illustrate how the technologist faced with the flow sheet of a powder system should always discuss, before proceeding to the actual mixing stage, whether modification of any or all of the ingredients to be mixed can expedite the mixing strategies to be adopted in a given process.

Sometimes modifying the powder before proceeding to create a mixture can improve the product, or one can sometimes make virtue out of necessity. For example, one food manufacturer was making a product to which he was having to add an artificial sweetener. This artificial sweetener had only low solubility in water. Therefore to make it soluble within a relatively short time, the sweetener had to be a very fine powder. This caused problems when the powder was used in that the final product caused annoyance to customers who disliked the sensation of inhaling a fugitive sweetener aerosol when they opened the product before placing it in a container. The problem was solved by dissolving the sweetener in water and then encapsulating the solution in gelatin to produce a free flowing powder which was much easier to mix into the other ingredients to produce the final product. In such a situation one could make virtue out of necessity by claiming in one's advertizing that

> New use of magic microcapsules enhances the flavor.

Sometimes microencapsulation of minor ingredients not only makes it easier to produce a powder mixture but can enhance shelf life and quality of the product. Thus the widespread use of microcapsules, sometimes called flavor buds, in gelatin desserts, not only makes it easier to produce the product but the flavor trapped inside the capsule has a much longer shelf life [1]. Again, if aspirin is microencapsulated before being added to a drug powder being used to produce medication, its shelf life is increased, and also one can hide the flavor of the drug by delaying the dissolution of the medication part of the tablet until it is in the desired part of the body. In this chapter we will look at various strategies which can be used to modify a powder ingredient to make it more usable. It is difficult to give general guidelines, and the intent of this chapter is not to solve everybody's problems, but to stimulate ideas into different ways of modifying the powder prior to its use in a mixing process.

4.2 MODIFYING THE SIZE DISTRIBUTION OF THE POWDER INGREDIENTS

In general, as we discussed briefly in Chapter 3, the fewer the fines in a powder, the easier it is to control the flow and to meter the amounts of the powder entering a mixer. Also in general, the narrower the size range of the powder, the easier it is to mix it with another ingredient. For free flowing systems in which one can use a chaotically assembled mixture, it is better to have all ingredients approximately the same size. This fact is well known in the food industry where they try to avoid a range of sizes in the final product. For example, dog biscuits, or dry dog food, although not fine powders, when one mixes them by the ton, they behave as if they were powders. If you look at dry dog food, you will notice that the ingredients are all approximately of the same size.

Another example is the breakfast cereal described as Muslix® [2]. This particular type of product is prone to segregation if it is left to stand in a gently vibrating working environment or domestic environment [3, 4]. It is interesting to note that this particular product is often sold in rather small packages compared to those typically used for breakfast cereals. The manufacturers have tried to make all ingredients approximately the same size. This strategy is not always available to the manufacturer of powder products, since the customer may not accept all ingredients having the same size. Consider the ingredients in a dried soup mix; the customer likes to be able to distinguish familiar anticipated ingredients, and does not like the soup to contain carrot powder which is so small that one cannot see that it is carrot. I was once asked to consider this problem by a company who had reduced their onion flakes in a soup to a very small size, only to have the customers confuse the onion flakes with the appearance of a widely used soap powder, and they complained about that particular ingredient without tasting it.

In machines that dispense soup, the manufacturers have no choice but to make the powdered ingredients all the same size to avoid segregation. This fine pulverization of all ingredients often leads to an anemic look to the soup when the powder is mixed with hot water and dispensed by the machine. The segregation of the ingredients in powdered soup is one of the reasons why commercial soup powders are usually individually packaged rather than sold in relatively large canisters from which one can make several servings of a soup. The same problem of size differential of ingredients occurs in soft drink mixes. Again, the vendors sell relatively small packages of the ingredients to avoid segregation which would occur in a large canister of material, which is unavoidably subjected to environmental vibration in storage places such as a domestic pantry. One particular manufacturer of soft drinks developed a package in the early 1990s which indicated that the package contained enough material for a relatively large quantity of drinks. My first assumption was that the manufacturer must have overcome the problems of segregation, only to find when I opened up the canister that it contained five subcanisters, each with a small amount of the soft drink mixture! In general, segregation problems from environmental vibration in an assembled mixture are minimized if the ingredient size range is not greater than three to one. They become severe, in the absence of electrostatic forces, if the size range is greater than nine to one.

If one is producing an assembled mixture, as in the ceramics industry, then one has little choice as to the size ratio of the different ingredients, and if one is going to exploit triboelectric mixing one usually needs the ingredient that will coat the other to be much smaller than the primary powder. In this section we will briefly discuss two or three ways of modifying the size range of a powder to obtain better flow and hopefully better mixing strategies. Several of the ideas presented here are from actual case studies, but obviously the examples have been changed a little so as not to identify the actual situation in which the ideas were discussed.

In one instance, a company was having trouble with a particular additive, which was a fine powder coating another. They were also having trouble with a small percentage of large fineparticles present in the feed powder. It so happened that the powder was being produced by means of a **Trost mill** [5]. The operating principles of this device are illustrated in Figure 4.1. The system is described as an **autogenous** grinding system. This word means **self-grinding**, but the origin of the term is pronounced 'Au-to-gen-ous'. Autogenous grinding or pulverization is any system in which a set of fineparticles are reduced in size without using such items as balls in a ballmill or hammers in a hammermill. The coarse-powdered material to be pulverized to a much smaller size is placed in a bin just above a Venturi throat feeding the mill element itself. A **Venturi throat** is a constriction of a pipe in which there is flowing liquid or gas. When such a constriction to the flow occurs, the internal pressure of the liquid falls (a fact known as **Bernoulli's principle**). Because of this lower pressure in the Venturi throat the coarse powder is sucked into the moving airstream. An opposing airstream is fed in at the other side of the mill. For the initial few minutes of operation the coarse granular material is moved by this opposing air flow up into a circular chamber which is actually a **centrifugal air classifier** for fractionating powder. The flow of air through the entire device is such that the larger granules being fed into the machine are not able to spiral into the central exit of this air classifier supported by the velocity of the air (or gas). Therefore they trundle around the outer rim of the classifier and fall back into the opposing airstream. Once these larger granules fall into the opposing airstream they start to collide with the primary stream in the collision zone. After being fragmented or deagglomerated in this zone, the smaller fineparticles are swept up into the classifying chamber and out of the chamber by the spiraling air stream moving into the exit at the center of the chamber. This equipment can be used to deagglomerate powders with the fines moving directly into a fluidized bed system or other mixing device. One of the main advantages of this type of system is that the deagglomerating or size reduction taking place is virtually contamination free, whereas ballmilling with steel balls etc. can contaminate the powder. It can also be operated with inert gas such as nitrogen or carbon dioxide. In some situations, the nitrogen and the carbon dioxide can come from cryogenic materials and can be used to cool the material before grinding it, a process which makes it brittle and easier to reduce in size. In the particular instance of powder mixing that was being considered, the group producing the powder did not realize that the group mixing the powders was having problems. The situation was quickly fixed by simply increasing the flow of gas through the trost mill. The increase in the air speed resulted in a smaller size of material being centrifuged out to the wall of the classifier and being trundled back into the feed. Thus the larger grains that were disturbing the mixing process were eliminated from the powder supply.

In another situation where a company was having difficulty mixing and stabilizing several ingredients to produce a mixture without fines which then flowed freely into some packaging machinery, I suggested the possibility of using the system shown in Figure 4.2. It operated on the same principles as the

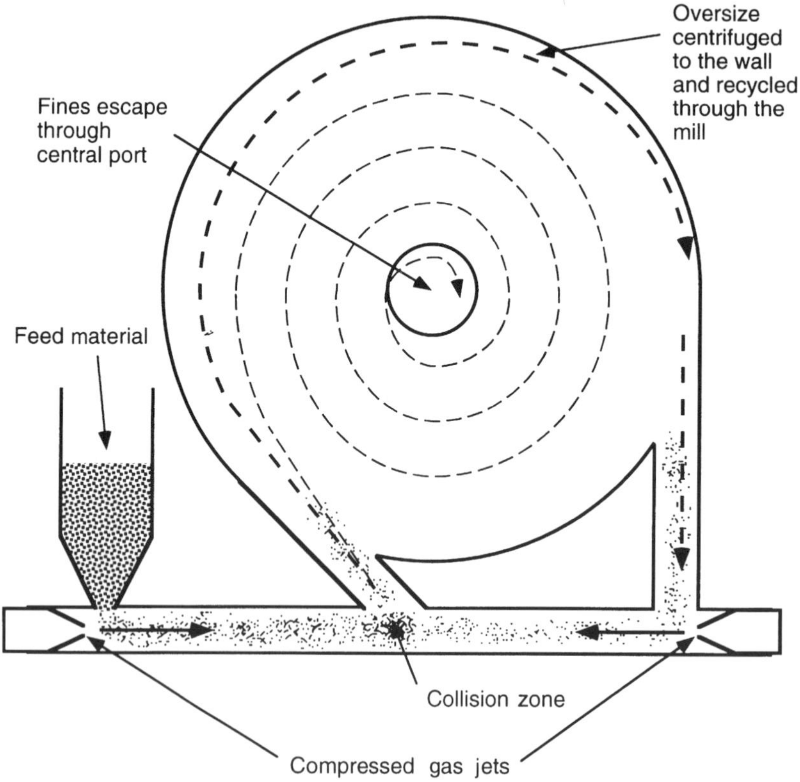

Figure 4.1 Fines can be reduced below a specified size by adjusting the flow. Schematic representation of the operation of a Trost mill.

trost mill in which the ingredients were fed into a collision using Venturi throats, as illustrated. The collisions in a mixing zone would deagglomerate the ingredients and then as they were swept out of this zone a mist could be sprayed into the mixture to produce granules which then move through a cyclone. The cyclone would throw the larger fineparticles to the wall and out to a delivery system, whereas the fines would be recirculated back into the mixing system. One could even provide corona discharge points to produce either neutralizing charge or charge to enhance triboelectric coating as the requirements demand. Such an enclosed system could also operate with nitrogen or other inert gas.

In another situation, a company was using a spray-dried product as an ingredient in an overall mixture. They were having problems with the spray-dried product because it had a wide range of sizes. Investigation established that the spray drying was taking place with the spray being generated at the top of a

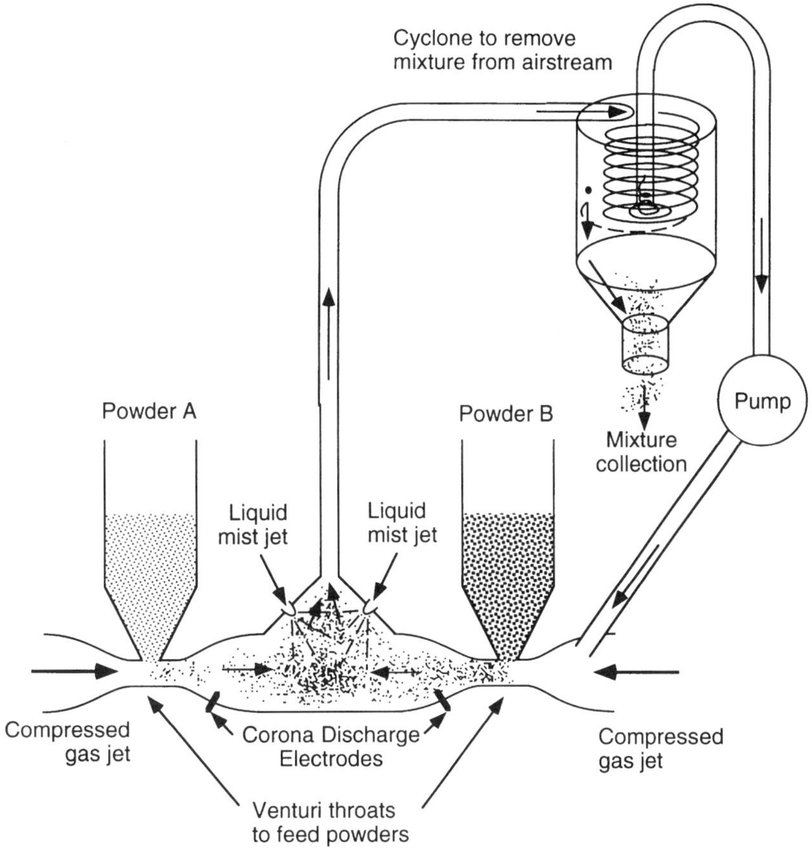

Figure 4.2 Suggested design of a turbozone mixer which could operate on principles similar to that of a Trost mill. By permission of Garlock Inc., Plastomer Products, 23 Friends Lane, Newton, PA 18940, USA.

tower by the breakup of a jet of slurry. The slurry droplets were being dried as they fell down the tower through rising hot air. If one takes an ordinary jet and breaks it up by directing it on a target plate or some other method, one always obtains a relatively wide range of sizes in the drops being produced by the breakup of the jet. One can make a much more monosized set of fineparticles in a spray drying system by using a spinning disk nebulizer rather than a random jet breakup. In this device the slurry to be turned into a spray flows onto a spinning disk and the droplets are produced by the centrifugal forces at the edge of the spinning disk. In this situation the mixing problem disappeared when the spray drying process was switched to a spinning disk spray generator [6].

The dramatic way in which the rheology of powder system, and hence the progress of a mixing procedure, can be changed by the removal of the fines, is demonstrated by the data summarized in Figure 4.3. A plastic powder was being mixed with a pigmentation powder and the fines in the plastic powder were

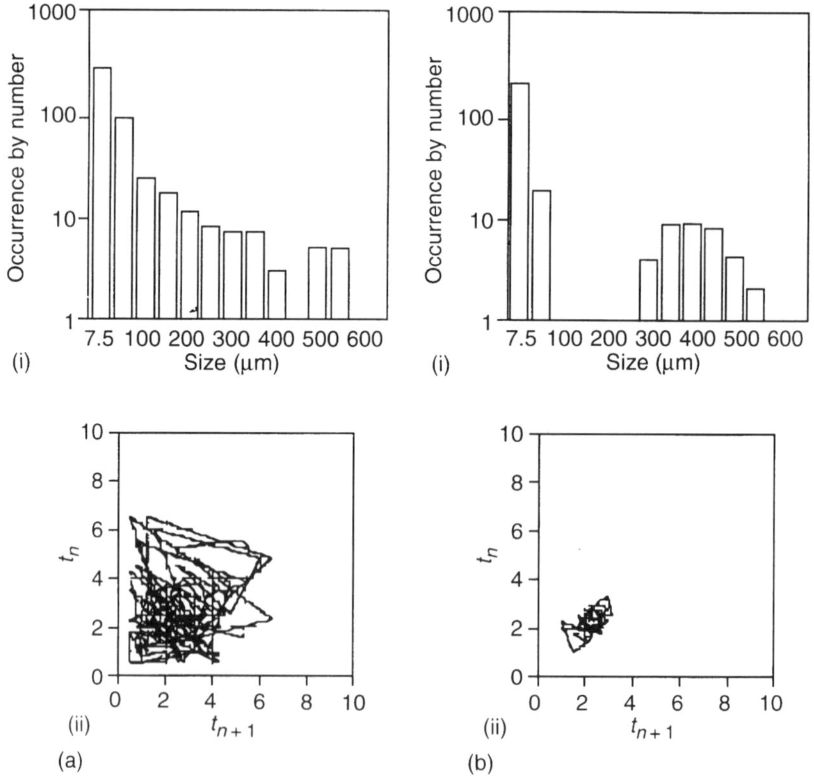

Figure 4.3 The removal of fines from a powder can drastically alter the flow behavior [7]. (a) (i) Size distribution of the original powder; (ii) strange attractor summary of the avalanching of the original powder. (b) (i) Size distribution of the sieved powder; (ii) strange attractor summary of the avalanching of the sieved powder.

causing problems. To solve this problem the powder was sieved on a 48 mesh Tyler sieve (nominal mesh size 297 μm). In Figure 4.3 the size distribution of the plastic powder before and after sieving is shown. It should be noted that these graphs were generated by image analysis and that the ordinate represents the number of fineparticles in a size group. In the sieved powder there appears to be some fine material which is relatively large by number, but which on a weight basis would be insignificant. Notice too that, on a number basis, the size distribution of the sieve fractionated powder is Gaussian. The fines in this sieved fraction fail to pass through the surface because they cling by electrostatic forces to the the larger grains of powder as the material is sieved. In Chapter 3 we discussed how the avalanching behavior of a powder could be studied using a ramp onto which the powder was fed, and we also discussed how the avalanching

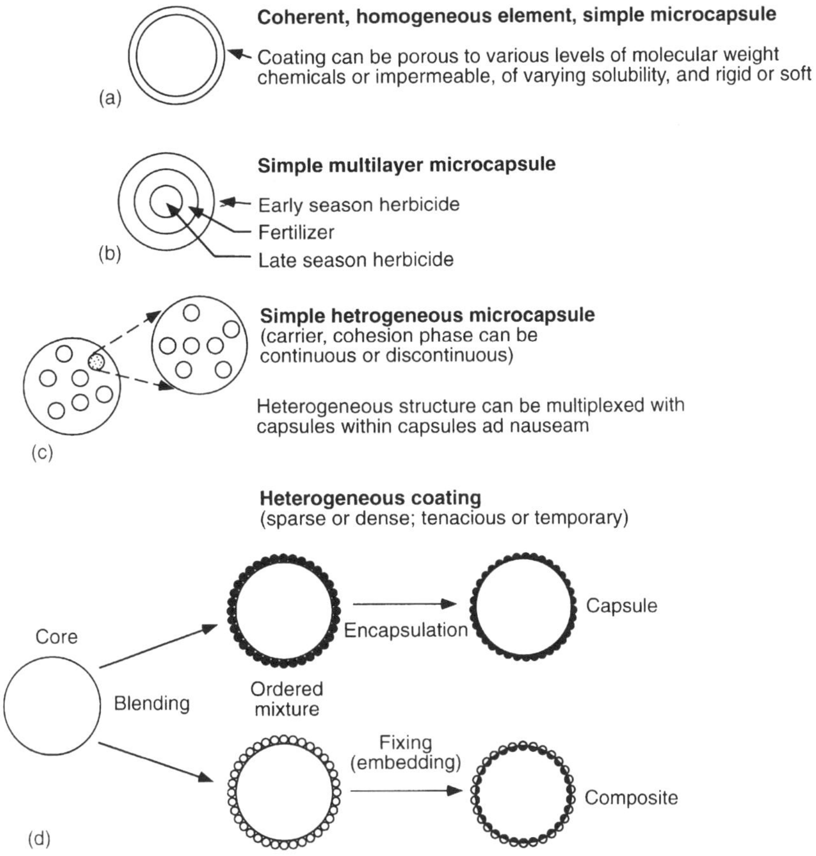

Figure 4.4 The structure of a microcapsule can have various levels of complexity. (From Kaye [1], reproduced by permission of the Council of Powder Technology and the Hosokawa Powder Technology Foundation, Osaka, Japan. © 1995 KONA. All rights reserved.)

data could be summarized using strage attractor diagrams. A piece of equipment has been developed in which the avalanching of the powder is studied in a rotating disk rather than down an inclined ramp. In this equipment a partially filled disk has powder building up as the disk is rotated until it cascades down in an avalanche; with subsequent rotation of the disk building up the powder until another avalanche occurs. The strange attractors summarizing the avalanching behavior of the two powders, with and without fines, are shown in Figure 4.3. These illustrate the dramatic change in flow behavior caused by the removal of fines. The fines removed from the powder were granulated by the addition of a small amount of liquid and used in another process. A discussion of granulation techniques for upgrading the sizes of smaller grains in a powder is deferred until Chapter 9.

4.3 MICROENCAPSULATION OF INGREDIENTS

In Figure 4.4 some of the vocabulary now used in the description of microcapsules is illustrated. The simplest form of microcapsule is known as a coherent, homogeneous element, simple microcapsule in short form, a simple microcapsule shown in Figure 4.4(a). This is the type of microcapsule used to encapsulate a liquid with a thin gelatin wall so that the liquid represents 99% of the volume of the microcapsule. The basic simple microcapsule can be increased in complexity by adding more coherent layers to the original simple microcapsule.

In a heterogeneous simple microencapsulation, such as that shown in Figure 4.4(c), a microcapsule containing dispersed smaller droplets or fineparticles can be prepared at any level of complexity of dispersed material within the microcapsule. Obviously one can make composite microcapsules which have a heterogeneously structured core with a coherent coating about the outside.

4.4 TECHNOLOGIES FOR PRODUCING MICROCAPSULES

In one review of the technology used to create microcapsules it is estimated that there are up to 25 different ways described in the scientific literature for making microcapsules [8]. In this review we will concentrate on a few of the major methods which should acquaint the reader with the basic concepts of the technology and introduce the methodologies available.

One of the most widely used methods of microencapsulation was developed by Wurster at the University of Wisconsin in Madison [9]. The patents for this process are now vested in a corporation known as The Coating Place Inc. [10].

The Wurster technique for creating microcapsules is essentially a spouted fluidized bed system in which the moving fineparticles are coated with the appropriate material. The basic equipment used in the process is illustrated in Figure 4.5. The powder grains to be coated are air fluidized. As they move up through the center of the fluidized bed they are sprayed with the coating material. At the top of the central cylinder the coated fineparticles fall back down the outside to recirculate for a second coating to be applied. One of the advantages of this procedure is that coatings of any given thickness can be built up on the powder grains by repeated circulation of the material. If one attempts to use this technique with very fine powders, the most difficult part of the process is to fluidize the powder grains. The circulating action of the fluidized

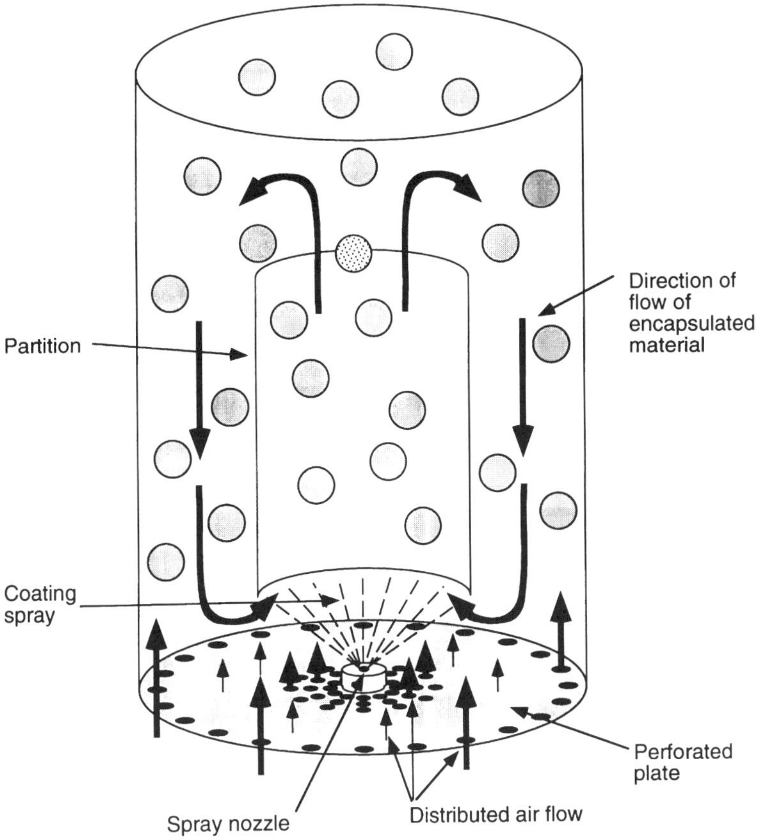

Figure 4.5 The Wurster technique for creating microcapsules: the powder to be coated is fluidized in a circulating spouted bed and the coating is sprayed onto the suspended powder grains. (From Kaye [1], reproduced by permission of the Council of Powder Technology and the Hosokawa Powder Technology Foundation, Osaka, Japan. © 1995 KONA. All rights reserved.)

bed also results in each increment of the coating being dried in the downward portion of the circulation trajectory. One has to be cautious with some materials to avoid explosions, since a dry circulating system can build up an electrostatic charge, and there is a danger of explosion in some situations. Other manufacturers use similar air circulation systems in fluidized bed equipment. Thus the two systems of Figure 4.6(a) and (b) are described in the trade literature of the Glatt corporation [11].

The basic system developed at the South West Research Institute for creating microcapsules is shown in Figure 4.7 [12]. The core material is fed to a spinning disk at the center of the equipment. Core droplets, or grains of the powder to be encapsulated, are thrown off the disk towards an outer cylinder containing many small holes around its periphery. The coating material is fed to this cylinder in liquid form so that it coats the holes in the cylinder with a thin film of the

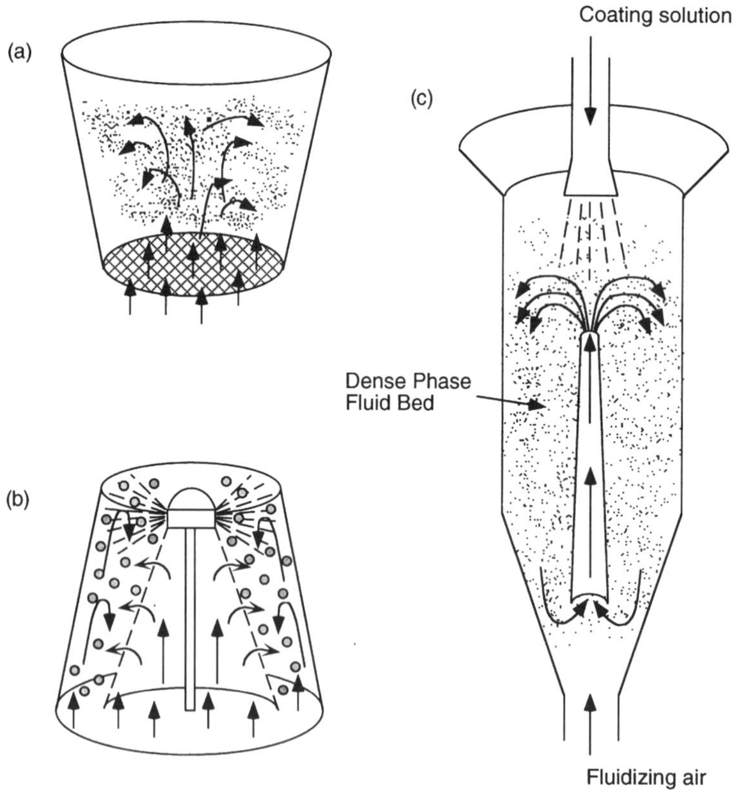

Figure 4.6 Variations in the configuration of air-fluidized spray coating systems have been implemented by various manufacturers. (From Kaye [1], reproduced by permission of the Council of Powder Technology and the Hosokawa Powder Technology Foundation, Osaka, Japan. © 1995 KONA. All rights reserved.)

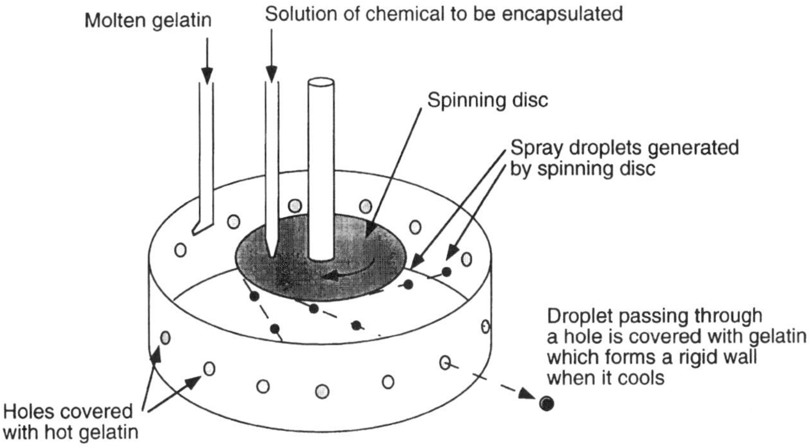

Figure 4.7 In the SWRI technique for creating microcapsules, a spinning disc is used to fling droplets or fineparticles through a film of coating material. (From Kaye [1], reproduced by permission of the Council of Powder Technology and the Hosokawa Powder Technology Foundation, Osaka, Japan. © 1995 KONA. All rights reserved.)

coating material. The core droplets or fineparticles pass through this film, which is constantly renewed, and are coated as they fall out into the outer container where they can be hardened by physical or chemical methods.

Several techniques have been developed for mixing core and coating fineparticles which have been given opposite electrostatic charges to attract them to each other. In this section we will describe the basic system developed by Leiberman and co-workers at the Illinois Institute of Technology Research Institute (IITRI), Chicago [13], illustrated in Figure 4.8. In this technique the core material can be either solid or liquid. The usual choice of coating is a liquid for ease in subsequent handling of encapsulated fineparticles. Typical coating materials used by Leiberman and co-workers include a wax that solidifies on cooling, a dissolved polymer resin that forms a skin upon the evaporation of a solvent, a polymeric skin formed by interfacial action between a component in the core material and a component in the coating material, and liquid coatings that solidify upon exposure to a suitable gas phase. The two components to be turned into a microencapsulated system are fed into a reaction chamber which can be heated. The aerosol fineparticles are given an ionic charge of the appropriate sign using subcorona discharge systems. To achieve sufficient encapsulation, the system must be designed to achieve a high rate of collision between the two types of aerosol in a turbulent supportive air or gas system. If there is any danger of explosion, appropriate supportive gas systems such as nitrogen must be used. In this process, one must choose the coating substance so that it will wet, i.e. spread out on, the core material and cover it completely. If the core is also liquid the coating must have a lower surface tension otherwise the desired encapsulation will not be achieved. Figure 4.9 an encapsulated sodium chloride crystal system coated by this procedure is shown.

A process known as **coacervation** was developed by colloid chemists Kruyt and colleagues in the 1930s and remained a technology without an application until it was further developed by scientists at the National Cash Register company to create microcapsules for use in carbonless copy paper [9, 12]. The basic technique is illustrated in Figure 4.10. The core material to be encapsulated is placed in an immiscible liquid to form liquid droplets. The coating material is also suspended in the liquid medium or actually dissolved in this support material. To induce the process known as coacervation the temperature, pH or other environmental conditions are changed in such a way that the wall material

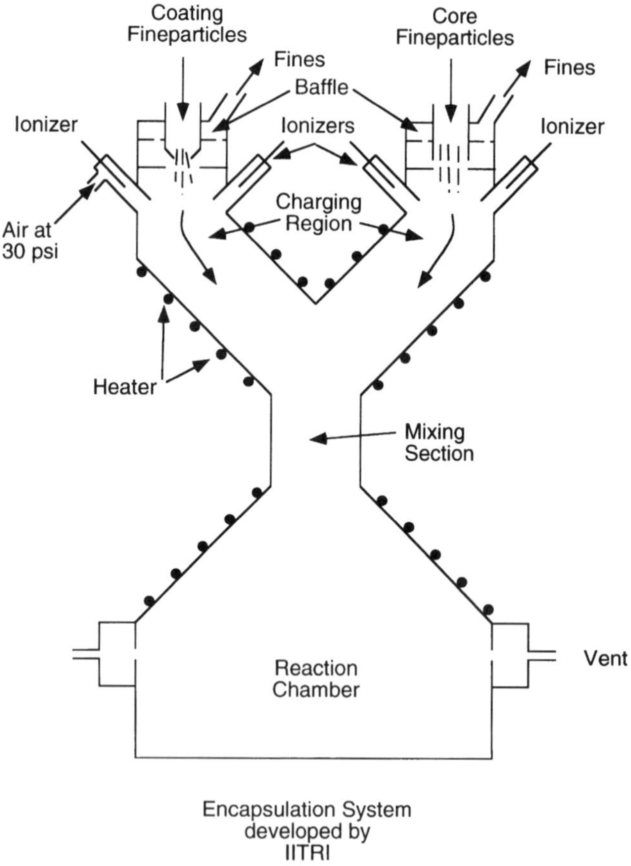

Figure 4.8 In electrostatic, aerosol based, microencapsulation technologies the ingredients are attracted to each other by electrostatic forces in a reacting chamber. (From Kaye [1], reproduced by permission of the Council of Powder Technology and the Hosokawa Powder Technology Foundation, Osaka, Japan. © 1995 KONA. All rights reserved.)

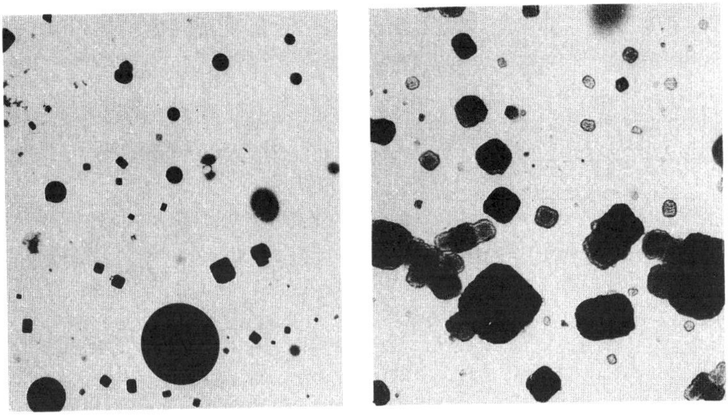

Figure 4.9 Electron micrographs of sodium chloride crystals, magnified 30 000 times, (a) without and (b) with carboxymethylcellulose encapsulation achieved using electrostatically charged aerosol system. (From Kaye [1], reproduced by permission of the Council of Powder Technology and the Hosokawa Powder Technology Foundation, Osaka, Japan. © 1995 KONA. All rights reserved.)

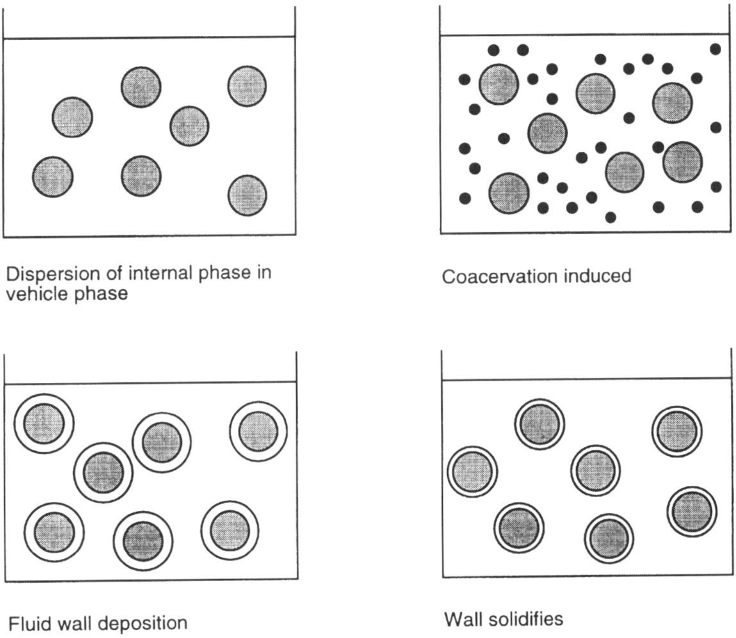

Figure 4.10 The coacervation process starts with an emulsion of two immiscible liquids. A wall is formed around the droplets to be encapsulated by changing the physical and chemical properties of the supporting liquid. (From Kaye [1], reproduced by permission of the Council of Powder Technology and the Hosokawa Powder Technology Foundation, Osaka, Japan. © 1995 KONA. All rights reserved.)

comes out of the solution and aggregates around a core droplet to form continuous encapsulating walls. Then in the final stage of the process these capsules are hardened.

An alternative technique for creating microcapsules out of the suspension of coacervated droplets is to spray dry the resulting slurry to create heterogeneously encapsulated composite microcapsules. A third avenue for creating microcapsules with coacervated suspensions is to add another ingredient that gels the support liquid to create a solid form which can then be crumbled to create composite microcapsules. Another way of forming the walls around the droplets suspended in the supporting liquid is to use a monomer dissolved in the support liquid, which can then be directly polymerized onto the droplet suspended in the liquid. The volume of the core material in the finished microcapsule can be varied anywhere from 20% to 99% of the volume of the microcapsule. By controlling the wall formation process, the wall can either be made solid or porous to any required level.

NOTES

1. Kaye, B.H. (1992) Microencapsulation: The creation of synthetic fineparticles with specified properties. *KONA,* **10**, 65–82.
2. Barker, G. and Grimson, M. (1990). The physics of museli. *New Scientist* (26 May) 37–40.
3. Barker, G.C. (1994) Computer simulations of granular materials in *Granular Matter, an Interdisciplinary Approach* (ed. A. Mehta), Springer-Verlag, New York, pp. 35–83. In this communication, segregation in a powder mixture is simulated on a computer.
4. Barker, G.C. and Mehta, A. (1993) Size segregation powders. *Nature*, **361** (28 Jan.), 308.
5. Trost mill manufactured by Garlock Inc., Plastome Products, 23 Friends Lane, Newton PA, USA, 18940–9990.
6. Beddow, J.K. (1980) *Particulate Science and Technology*, Chemical Publishing Co., Inc., New York.
7. Bak, P., Tang, C. and Weisenfeld, K. (1988) Self-organized criticality. *Physical Review* A, **38**(1), 364–74.
8. Anon (1984) New life for microcapsules. *Chemical Engineering* (1 Oct 1984), 22–25
9. Watson, A. C. (1970) Microencapsulation. *Science Journal* (Feb.), 57–60.
10. The Coating Place Inc., Box 248 Verona, WI 53593, USA, Application literature describing many uses of microencapsulation products are available from this corporation.
11. Glatt Air Techniques Inc., 520 Livingston Street, Norwood, NJ 07648, USA.
12. Mattson, H.W. (1965) Miniature capsules. *International Science and Technology* (April), 66–76.
13. The electrostatic encapsulation technique was developed by the fineparticle section of the IIT Research Institute, 10 West 35th Street, Chicago, IL, USA.

5

Monitoring mixers and mixtures

5.1 DISTINGUISHING BETWEEN CHAOS CREATING OPERATIONS AND DISPERSION MECHANISMS

It is the philosophy of this book that, when looking at the performance of powder mixing systems, it is necessary to distinguish between the powder of the mixer to create chaotic conditions inside the mixer and the ability of the mixing mechanisms to disperse aggregates of the constituent ingredients within the powder mixture. It is useful to distinguish the two aspects of mixing as (a) dispensing the ingredients and (b) increasing the intimacy of the mixer. For example, when mixing cocoa and flour, a random mix of the ingredients may be readily achieved by mixing in a device such as a Y-mixer, but to create an intimate mix one may have to dry grind the randomized mixture (see discussion later in this chapter of the optical properties of a mixer). All too often, technologists demand too much of their mixers. The mixer which is very efficient at distributing the ingredients chaotically to form a chaotic mixture often contains no working mechanisms to break down aggregates within the overall mixer structure. In many situations, it may be more efficient to create a chaotic mixture of the aggregated constituents, and then pass that first-stage mixture through a device with high shear zones to create the dispersion of the aggregates in the overall mix. One therefore needs to monitor the performance of a system with regard to these two mechanisms. In the past, the two have usually been monitored together and the mixer pronounced good or bad solely on the quality of the resulting mixture. It is my opinion that, very often, the 'fail' grade is awarded to a mixer when the mixer is working extremely efficiently to create chaotic conditions, but the intimate mixture is not being created because of the lack of shear zone dispersion. Scientists have attempted to explore the dynamics of mixers by using free flowing sand of well-defined sizes with one of those sizes being given a different color (see, for example, the discussion of the work of van den Bergh on the performance of the Nauta mixer in section 5.3).

What their investigations have been exploring is the chaos producing system of the mixer, and when one then pronounces the mixer at satisfactory from that perspective, one receives very little guidance as to how a wide-size-range powder or a cohesive powder will disperse within that same powder mixer. In this overview of techniques for monitoring the performance of mixers and exploring the structure of mixtures we will attempt to delineate exactly what was achieved in any particular study. In the next section, a new technique for monitoring the chaotic producing behavior of a mixer called 'Poisson tracking' will be described.

5.2 POISSON TRACKING AS A TECHNIQUE FOR STUDYING CHAOTIC CONDITIONS IN A POWDER MIXER

In this book, we distinguish between tracers and trackers. A **tracer** is an ingredient that is meant to trace out movements within a powder mixing system in the same way that tracer bullets make the missiles of a machine visible to the gunner. One uses tracers to find out directions and pathways within a powder mixer. When using **trackers**, we have no concern at all as to how a fineparticle designated as a tracker gets to a certain location, we just monitor how frequently trackers occur in a sample from a specified location. We check if the population fluctuation of the tracker fineparticles in the sample are consistent with the existence and persistence of chaotic conditions in the powder mixer. Ideally, tracker fineparticles should be of the same density as the fineparticles being mixed in the system. They can be distinguished from the other fineparticles either by color or by some internal ingredient which enables them to be abstracted from the powder sample. Thus one could probably create very useful trackers by agglomerating some of the finer powder to be mixed in a mixer system. One can either add some magnetic powder to the powder being agglomerated to create the tracker so that it can be abstracted from a sample using a magnet or, if one is going to work with a tracker which is somewhat larger than the powders to be mixed, one does not even need to add a distinguishing color, since they can be easily separated from the rest of the powder by using a sieve. In fact I have suggested that sometimes a sample for monitoring chaotic conditions within a powder mixer could be collected in a cup with a sieve bottom mounted on a chaos inducing bar placed across the mixer, so that powder being mixed would pass through the sampling cup and the trackers would remain on the sieve mesh of the bottom of the cup. If these trackers were something like a bright red color, they could probably be viewed directly through a viewing port on the exterior of the mixing chamber. Their presence could be recorded by a simple television camera so that almost in real time the chaotic conditions in the mixer can be monitored. One could also color-code trackers being introduced into a mixer via different ingredient additions to the mixing system so that the chaotic distribution of different ingredients can be monitored simultaneously. Consider, for example, the simple case of a three ingredient powder mix. If the proportions of the ingredients were 1 : 2 : 3 and if trackers were added to the different ingredients such that one tracker could be expected for every 50 cm^3 of that ingredient then in a sampling cup one would expect to obtain red to green to yellow trackers in the proportions 1 : 2 : 3 with all three of them fluctuating according to the appropriate stochastic relationship. Note that open sieve bottom containers for samplers would capture much larger samples than a closed bottom cup and they would have to be calibrated to determine the expected populations in the sampling cup from some empirical investigations. Investigations of multicomponent ingredient mixing using sieved bottom tracker capture cups are underway and it is hoped to report on this system in the not too distant future.

Currently, the only investigation which has been reported in the scientific literature using Poisson tracking to monitor chaotic conditions is the work reported at the Nuremberg Conference on Particle Characterization in 1989 [1]. In this investigation, which was carried out using a free-fall tumble mixer of the type shown in Figure 1.6, the number of trackers expected in the sampling cup was a small number. Therefore the expected fluctuations in the population of a set of trackers in sequential samples taken from inside the mixer would be expected to follow the stochastic function known as the 'Poisson probability distribution'. The mathematical expression for this distribution, and its physical significance, is discussed in many textbooks on statistical analysis [1–3]. POMM, our expert system discussed earlier, can work directly with the mathematical form of the Poisson distribution, but in a working environment, the technologist will probably make use of a special graph paper developed for use when fluctuating populations of a system such as the numbers of tracker spheres in a sample cup are expected to follow the Poisson distribution. This graph paper is also the basis of a visual display that is presented by POMM. For this reason, we will concentrate on the graphical use of data to demonstrate the use of Poisson trackers to monitor chaotic conditions.

To understand the procedure involved in using Poisson trackers to monitor chaotic conditions in a mixer consider the data summarized in Tables 5.1 and 5.2. These data were generated during an investigation to see if there was any difference in the mixing capacity of tumbling chambers of different shape used in the free-fall tumbling mixer. The three chambers used in the investigation were a cylindrical, a cubic and an icosahedral chamber of equal volume. In this experiment, the aim of the investigation was to study the homogenization of a stored powder having a wide range of sizes. The powder used was a commercially available sugar powder with individual grains being of the order of 500 μm in diameter. The tracker spheres used in the experiment were plastic spheres 1–2 mm in diameter; 350 trackers were placed in each chamber. The volume of the containers was 140 cm^3 and they contained 35 cm^3 of powder. The rolling drum had a diameter of 30 cm. When the drum was rotated at a speed of four revolutions per minute, the chambers were picked up by the foam rubber lining of the cylinder and elevated until they fell down at random time intervals, tumbling in chaotic order down to the bottom of the chamber. Again they were picked up by the rotation of the drum and underwent subsequent tumbles. It was

Table 5.1 Numbers of tracker spheres observed in free-fall mixer experiments

Sample time (min)	Number of trackers found in sample		
	Icosahedron	Cube	Cylinder
1	12	11	11
2	7	10	11
3	8	8	14
4	7	11	8
5	15	7	7
6	15	8	12
7	3	5	11
8	6	5	5
9	14	8	6
10	9	11	6
11	6	10	8
12	12	12	9
13	4	10	6
14	9	13	5
15	5	8	11
16	10	5	9
17	8	12	6
18	5	8	9
19	6	7	9
20	4	5	6
21	10	14	5
22	12	7	9
23	3	10	6
24	13	7	4
25	8	11	10
26	6	8	6
27	13	11	3
28	4	11	7
29	6	8	13
30	13	2	10
31	9	8	11
32	6	13	5
33	8	11	5
34	13	10	10
35	12	11	6
36	11	6	4
37	7	10	6
38	10	9	9
39	8	6	8
40	6	11	8

Table 5.2 Cumulative occurrence of observed population of trackers

Observed population	Icosahedron			Cube			Cylinder		
	f	$\sum f$	%	f	$\sum f$	%	f	$\sum f$	%
1	–	–	–	–	–	–	–	–	–
2	–	–	1	40	100	–	–	–	–
3	2	40	100	–	–	–	1	40	100
4	3	38	95.0	–	–	–	2	39	97.5
5	2	35	87.5	4	39	97.5	5	37	92.5
6	7	33	82.5	2	35	87.5	9	32	80.0
7	3	26	65.0	4	33	82.5	2	23	57.5
8	5	23	57.5	8	29	72.5	4	21	52.5
9	3	18	45.0	1	21	52.5	6	17	42.5
10	3	15	37.5	6	20	50.0	3	11	27.5
11	1	12	30.0	9	14	35.0	5	8	20.0
12	4	11	27.5	2	5	12.5	1	3	7.5
13	4	7	17.5	2	3	7.5	1	2	5.0
14	1	3	7.5	1	1	2.5	1	1	2.5
15	2	2	5.0	–	–	–	–	–	–

f = frequency of population. $\sum f$ = cumulative frequency.

observed that a 1 min tumbling period resulted in approximately 12 free-fall tumbles of the mixing chamber. A 1 cm^3 sampling cup was placed at the end of an arm 5 cm long in the lids of the containers. This meant that when the mixing chamber was at rest, and sat on its base, the sampling cup was 0.5 cm below the free surface of the powder in the container. After the container had been tumbled for 1 min, the sample was withdrawn and the tracker spheres separated from the fine powder on a 35 mesh sieve. After the number of tracker spheres was counted, both the tracker spheres and the fine powder were returned to the mixing chamber, which was again tumbled for 1 min and the sample withdrawing experiment repeated. The observed fluctuating number of tracker spheres from a set of experiments are as recorded in Table 5.1. To be able to plot these data on Poisson probability paper, one needs to convert the frequency data into the cumulative occurrence of observed population of trackers, as shown in Table 5.2. In Figure 5.1(a) the structure of Poisson probability graph paper is shown.

There are three sets of lines on the Poisson graph paper. The abscissa is a logarithmic scale, the ordinate a probability scale structured according to Gaussian probability function. The grid formed by the logarithmic abscissa and the probability ordinate is crossed by a set of curves which are labelled 1, 5, 10 etc. These curves are the observed population densities which are fluctuating according to the Poisson probability function. Thus point A on curve 1 indicates that, in an experiment, 90% of the observed populations (0.9 on the probability scale) had a population of one or more. Point B represents the fact that in the same experiment populations of two or more were observed with a probability of 0.72.

The technique for plotting points such as A and B is to locate the observed population curve, and slide down it until it crosses the observed probability line; at that intersection the datapoint is plotted. Thus to plot the datapoint F which represents the observed frequency of populations of six or more, one slides down curve 6 (not shown) until it meets a probability of 0.03. If an observed set

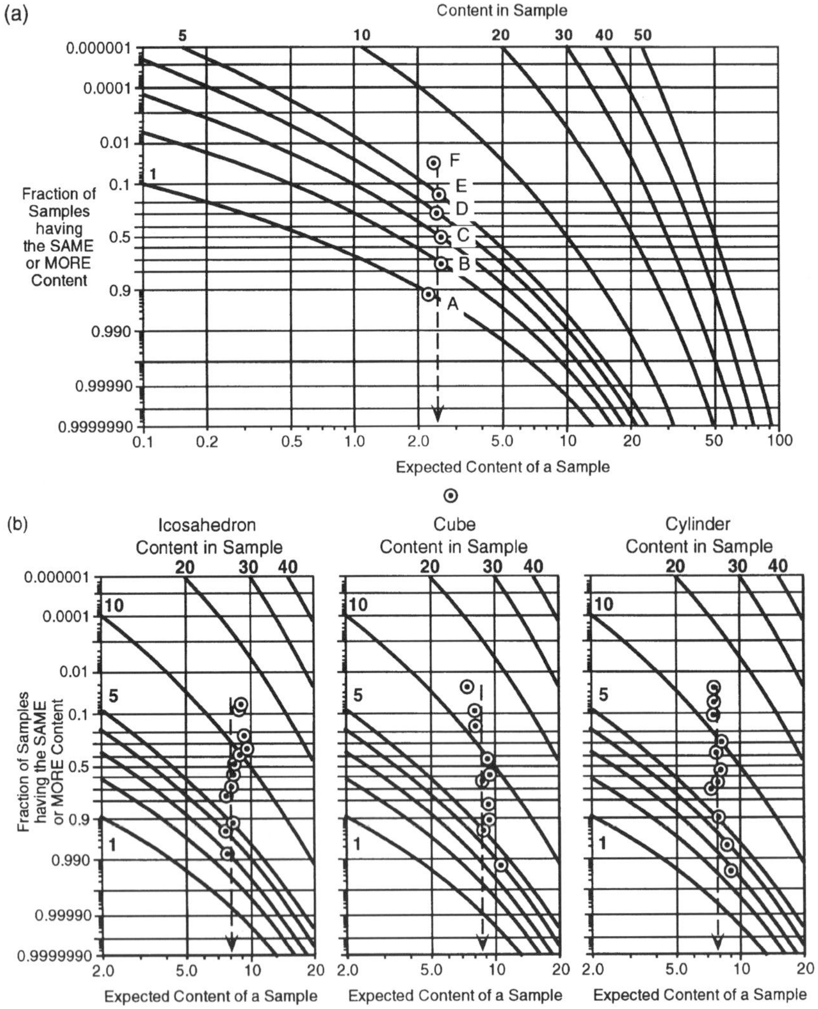

Figure 5.1 If a fluctuating population, such as the observed population of tracker beads in a sampling cup, can be described by the Poisson distribution, then the data for the observed fluctuations generate a straight line on Poisson probability graph paper, intersecting the abscissa at the expected average population in the sample cup [3]. (a) Typical data presented on Poisson probability graph paper. (b) Data from Table 5.2 plotted on Poisson probability graph paper.

of populations are following a Poisson distribution, then the datapoints such as A to F on the graph define a straight line which cuts the abscissa at the expected population [3]. Thus in the case of Figure 5.1(a) the datapoints cluster along the straight line which can be drawn to cut the expected population axis at the value of 2.5. Thus the observed frequencies of populations ranging from one to six in this case were consistent with an expected average population of 2.5. In Figure 5.1(b) the fluctuating data of Table 5.2 are plotted on Poisson graph paper. It can be seen that all three mixing chambers gave populations that could fluctuate with an expected population consistent with the hypothesis that the mixing device was producing chaotic conditions with all mixing chambers. A worker new to the use of this type of graph paper might be a little disturbed at the expected population for all three graphs of Figure 5.1(b) which was theoretically 10. The lines on the graph paper differ slightly from 10. This is because in all stochastic systems there is going to be a variation in the actual number generated by such data sets. One of the advantages of POMM is that it can very rapidly generate sets of simulated population sampling to educate the worker as to what is a reasonable fluctuation in a set of Poisson tracker populations for a given situation. Thus Figure 5.2 demonstrates a simulated output produced by POMM, showing the variations one could expect on 50 items of information (from 50 sequential samples for several different expected populations). Graph (a) represents the simulated variation in sets of 50 sequential samples for expected populations of 2.5, 10 and 22.5. The scatter of these datapoints about the expected population lines and the variations in the actual population lines from anticipated populations are not due to experimental error, but are inherent stochastic variations expected by pure chance in a system which is ideally chaotic. It should be noted that Poisson graph paper becomes indistinguishable from Gaussian graph paper for expected populations greater than about 20. The simulated data of Figure 5.2 emphasize that when using this type of graph paper and other probability-based graph papers with cumulative versions of distribution data, one should always give more weight to the datapoints in the central regions of the graph. One must not put too much emphasis on the tails of the distribution unless they are of vital significance of the given situation, since they represent rare events in the statistical fluctuations of the populations. A few experiments with people new to the field will soon demonstrate that the type of fluctuations shown in Figure 5.2 are surprising to the uninitiated.

The power of Poisson trackers to demonstrate the existence or absence of chaotic mixing conditions is illustrated by the data of Figure 5.3. An important variable governing the performance of a free-fall tumbling mixer is the volume fraction of the chamber occupied by powder. The higher this fraction, the less free are the fineparticles in the mixer to move about at random. The data for Figure 5.3 shows the variations in chaos tracker populations in the three mixing chambers illustrated in Figure 5.1 when the chambers were half full of powder. It can be seen that the population fluctuations in the case of the cylindrical and cubic chamber appear to be basically Poisson, but that the chaotic conditions are apparently not persisting in the icosahedral chamber for such high powder volume operation. At first sight it might appear surprising that the icosahedral chamber with its many faceted surfaces is not as efficient as the other simple geometric shapes. This failure of the icosahedral mixing chamber at high volume

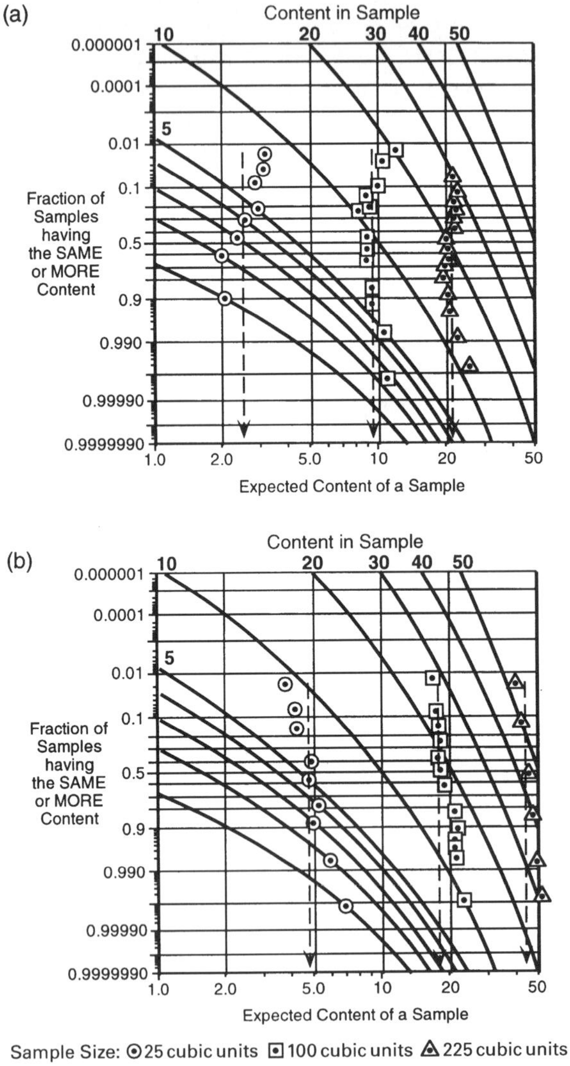

Figure 5.2 The expert system, POMM (Powder Mixing Monitor), can simulate fluctuation in expected populations and the effect of sample size on the expected results. Thus, for the data displayed above, fluctuations in 50 sequential samples of various sizes for two different sample concentrations are displayed. (a) Fluctuations in the expected population when the simulated concentration is 10%. (b) Fluctuations in the expected concentration when the simulated concentration is 20%.

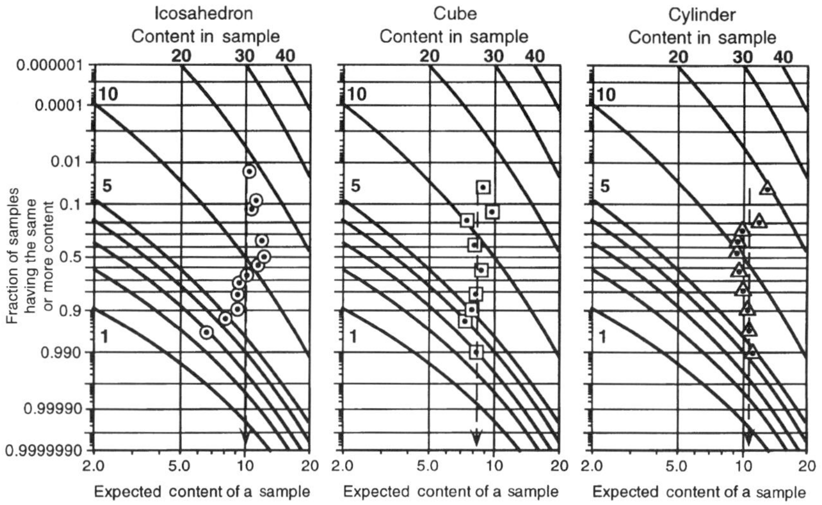

Figure 5.3 Actual results for three different shapes of mixing chambers used in the AeroKaye® mixer/sampler, shows the actual population of tracker beads present in the sample cup when samples were examined periodically during the mixing process. The mixture was deliberately created with an expected population of ten tracker beads to the equivalent volume of the sample cup.

fractions appears to be a function of the fact that the internal volume of the icosahedron is more restricted in possible spatial leaps and that, although there is more tumbling action with the icosahedron, there is less internal movement as the powder charge fraction increases. It must be emphasized that tracker spheres can only establish chaotic conditions and not the quality of the mixture. If experiments established by means of chaos trackers that the mixer is performing as well as can be expected from the laws of physics, any inadequacy of the resultant powder mixture must lie in the failure of the mixer to disperse agglomerates of the ingredients or to create intimate intermingling on the mixture. In such situations one should seek to introduce a shear mechanism into the mixer to process the first-order mixture created by chaotic conditions with dispersion equipment such as a pinmill or other high-shear equipment. It should be noted that a similar technology to the use of Poisson trackers to assess the performance of powder mixing equipment was developed by Eisenberg [4, 5].

5.3 USING RADIOACTIVE TRACERS TO FOLLOW POWDER DISPERSION IN POWDER MIXING EQUIPMENT

In the academic world, in-depth studies of the movement of powder in mixing systems have been carried out by several workers. For example, van den Bergh has used radioactive-labeled sand to explore powder movements in a Nauta mixer [6]. He has also developed a computer simulation of a Nauta mixer [6–8]. The basic structure and internal movements of the Nauta mixer are illustrated in Figure 5.4(a). A conical bin is equipped with an elevating screw which can be

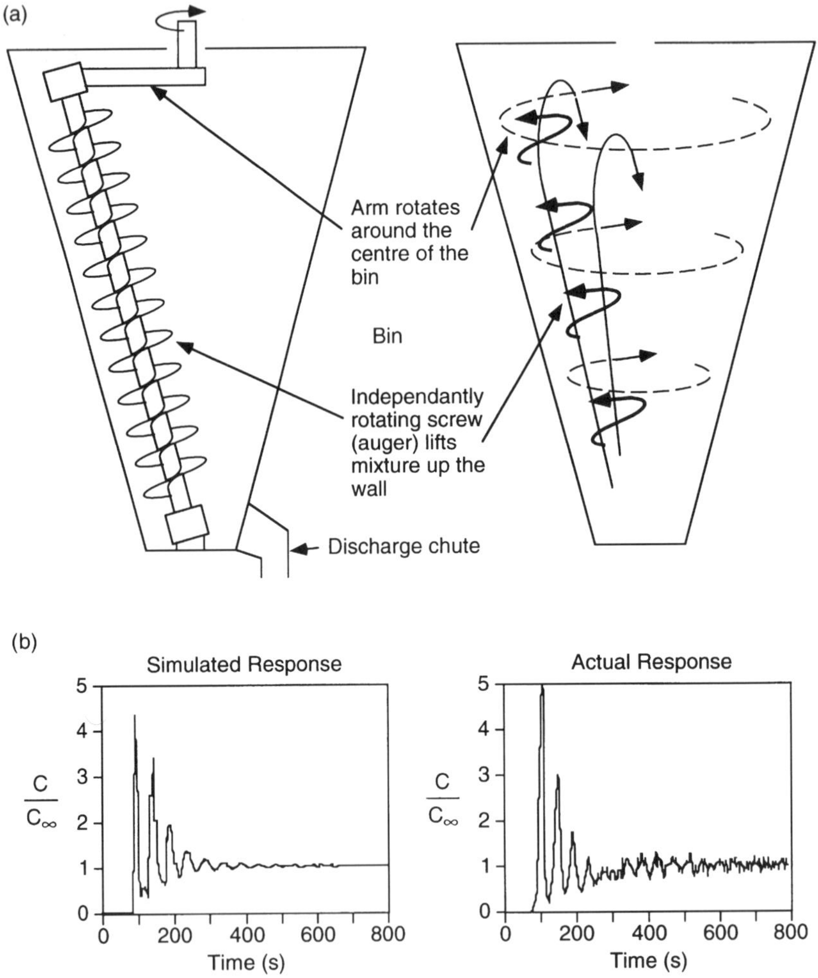

Figure 5.4 van den Bergh has carried out exclusive studies of the mixing action within a Nauta mixer, using radioactively labeled sand. He also developed a computer model of the mixing action. (a) Construction and mixing actions of the Nauta mixer. (b) Simulated and measured response curves when the angular velocity of the screw around the bin was 0.131 radians per second and the angular velocity of the the screw was 5.76 radians per second. C = radioactivity measured at the detector at time t. C_∞ = radioactivity measured for a completely mixed system (after infinite mixing time). (Reprinted from van den Bergh *et al.* [8], with kind permission from Elsevier Science, S.A., Lausanne, Switzerland).

rotated around the periphery of the bin. As it is rotated around the bin, the screw turns, lifting powder from the bottom of the bin and pouring it into the center of the mixer. To investigate the dynamics of powder movement in such a bin, van den Bergh carried out experiments with the silica sand fractionated so that the range of diameters in the sand was 600–850 μm. Van den Bergh notes that this narrow size range avoids significant segregation and that for each experiment the mixer was loaded with 100 kg of sand. Then 0.05 kg of sand labeled with radioactive iodine was poured into the center of the free material surface, and the mixing operation was initiated. The radioactive sand was monitored by a scintillation detector fitted into a thick-walled cylindrical collimator of lead placed close to the bin wall on a fixed support. It was assumed that the measured radioactivity at the detector was proportional to the amount of radioactive sand in the vicinity of the monitor. Typical curves generated by van den Bergh in his study are shown in Figure 5.4(b). It can be seen that the dispersal of the radioactive sand in the mixer resulted in a good mixture after 800 seconds and that the progress of the mixing matched the predictions of the computer simulation. It should, however, be noted that this experiment only establishes that, with a free flowing powder not subjected to segregation, the mixer works well. This reinforces the comments made earlier that, if in fact one was to use a Nauta mixer with a real mixture and if the mixture was unsatisfactory after 1000 seconds of operation, then the problem of failure to mix lies with the physical properties of the powders being used and cannot be attributed to the randomizing mechanisms of the specific mixer.

Similar radioactive trace experiments for a fluidized bed mixing system have been carried out by Moselmian *et al.* [9, 10]. The equipment they used, and a typical set of results, are shown in Figure 5.5. The glass beads used contained about 10% by weight of sodium. Some of these beads were activated in a nuclear reactor to convert sodium into its radioactive isotope. Then the radioactive beads were placed in the powder bed. These tracers have a relatively short halflife of 15 hours. Therefore the radioactive tracers decayed rapidly to their natural state overnight and so did not need to be separated from the original bed particles between test runs, which were scheduled longer than 15 hours apart. Again, it is difficult to extrapolate the implications of such idealized experiments to the performance of mixers with real powders of interest to industrialists.

5.4 MONITORING MIXTURE STRUCTURE BY MEANS OF OPTICAL REFLECTANCE MEASUREMENTS

One of the most difficult tasks facing the powder technologist is the consistent production of mixtures of two different colored powders to produce a uniform product for coloring paints, filled plastics, composite materials, cosmetics etc.

It soon became obvious to pioneer workers, when carrying out a study of the optical properties of powder mixtures, that the optical reflectance of the powder mixture could form the basis of a method of assessing the progress of the mixing process and the operational acceptability of the product. The difficulty of intermingling two or more powders of different colors varies with the physical

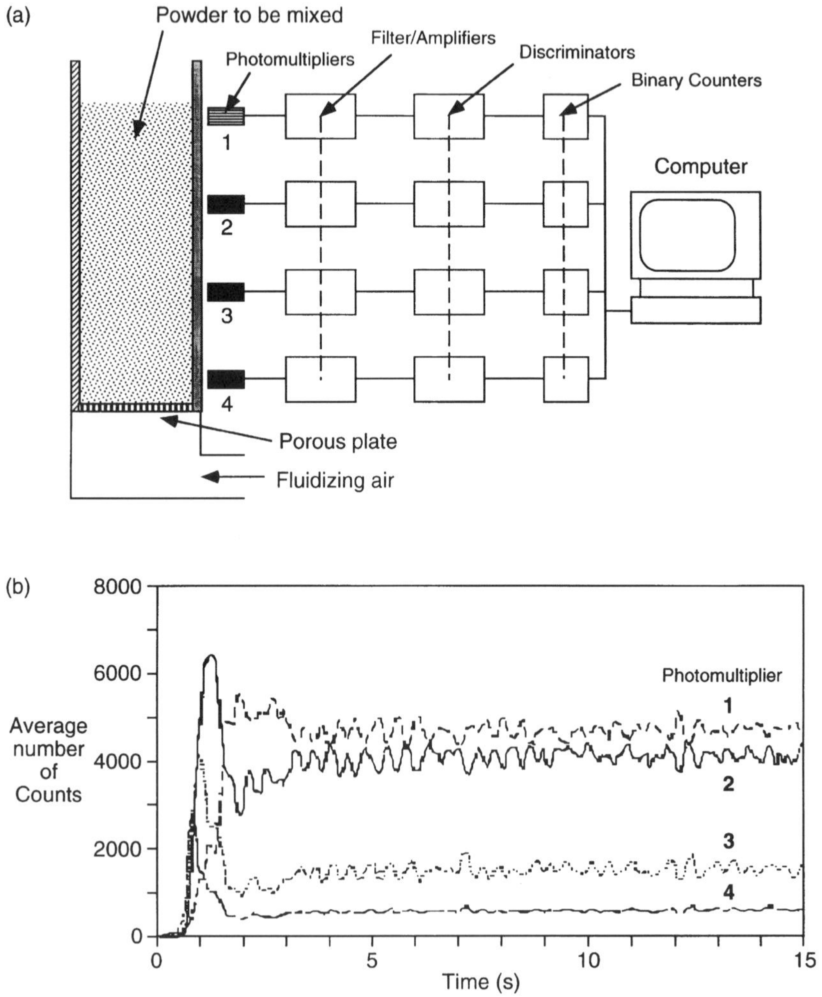

Figure 5.5 Moselmian and colleagues have used radioactive tracer spheres to explore the dynamics of powder in a mixing system. (a) Schematic diagram of the fluidized bed, data acquisition and data processing system used by Moselmian *et al.* (b) Typical data obtained using 600 μm tracer spheres dispersed in glass beads in the size range 425–600 μm. (Courtesy of the authors D. Moselmian, M.M. Chen and B.T. Chao, [9].)

properties of the ingredients and becomes more difficult as the powders become finer in constitution and cohesive in their physical flow behavior. In the past, there has been some hope that one might be able to generate calibration curves for the operation of mixers blending colored ingredients. The prediction of the optical properties of powder mixtures is, however, a complex subject, and the effective reflectance and color of a powder mixture is a complex function of the structure of the powder mixture. The various factors that can affect these properties in a mixture of two powders are illustrated in Figure 5.6. Many powder technologists are interested in multicomponent mixtures, and most of what is said here applies to more complicated mixtures than a two ingredient system, but it is easier to discuss the contributing factors and their interaction with respect to a two-component mixture for the purpose of describing concepts and measurement technology. Most complications that arise from the presence of more ingredients than the two basic ones being discussed are essentially a relatively easy extrapolation of the problems and solutions encounters when studying two component systems.

The first attempt to develop a probe for investigating the internal structure of a mixture and the performance of a mixer using an optical device was reported by J.B. Gray [12]. Gray's probe consisted of a white light source and a photocell behind a small glass window at the end of the probe. The probe measured the intensity of light reflected from a layer of powder immediately outside the glass window. The readings of a photocell meter were then related to the composition of the mixture. Empirical calibration curves were prepared to interpret the data generated with this type of probe. A similar type of equipment was developed by

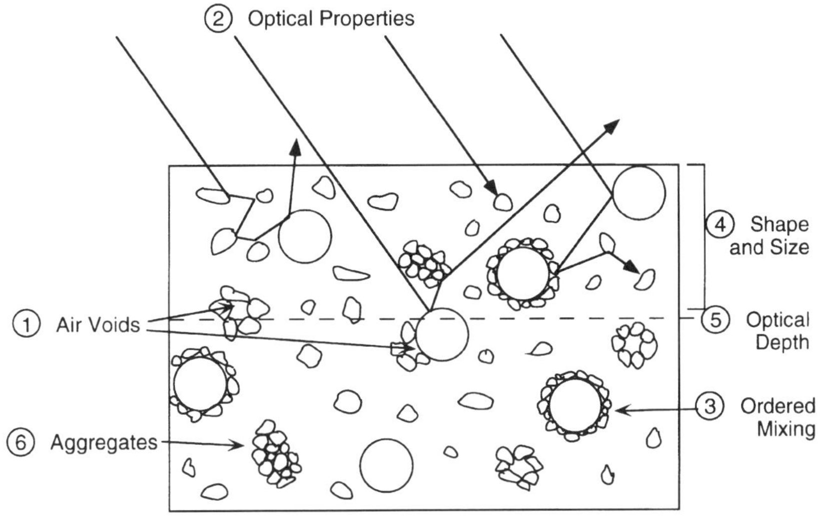

Figure 5.6 Many factors determine the effective reflectance and color of a powder mixture [4,11(b)].

Schofield. The probe used by Schofield and the mixing equipment that he investigated using the probe are shown in Figure 5.7. Typical results obtained by Schofield and colleagues when investigating the performance of the mixers with one-to-one mixtures of 150 μm red and white sand are shown in Figure 5.8. Note that 1 on the mixing index axis of these graphs is equivalent to the best structure that can be achieved with a chaotically assembled mixtures. The studies carried out by Schofield suggest that all of the mixers of Figure 5.7 are equally proficient at creating chaotically structured mixtures. Perhaps the main importance of Schofield's work in current approaches to powder mixing is to indicate that any failure of these standard mixers to achieve mixing in a specific industrial situation lies not in the chaos creating capacity of the mixer but in the physical properties of the powders.

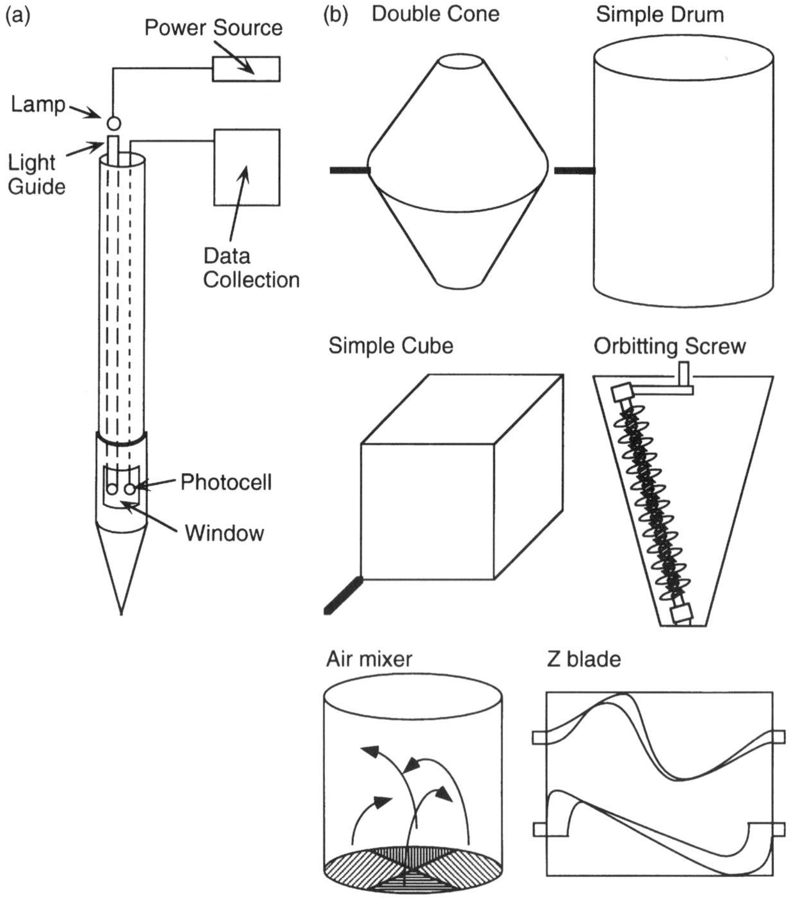

Figure 5.7 Optical probe equipment used by Schofield *et al.* to explore variations in powder mixture structures in free flowing powders [12].

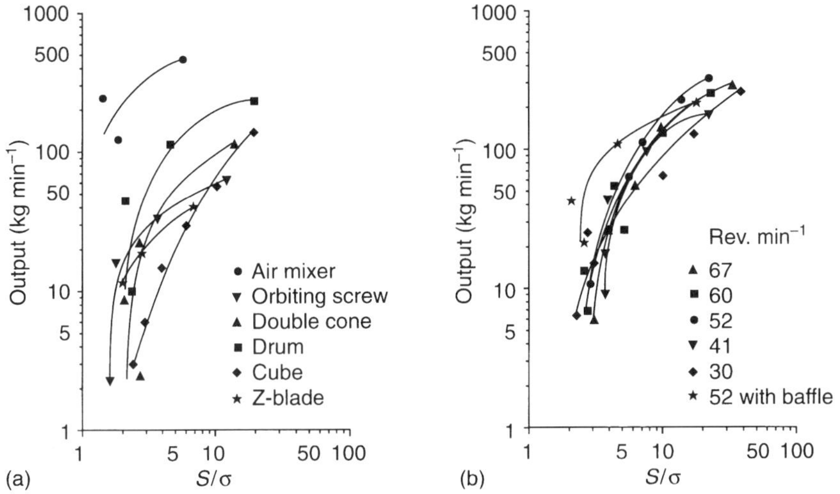

Figure 5.8 Typical data for the variations in powder mixture generated by Schofield *et al.* in studies of free flowing powders [12]. (a) Comparison of the performance of various mixers. (b) Results for a drum mixer run at various rotational speeds. S = sample composition standard deviation; σ = standard deviation for the same number of completely random samples.

To workers in the powder mixture field the advent of fiberoptics obviously gave greater flexibility and capacity for sophistication in the design of optical probes. One of the first reports of the use of fiberoptic probes to explore powder mixtures was published by Kaye *et al.* [14]. They measured the forces involved in pushing a fiberoptic probe into a loosely compacted powder bed of titanium dioxide mixed with cabonblack. Typical curves of reflectance versus position of the probe obtained by these workers are shown in Figure 5.9. Further development of the basic system developed by Kaye and co-workers was undertaken by Harwood *et al.* They used ultraviolet light to explore the reflectance of mixtures which were optically white in the visible spectrum. The system used by Harwood and co-workers is shown in Figure 5.10. In their equipment, to overcome the problems of exploring the mixture structure while the mixer was still operating and also to avoid problems due to powder coating the probe tip, the tip of the optical probe and the probe monitor were mounted in the wall of the mixing vessel. This enabled the probe to be kept clean by the movement of the internal mixing arm as it swept past. The equipment was quite successful for the powders that they studied.

In general, the problem of looking at powder mixtures by means of a probe inserted into a mixture tended to be complicated by the fact that every powder mixture had to be individually calibrated, and one could never be sure of the effect of compression by the probe on the powder mixture, altering the reflectance of the packed powder ahead of the probe. The use of fiberoptic probes to monitor mixture quality and mixing process has been reported by Japanese scientists. At a fineparticle meeting in Miami in 1985, Saito and Kamiwano of the Department of Chemical Engineering at Yokohama National University reported on an optical method for measuring flow velocity and composition of a mixture of powders continuously [16]. They used their probe to look at the constitution and individual fineparticle dynamics in a fluidized bed. The study was essentially limited to free flowing powders of two very different optical

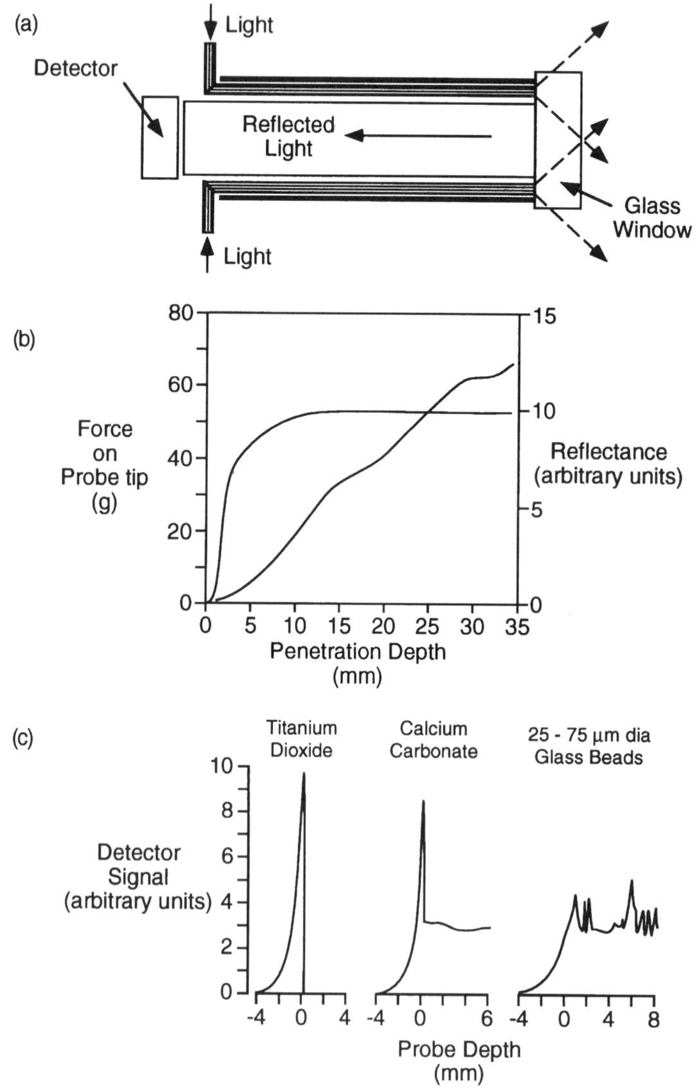

Figure 5.9 Typical data curves reported by Kaye *et al.* for the study of powder systems using a fiber optic probe [14].

reflectances (they used, for example, silver-coated polystyrene particles and free flowing sand).

Alonso and co-workers at the University of Osaka have recently published a comprehensive study of powder mixing and powder coating using fiberoptic probes [17]. They fitted several fiberoptic probes into different types of mixer and studied the mixing process. The optical system that they used is shown in Figure 5.11(a). Typical calibration curves and performance curves generated with this equipment are shown in Figure 5.11(b), (c). The effect of varying mixing speeds on the efficiency of powder mixing is shown for a mixture of magnetite and polymethyl methacrylate spheres in Figure 5.12(a). In the first part of the study the effect of agitation speed in the mixer was followed. The powder mixture obtained by ordinary high-speed stirring was then subjected to a period of dispersion in the Angmill system. This is a system which achieves powder mixing by dry grinding and results in the heterogeneous coating of one of the components by the other. The process is known by the trade name of Mechanofusion™ when the process is achieved in the Angmill manufactured by Hosokawa (see discussion at end of Chapter 1). A similar process has been developed by the Nara Machine Company who term the process Hybridization® of the powder system. (The process of heterogeneous encapsulation occurring in the hybridization system is illustrated in Figure 1.21.) It can be seen in Figure

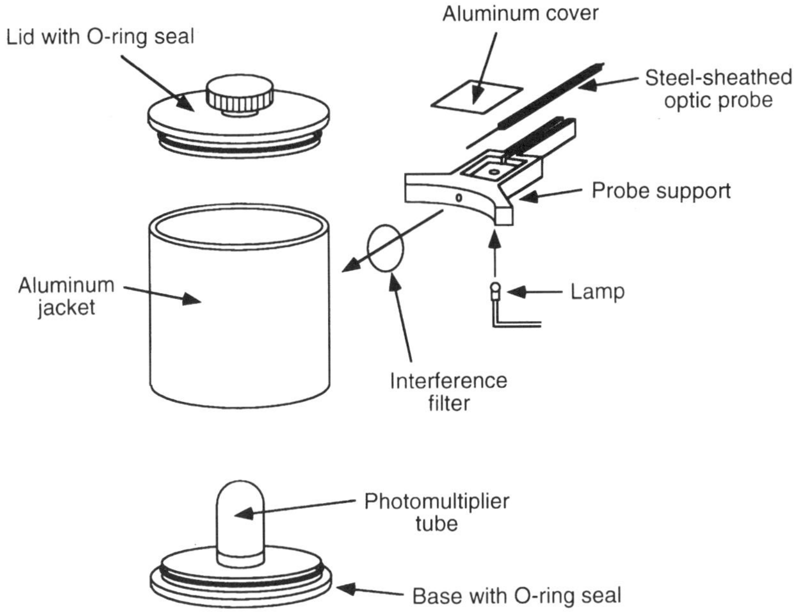

Figure 5.10 System developed by Harwood *et al.* to study powder mixtures (From Harwood *et al.* [15], with kind permission from Elsevier Science, SA, Lausanne, Switzerland).

5.12(b) that a much more intimate mix is obtained when the best that can be achieved by high-speed stirring is subjected to a Mechanofusion process. The structured mix produced, with one powder coating the other, is a very efficient optical scattering system. However, the optical power changes as the fine-particles coated on to the core fineparticles begin to penetrate into the system and the effective refractive index of the scattering centers formed by the smaller fineparticles decreases because of the change in interface refractive index. This latter example illustrates the complexity of the refractive index. This latter example illustrates the complexity of the problem of interpreting the optical

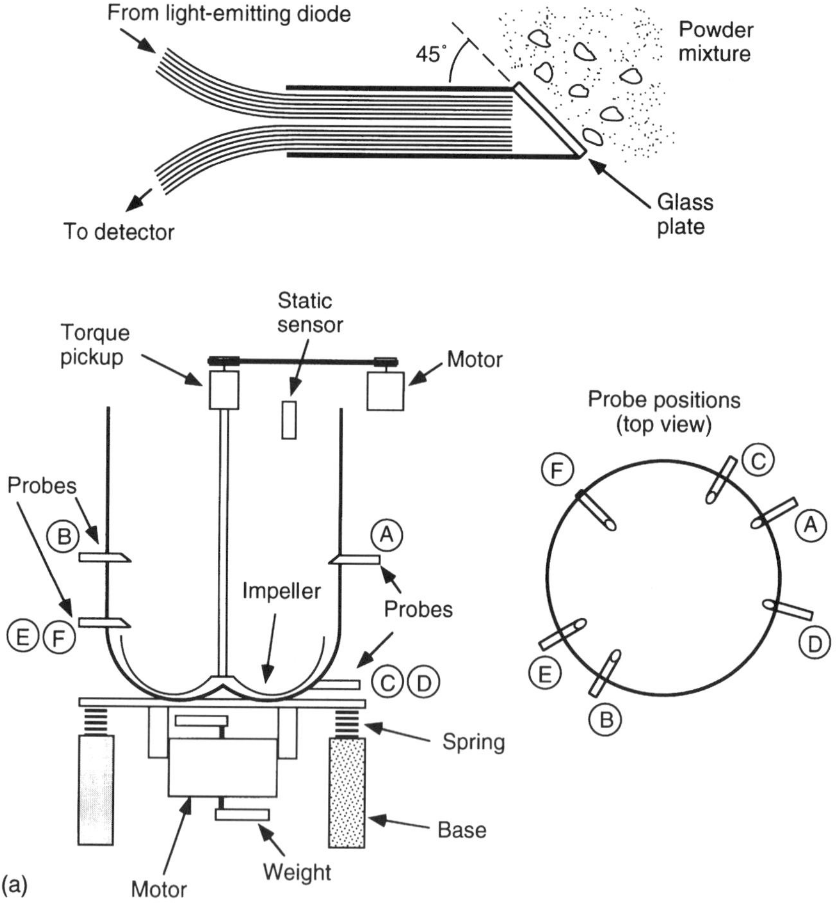

Figure 5.11 Alonso *et al.* used a fiberoptic probe to explore the progress of powder mixing. (a) Optical probe and mixing equipment used in the studies. (b) Calibration curves for various powder mixtures. E = intensity of reflected signal; E_0 = intensity of inclined signal. © Performance curves for two mixtures. (From Alonso *et al.* [17], reproduced by permission of the Council of Powder Technology and the Hosokawa Powder Foundation, Osaka, Japan. © 1995 KONA All rights reserved).

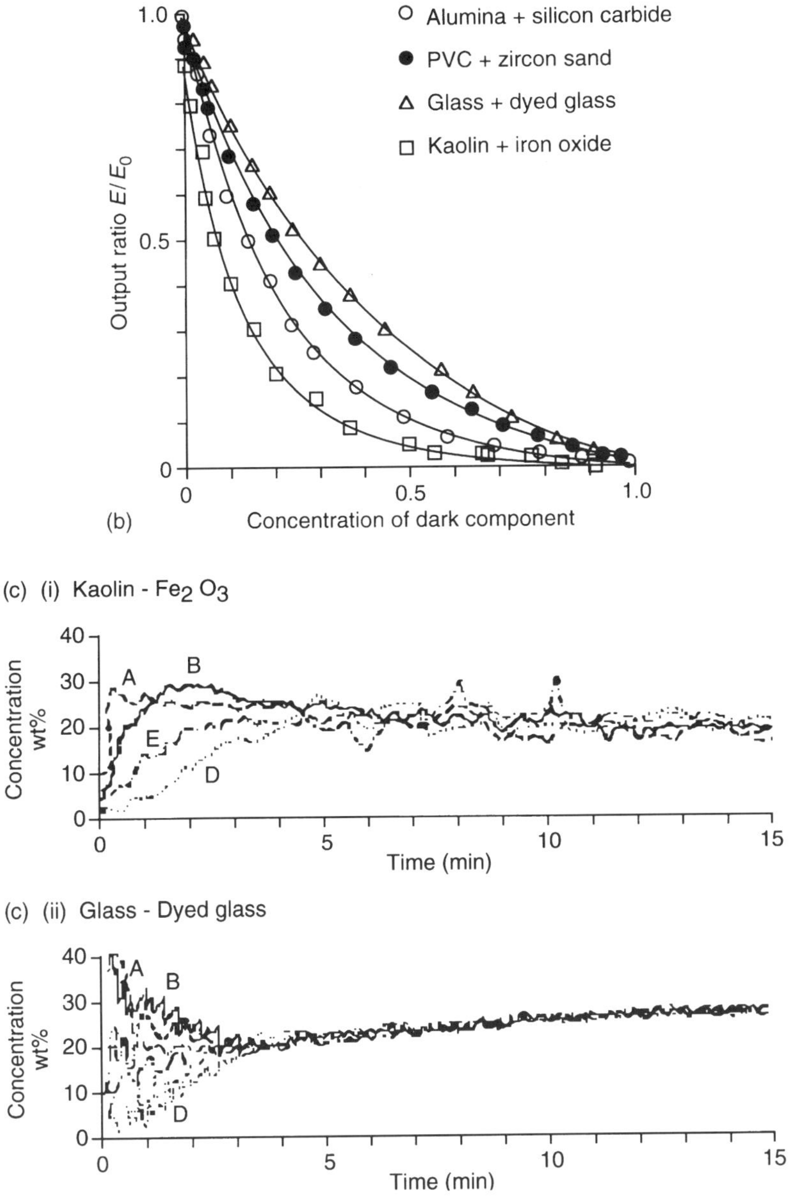

Figure 5.11 *Continued*

reflectance of powder mixtures. Many other powder mixtures were studied by Alonso *et al.*

Recently, Kaye has suggested that new innovations in the sampling probes could greatly improve the ability of technologists to follow mixing procedures in powder mixers by optical probes. First of all it was suggested that a pneumatic lance be modified to carry the optical probe, thus simplifying the problems of inserting and withdrawing the probe. The design is shown in Figure 5.13(a). The tip is made of Teflon® or some other nonstick surface. When the location within the powder mixture to be characterized is reached, the air flow is reversed, gradually drawing the powder sample into the sampling chamber. The sampling

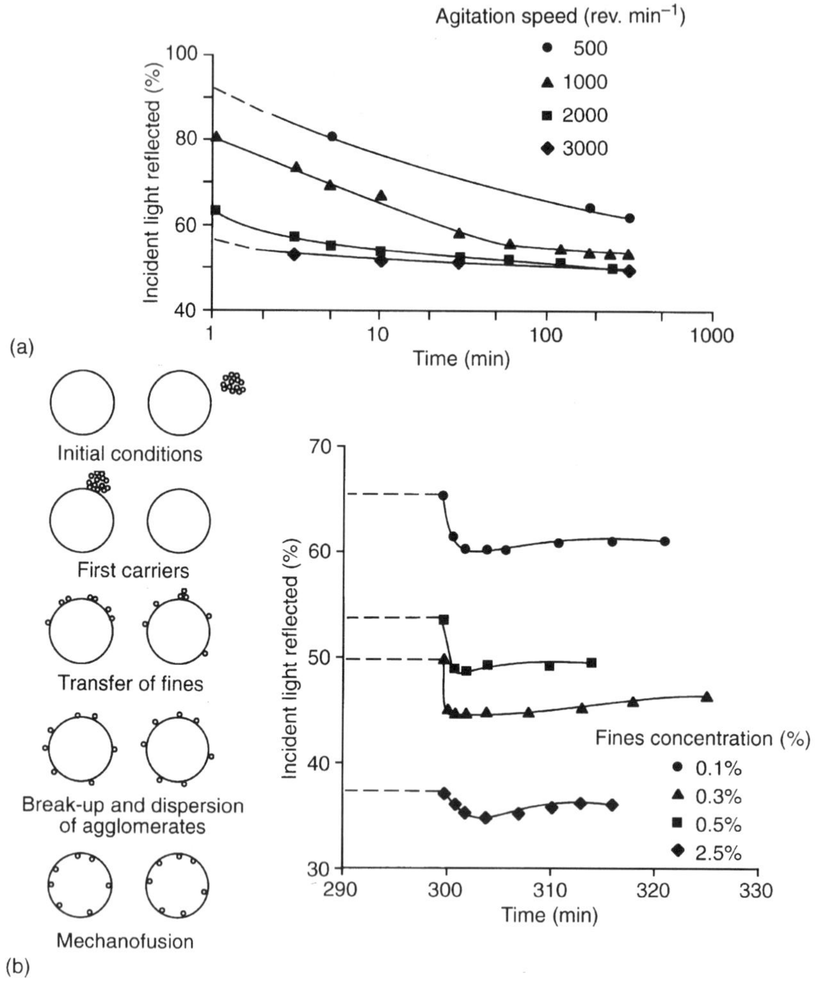

Figure 5.12 Performance studies of the dispersion of magnetite into and onto polymethyl methacrylate (PMMA) spheres. (a) Progress curve for optical monitoring of the dispersion of fine magnetite (0.3%) into the PMMA matrix. (b) The effect of Mechanofusion™ treatment (2600 rev. min^{-1}) of PMMA–magnetite mixture. (From Alonso *et al.* [17], reproduced by permission of the Council of Powder Technology and the Hosokawa Powder Foundation, Osaka, Japan. © 1995 KONA. All rights reserved.)

chamber within the tip has a well-defined geometry and can be compacted to a known amount by increasing the pneumatic pressure after the powder has entered the probe. In some cases this might result in the aspiration of fines from the mixture outside the probe, so that one would have to investigate the potential usefulness of this operational feature with a particular mixture. After the reflectance of the well-defined sample has been obtained, the powder sample can be expelled from the probe by reversing the air pressure. The probe is then moved to a second location.

A lower-cost probe which could be used to explore the performance of mixers such as drum mixers or Y-mixers etc. is shown in Figure 5.13(b). The chaos inducing bar placed across the mixer (called an intensifier bar in some commercial equipment manufacturers) can carry sampling cups with highly reflective internal surfaces. A fiberoptic probe can then be inserted into the cup and

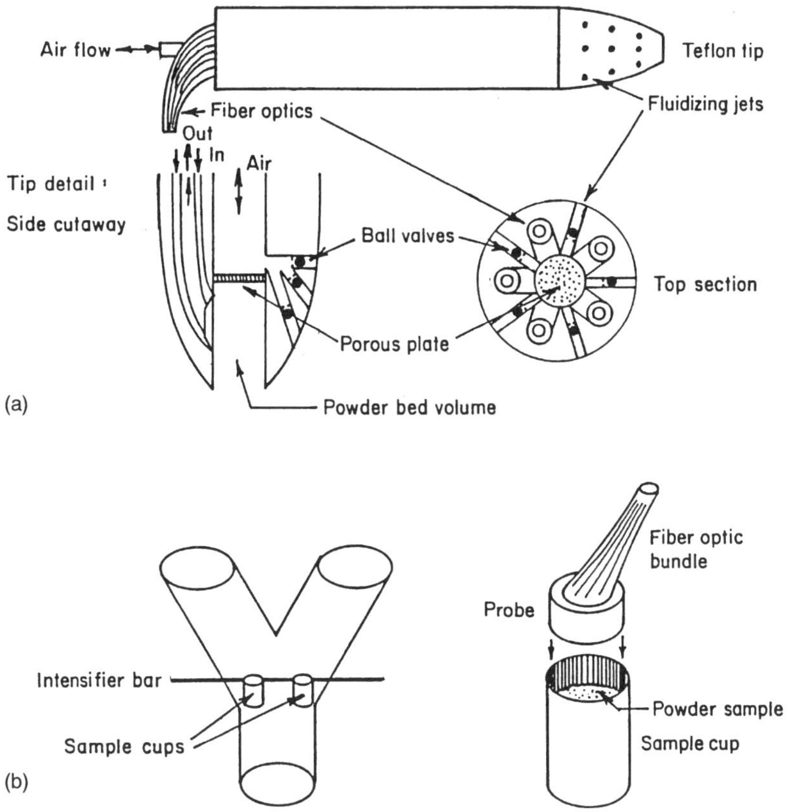

Figure 5.13 Two suggested innovations in the design of fiberoptic probes may improve the efficiency of equipment for monitoring the mixing of cohesive powders. (a) Pneumatically assisted fiberoptic sampling probe suggested by Plessis and Kaye [11(a)]. (b) Sample cup fiberoptic probe for monitoring powder mixtures described by Kaye [11(b)].

pressed down to within a prearranged pressure to look at the optical reflectance of the compacted powder sample contained there. Sampling cups can be also placed lower in the mixer so that a post-mixing checkup on a mixer can be achieved by looking at various sample cups after the mixer has been emptied. To follow the process of mixing, one would look at the samples in the cup on the chaos inducing bar, and then the next operation of the mixer would empty the cups ready for further samples to be obtained at the next prearranged sampling time [11(b)]. Studies of fiberoptic testing of powder mixtures were described by Weinekbötter and Reh [18]. The experimental set-up consisted of an argon laser (514 nm), a photometer and three different fiberoptic probes (Figure 5.14),

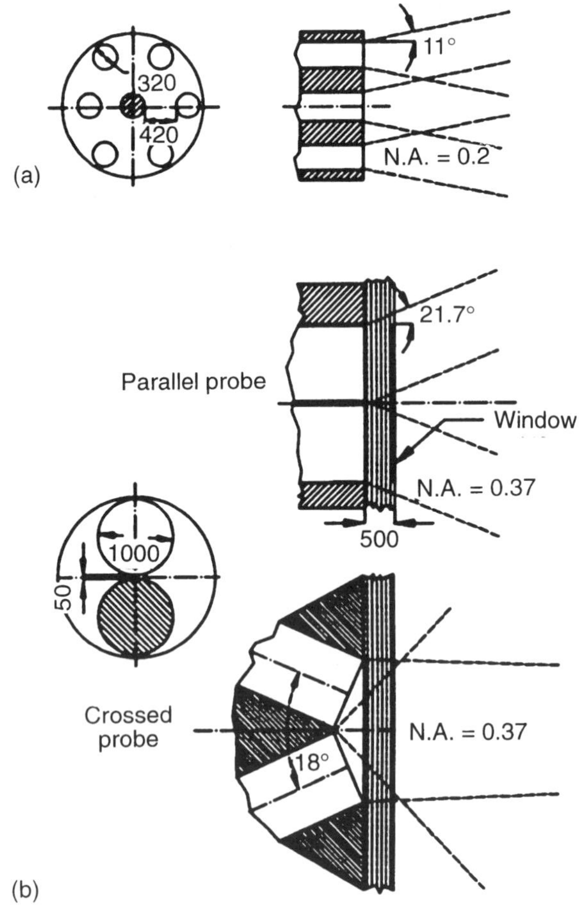

Figure 5.14 The three fiberoptic probe configurations for reflectance measurements used by Weinekötter and Reh [18]. (a) Coaxial bundle probe; (b) pair probes. (Dimensions μm.) (Reprinted with permission of the authors.)

viewing angles which are given by

$$\text{N.A.} = n \sin \theta$$

where N.A. is numerical aperture; θ is refraction angle; n is refraction index. The coaxial bundle probe directed light out through the peripheral or central fibers, and received the signal through the other fibers. The distance between the fibers was wide and the numerical aperture small, therefore creating a small sampling volume for inspection by the probe. The parallel probe of Figure 5.14(b) used a pair of fiberoptic cables separated by a distance of 50 μm. This geometry created a larger sampling volume than that of the bundle probe. In the third probe configuration used, two crossed probes with axes oriented at 48° to each other created the largest sampling volume used with their experiments. In this third probe design the emitting and acceptance light cones were very close to the surface of the probe. The utilities of the three probe arrangements were explored using two mixtures containing aluminum hydroxide (white), one with silicon carbide (black) and the other with irgalite (green). Samples of different constitutions of the test powders were intensively mixed in order to assure good quality mixtures. Results obtained by Weinekötter and Reh for the two mixtures using the three probe arrangments are summarized in Figure 5.15. The reflectances (R) were normalized with the mean signal of the aluminum hydroxide.

In their study, Weinekötter and Reh compared their experimental results with predictions from Kubelka and Munk's theory of light scattering by paint films [19]. Weinekötter and Reh found their results to be in good agreement with this theory for light colored mixtures. It did not apply to the diffuse reflection of highly absorbing media, and cannot be used for mixtures containing higher concentrations of the dark component. Weinekötter and Reh studied the effects of particle size, incident light intensity, sample weight, and of sample size on the reflectance of light from a powder, while using their three probe configurations. They also studied the possibility of using their system for on-line measurements of the performance of continuous powder mixing systems.

Recently, Gratton-Liimatainen has made a study of the changes in reflectance of cohesive powder mixtures as a possible first step to creating feedback loops on industrial mixers. The original aim of these experiements was to test the possibility of developing fiberoptics inspection systems of the type illustrated in Figure 5.13(b) [20]. Preliminary experiments established the fact that mixing the powders in a Y-mixer such as that of Figure 5.13(b) did not lead to an intimate contact of the two different colored ingredients, even when the powders were as well mixed as one can achieve with this type of mixer. To obtain color-consistent results, it was found necessary to take the mixture sample out of the powder cup mounted on the intensifier bar, grind the material in a pestle and mortar for 2 min to increase the intimacy of the mix and achieve consistent color-reflectant mixtures (after 2 min, further grinding did not change the color reflectance signal). The equipment used by Gratton-Liimatainen to measure the reflectance of the powder mixture is shown schematically in Figure 5.16 [20]. In Figure 5.17 one of the graphs generated in these studies is shown. The food industry, in such operations as the creation of cake mixes, must mix cohesive powders such as wheat flour and cocoa. When carrying out experiements of this kind, it was found that any one mixture, say 20% by weight of cocoa, would have small differences in reflectance between independent samples. This is not an error, but an uncertainty due to the difficulty of ensuring an intimate mix.

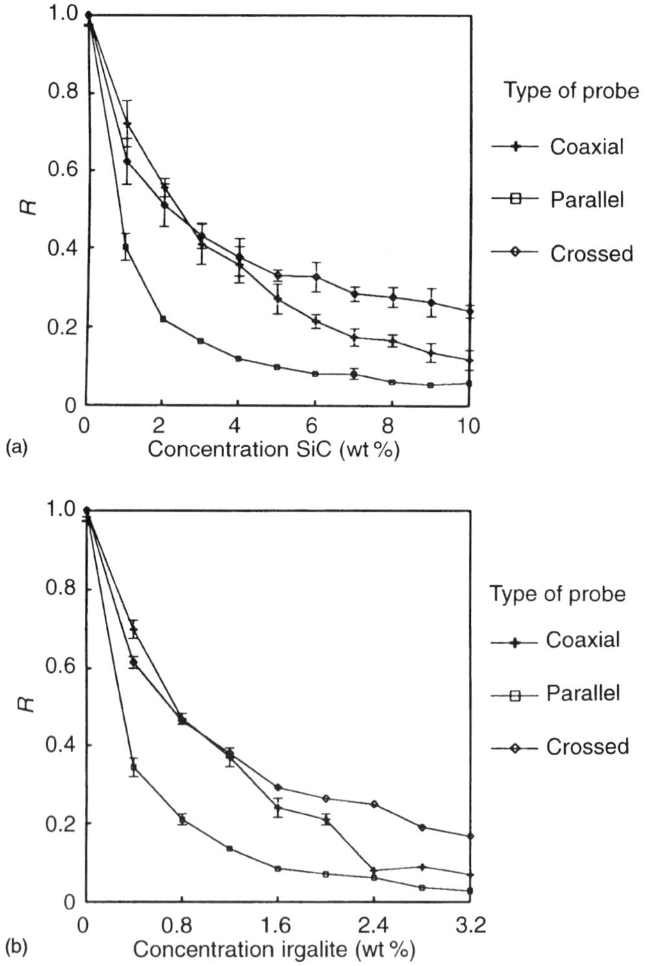

Figure 5.15 Reflectance curves for aluminum hydroxide mixed with silicon carbide and iragalite measured by Weinekötter and Reh, using the three probe types of Figure 5.14 [18]. (Bars represent 95% confidence intervals.) (Reprinted with permission of the authors.)

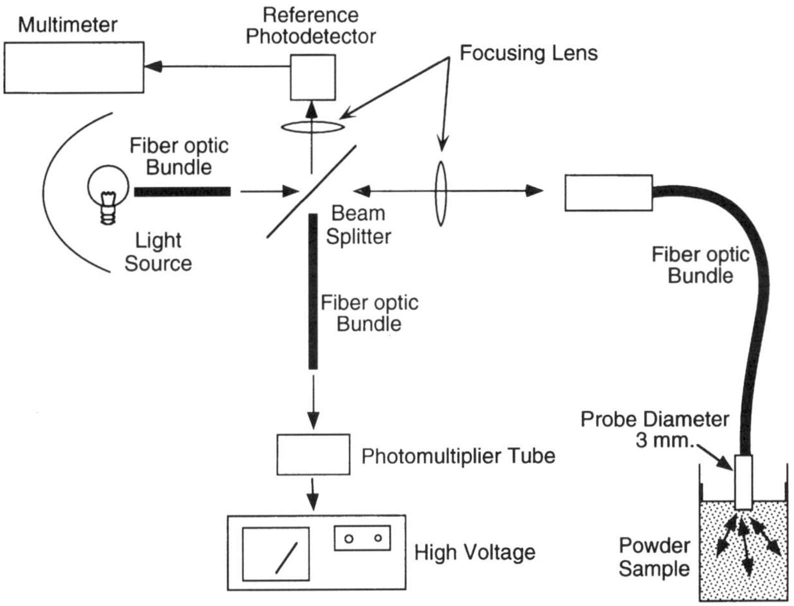

Figure 5.16 Configuration of the optical equipment used by Gratton-Liimatainen to study the reflectance of various powder mixtures.

(Studies involving taking color scanning electron microscope pictures of the different mixtures to demonstrate their intimacy, or their lack of it, are being carried out, in co-operation with Bonifazi of the University of Rome.) To allow for this uncertainty, the data summarized in Figure 5.17 were based upon

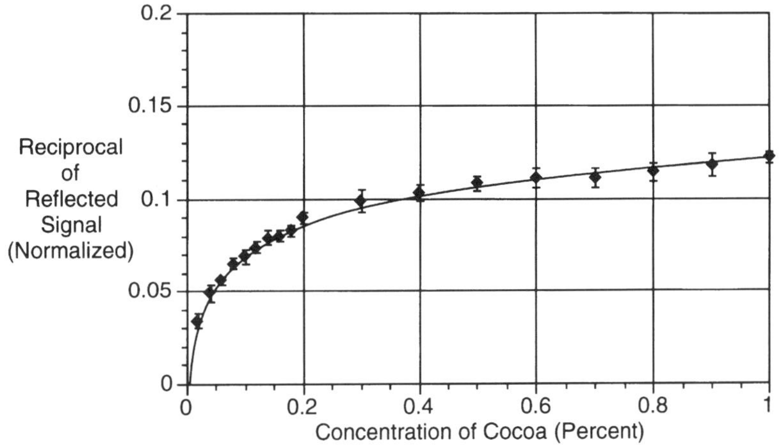

Figure 5.17 Reflectance curves for mixtures of flour and cocoa at various concentrations.

measurements carried out on five independent samples of a given mixture ratio. It can be seen that such reflectance measurements can be used to determine the richness of the mixture up to a concentration of about 0.3 (30% cocoa), whereas for the remainder of the graph, it would not be possible to determine the richness of the mixture from reflectance measurements within a useful range of certainty. However, many industrial procedures would be concerned with mixtures in the range of 80% and upwards of flour, so that the equipment of Figure 5.16 could indeed be used to create a feedback loop to assess the efficiency of this type of mixing process. Current studies are aimed at elucidating the potential of such feedback control of this type of mixture.

Another industry where the efficient mixing of cohesive powders is an industry problem is that of coloring calcium carbonate fillers used in the development of plastic objects. Figure 5.18 shows data generated on the mixing of pigment grade iron oxide with a calcium carbonate filler powder. Again, the utility of reflectance measurements for developing feedback control loops depends upon the mixture ratios. Up to a concentration of about 0.1 (10% iron oxide) the technique would be very efficient at assessing mixture ratios to within 1% of the nominal value. Less precise decisions could be taken for concentrations in the region of about 0.1–0.4, whereas for concentrations above this region, the method would not have any utility. However, bearing in mind again that the color pigment would be the expensive ingredient and that one is more likely to have additions of pigments of the order of less than 10% by weight, it appears that such reflectance measurements could be useful in the evaluation of the efficiency of a mixing process. Further studies on the use of such reflectance measurements to monitor mixing procedures are underway and it is anticipated that detailed studies will be published.

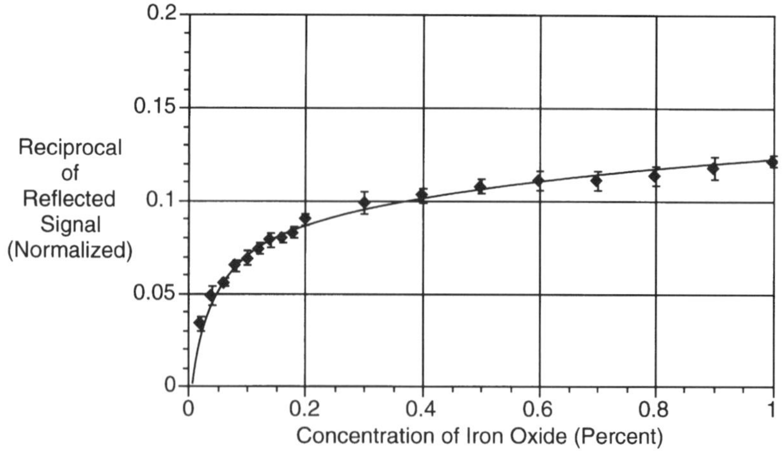

Figure 5.18 Reflectance curves for mixtures of pigment grade calcium carbonate and yellow iron oxide pigment.

5.5 FINGERPRINT SIZING OF POWDER MIXTURES TO MONITOR THE PERFORMANCE OF POWDER MIXING EQUIPMENT

Although the best way to make a mixture is to have ingredients all of the same size, in the real world most mixtures will have ingredients of different size distribution and/or shape. Theoretically, it has always been possible to assess the richness of a powder mixture by examining it under a microscope and carrying out a size distribution and relative proportion of two or more ingredients. However, I am not aware of any studies in which the variation in the constitution of a powder mixture has been assessed from image analysis. Recent developments in the field of particle size analysis have made it possible to implement a study of mixture richness from a size analysis of the ingredients. In Chapter 2 the basic operating principles of the Amherst Aerosizer® were presented. When discussing the sampling of powders, it was pointed out that, if one could achieve a good mixture of the different-sized ingredients of a powder, one could take a representative sample by taking any sample at random from within the volume of the mixer. The AeroKaye™ sampler was also described as an efficient mixing device and data were presented on how one could tackle a difficult problem such as the mixing of two fine cohesive powders in the AeroKaye™. The advantage of using the Aerosizer® for size analysis of mixtures is that it uses a small amount of powder, and the size distribution data from the aerosized sample are generated in a few minutes. It is feasible to consider using a pneumatic lance to take a small sample from within a powder supply and transferring this sample to an Aerosizer® to obtain information on a mixture within a few minutes. The power of this type of strategy to follow the rate of mixing is demonstrated by the data summarized in Figure 5.19. Two fine calcium carbonate powders used as fillers in the plastics industry, one nominally 6 μm and the other nominally 15 μm, were used in this study [21]. In Figure 5.19 the size distributions of the two ingredients as measured by the Amherst Aerosizer® are shown. A small icosolhedral mixing chamber was a quarter-filled with the 6 μm powder and then another equal amount of 15 μm powder was added. The mixture was tumbled for 1 min and a sample taken for size analysis. As the mixing proceeded further, the size distribution of the sample of the mixture was taken at a series of time intervals. The data generated demonstrated the progress of the mixing, as shown by the curves of Figure 5.19(b). After 20 min mixing there were no changes in the size distribution of the mixture because the ingredients were now well mixed. This is demonstrated by the data of Figure 5.19(c), in which the measured size distribution of the sample taken after 20 min is compared to the expected distribution of the mixture as calculated from the proportion of the ingredients and the known size distribution of the original ingredients.

In a more complex mixture there is probably one ingredient which can be tracked to fingerprint the complexity and richness of the mixture. Size

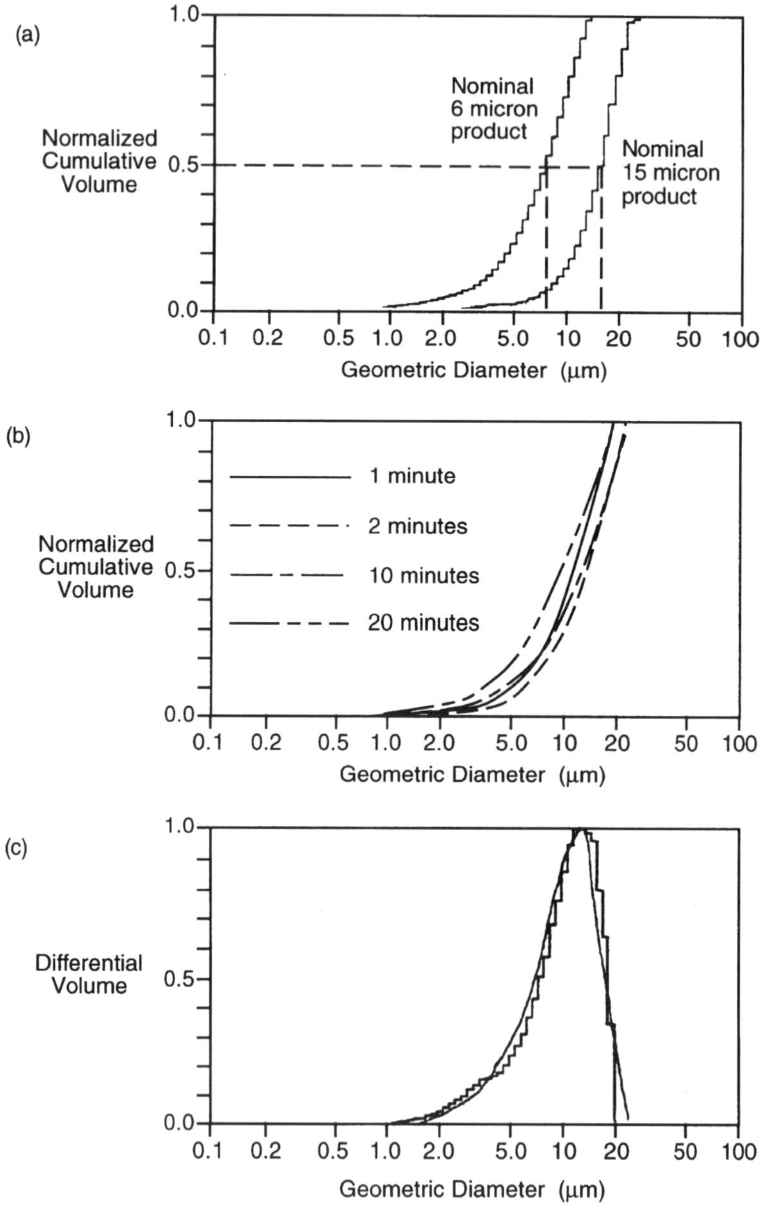

Figure 5.19 The Aerosizer® can be used to monitor the progress of a mixing process by 'fingerprinting' samples of the mixture at various times during the mixing process. (a) Size distibutions of two calcium carbonate powders to be mixed. (b) Size distributions of samples taken from the AeroKaye® at various times during the mixing process. (c) After 20 min of mixing, the size distribution of the sample matched the size distribution of the mixture from the size distributions of the two ingredients.

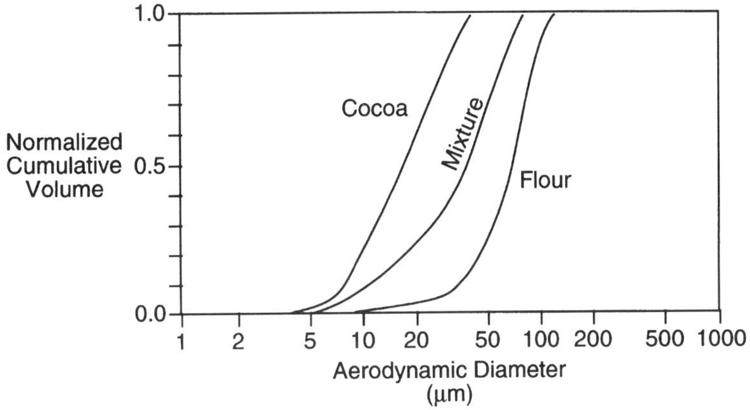

Figure 5.20 Since the size distribution as determined by the Aerosizer® is dependent on the density of the material being sized, when studying mixtures of different ingredients and use the resulting size distribution as a 'fingerprint' of the mixture, as illustrated for the mixture of cocoa and flour.

characterization of a mixture of ingredients is no longer a fundamental study of size, but rather a fingerprinting of the constitution of the mixture. Thus in Figure 5.20 a study of the mixture of cocoa and flour as fingerprinted using the Amherst Aerosizer® is shown. In the processing of the raw data generated by the actual measurements of the Aerosizer®, it is usually necessary to know the density of the different fineparticles. When studying a mixture, however, one assumes that the density of all ingredients is 1, with the subsequent display of the 'fingerprint size analysis curves', as shown in Figure 5.20.

5.6 CHARACTERIZING THE STRUCTURE OF CONSOLIDATED MIXTURES BY OPTICAL INSPECTION

Sometimes a powder mixture or a dispersion of powder in a supportive matrix has to be examined after consolidation to characterize/monitor the effectiveness of the mixing process and to characterize the stochastic nature of the dispersion of one or more ingredients. When dispersing the ingredients of a solid rocket fuel, inspection of the randomization of the ingredients involves a study of a section through the material. Most people are familiar with the appearance of sections through a concrete sample, in which the sections through gravel, sand and cement components are visible [22]. When viewing such a section through a mixture, any lack of uniform dispersion of the ingredients of the concrete is often visible to the observer. To consider the problems involved when attempting to characterize a consolidated mixture of powder by examining a section through a matrix, we will discuss the characterization of the system shown in Figure 5.21. The system of Figure 5.21(a) shows a magnified section through a toner

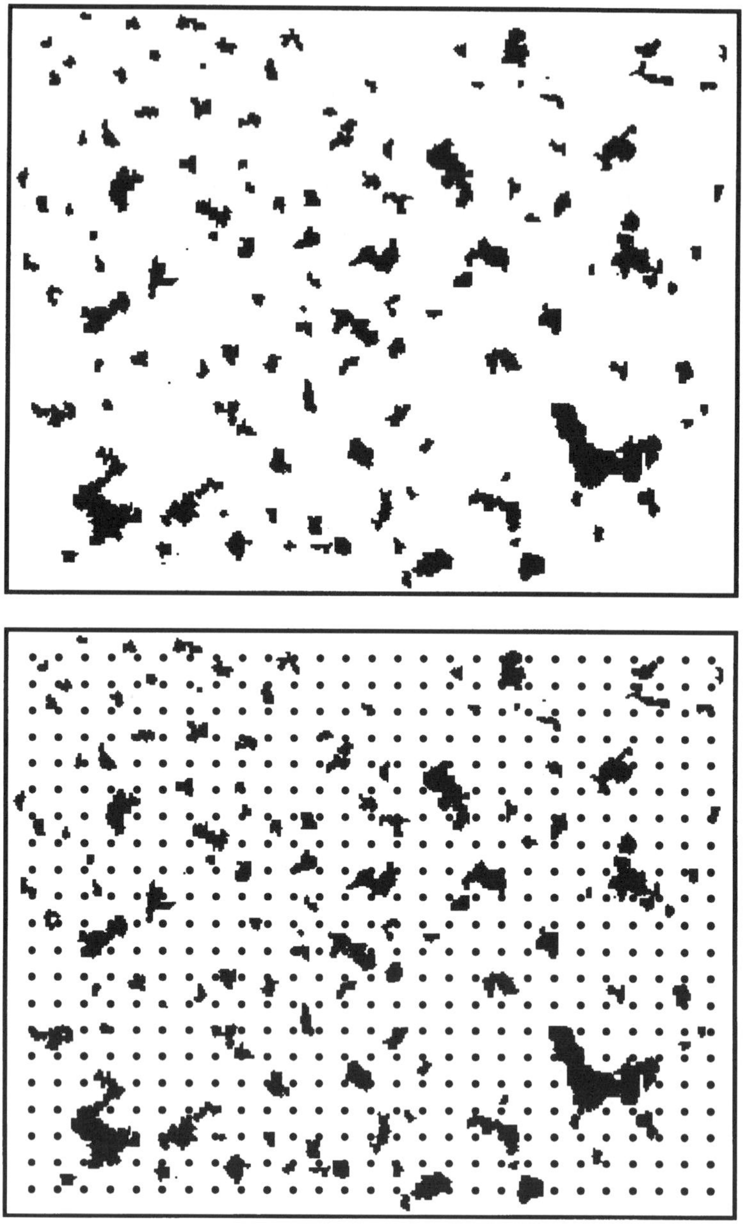

(a)

(b)

Figure 5.21 The percentage area of a pigment species in a section through a mixture can be evaluated using a dot counting technique, developed by Chayes. (a) A high magnification image of a section through a toner bead. (b) In Chayes' method, a dot grid is overlaid on the image and the number of dots falling on the carbonblack fineparticles is used to determine the fraction of the area covered by the carbonblack. (Original micrograph courtesy of D. Alliet, Xerox Corporation; used with permission.)

bead used in an electrostatic copying machine. In the electrostatic copier industry, which includes machines manufactured by the Xerox Corporation, Cannon Incorporated and other manufacturers, the dry powdered ink is sprinkled over the electrostatically charged paper to produce the copy of the document is described as the **toner**. Toner is usually made by dispersing carbonblack or other pigments in a clear plastic matrix. The system of Figure 5.21 is from a sample of black toner and it can be regarded as a simple paint in which black pigment has been dispersed in a clear matrix. Commerical paint systems often contain two or more pigments playing different roles. The system in Figure 5.21 can be regarded as a prototype of any composite material in which the carbonblack of Figure 5.21(a) represents one of the dispersed components. It is surprising that techniques for characterizing the structure of such systems have received very little attention in the technical literature of mixing.

The first characterization problem confronting the technologist when examining the structure of systems such as that of Figure 5.21(a) is to determine the percentage of the area of the section through the structure which is represented by the dispersed component, in this case the carbonblack pigment. Geologists often tackle similar problems, such as the evaluation of the richness of an ore body, which they must determine by examining a section taken through the body. To a geologist, many rock specimens are in fact mixtures of various ingredients, made by natural forces. There is no difference to the geologist in the appearance of a sectioned piece of rock and a section through a piece of concrete, which can be regarded as a synthetic rock. Geologists have developed a technique for measuring the percentage of a section represented by one type of mineral. The technique is sometimes referred to as **Chayes' dot counting procedure** or **modal analysis** [23]. The physical basis of the technique can be appreciated from Figure 5.21(b). A regular grid of dots is placed over the section to be characterized. The percentage of dots falling within the dispersed species is counted and expressed as a fraction of the total number of dots. Thus for the carbonblack dispersion of Figure 5.21, the dot counting technique yields an estimate of 8.03%. If one wishes to know if a field of view such as that of Figure 5.21(a) is a statistically acceptable variation in an expected mixture of the carbonblack in the matrix, one would carry out this measurement on a series of sections. The variation in the measured percentage of pigment in the matrix should have a Gaussian distribution about a mean value which should be the same as that of the expected percentage of the dispersed part of the mixture. Thus the measurement of 8.03% generated for Figure 5.21 might eventually prove to be one of a set of valid fluctuations in an expected population of 8% by volume pigment richness [24].

The mixing technologist can use another technique developed by the geologist for evaluating the percentage of exposed material in a rock section. This technique is known as the **Rosiwal intercept method** [23]. In this technique, the percentage of a line, drawn at random across the section, occupied by the dispersed material being examined, is measured. It can be shown that, provided that the section is a random representative section through a volume of the dispersed material, the measured fractional occupation of the line is the same as the volume percentage of that material in the sectioned dispersion.

In Figure 5.22 we show three methods for creating a set of random lines which could be used for the Rosiwal intercept method. Thus the first set of random lines were created by first selecting an angle θ from equally distributed values of θ using a random number table. In this procedure, a number between 0 and 360 is selected at random to determine the angle of a radius vector to be constructed from the center of the circle. Having selected the angle, the magnitude of the radius vector $\boldsymbol{r}$ to be drawn in this direction is also selected in a similar manner. Then a line is drawn at right angles to the selected radius vector spanning the circle. The next random line for this set is then selected by

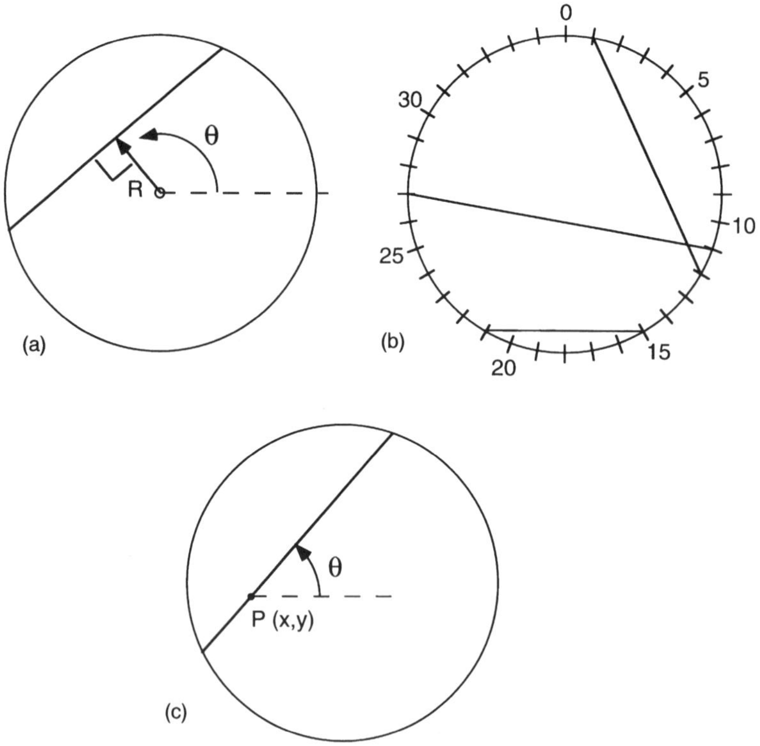

Figure 5.22 Drawing 'random' lines is not a simple task. Several different sets of lines can satisfy the definition of randomness depending on how 'random' is defined. (a) Length and angle can be used to construct lines on a circle (type 1 randomness). (b) Connecting randomly selected points on the circumference of the circle generated random chords (type 2 randomness). (c) Chords may be drawn by selecting an (x,y) location within the circle then choosing an angle at which to draw the line, all at random (type 3 randomness).

choosing another pair of values of θ and $\boldsymbol{r}$ with the subsequent drawing of the chord at right angles to the selected values.

In the second method, shown in Figure 5.22(b), called 'type 2 randomness', points along the perimeter of the circle are given a linear distributed set of numbers as shown, and then two numbers are selected at random and a line is drawn to join them. When lines are selected in this way, the resultant lines have a higher spatial density towards the perimeter of the circle used to construct the set of random lines.

The third method, shown in Figure 5.22(c) correspond to what many people feel is an intuitively acceptable definition of randomness. One imagines that the circle is located on a grid system and that a coordinate of a point through which a chord is to be drawn is selected by selecting y_1 and x_1 from a random number table. Then one selects a value of θ at random, as in the first set of lines, to create a line such as that shown in Figure 5.22(c). To create the set of lines the process is repeated. This type 3 random line population has the property that the probability of a line passing through a point in space is equal all over the spatial area of the space occupied by the lines. By using the circular field of view, one can carry out repeated independent measurements on the same field of view by rotating the set of random lines on top of the profiles to be evaluated.

If the pigment or other powder grain profiles are distributed randomly in the field of view, then a set of parallel lines constitutes the equivalent of a random search for pigment density in the field of view. Although it does not appear to have been used in the scientific literature, one could actually test whether the profile in a field of view such as that of Figure 5.21(a) is really randomly distributed by carrying out intercept measurements using a set of parallel lines of the same length as a set of random lines and comparing fluctuations obtained from the two line sets. If the standard deviation or the distribution of information obtained using parallel lines placed in a specified direction on the specimen differs from the data generated by interrogating the field of view with a set of random lines, then bias exists in the field of view. POMM can supervise and interpret a set of measurements of this kind and print out a decision as to the probability of a random dispersion or a biased dispersion in space. Note that a biased dispersion in this context would mean that the variation from area to area in space is not consistent with the hypothesis that the pigment has been dispersed chaotically. POMM can also interpret the probability that a set of dot counting estimates of the area of dispersed species is compatible with the hypothesis that the pigments are randomly (chaotically) dispersed in space.

It should be noted that the average length of the intercept made by a line crossing the pigment at random can be related to the average size of the dispersed pigments. A study of the efficiency and logic of the Rosiwal intercept method and Chayes' dot counting techniques are part of a branch of mathematics known as **geometrical probability** [25]. The general theorems of geometrical probability have undergone considerable development by scientists active in a subject known as **stereography**, which is defined as a study of three-dimensional structures reconstructed from information collected in one- and two-dimensional space from sections made through three-dimensional bodies. Stereography has been an active area of study for scientists in disciplines as varied as the study of size and spatial distribution of cancerous tumors in humans, the structure of porous metal, ceramic bodies and the structure of ore bodies being exploited by depth mining. Several books have been written on the subject of stereography and on the theorems of geometrical probability, most of which can be applied to the study of sectioned powder mixtures [25–27].

In our discussion so far, we have assumed that the sections made through a consolidated mixture such as that of Figure 5.21(a) are physical sections made directly through the structure. In recent years, specialists in image technology have developed techniques for generating computer images of a two-dimensional section through a body created from multiple X-ray photographs taken at many angles around a two-dimensional examination plane of the system. In medical technology in North America, such techniques are known as **computerized axial tomography** (CAT) [28–30]. The word **tomography** meaning 'a drawing of a section' comes from the Greek root word *Temnein* meaning 'to cut'. This Greek word has given us the English words **tome** (a book, which was originally a section cut from a scroll) and **microtome** (a scientific instrument for making very thin slices through specimens to be examined in a microscope). Tomography is the art of making drawings with the use of multiple photography from many angles using penetrating radiation and a large computer [31]. The tomograph looks like a real cut through a system. The word 'axial' enters into the phrase given earlier because in the initial medical technology the patient being examined was rotated about an axis located orthogonally with respect to a fixed beam of X-rays. In modern instruments many beams are used and there may be rotation of the beams and/or the patient. In North America the technique is known as **CAT scanning**. CAT scanning is beginning to be used industrially and there have been reports of techniques for studying the structure of settled suspensions [32–35].

Tuzun and colleagues at the University of Surrey in Guildford, UK have reported on the use of a gamma-ray-based CAT scanning to study the movement of powders in fluidized beds and to study the movement of powder leaving a storage bin [36]. Thus Tuzun *et al.* show that, as powder leaves the bin, the voidage increases along the flow patterns because of the entrainment of air. Tuzun reports that with his gamma ray technique, the resolution is already better than half of a millimetre. We can expect increased usage of CAT scanning based on X-rays and other penetrating radiation, such as gamma rays, as the cost of the necessary computer processing continues to fall. Although there have not yet been any direct studies of mixtures of powders using CAT scan technology, Broadbent and colleagues have used tomographic techniques to follow the movement of a radioactive pellet in a Lodgie–Littleford type plough mixer [37].

A new way of characterizing the appearance of a field of view such as that of Figure 5.21(a) has developed through the use of fractal geometry to study such systems. The mathematical model used to characterize such systems is known to the mathematicians as the Sierpinski carpet [38]. We mentioned briefly the construction algorithm for a particular Sierpinski carpet in Chapter 1. In Figure 5.23, the first four stages in the construction of a particular Sierpinski carpet (an infinite number of carpets are possible depending on the size of the holes used to create the carpet) are illustrated. In this case, the initial square from which the carpet is to be constructed is divided into nine squares and the central hole removed. The process is then repeated with the residual squares and so on to create the systems of Figure 5.23(a). As already mentioned briefly in Chapter 1, the resultant system after this construction algorithm has been carried out an infinite number of times, is a carpet with no area constructed from an infinite number of infinitely thin threads. At this stage of its development, the Sierpinski carpet does not appear to resemble the carbonblack dispersion of Figure 5.21(a). However, one can create an alternative version of the carpet, known as a statistically self-similar carpet, by shuffling the holes in the carpet at random. The resultant self-similar session of the carpet at the third stage of construction is shown in Figure 5.23(c). It should be noted that an important difference between the idealized and self-similar versions of the Sierpinski carpet is that the holes of the ideal carpet are unconnected, whereas in a well-developed self-similar carpet the holes are interconnected.

Mathematicians characterize the structure of an ideal Sierpinski carpet by looking at the rate at which the area of the carpet disappears as one creates more and more holes in the original square. Figure 5.23(b) shows the rate of disappearance of the ideal Sierpinski carpet shown in part (a) against the size of the holes being created in the carpet. This log–log plot for the 'holiness of the carpet' is known, for historical reasons, at the Richardson plot which characterizes the carpet. It can be shown that the slope of the rate of the disappearance of the carpet on such graph paper, when subtracted from 2, results in a characteristic number known as the **Sierpinski fractal** of the porous body. The ideal carpet of Figure 5.23(a) has a mathematical fractal dimension of 1.89. The statistical self-similar version of the Sierpinski carpet does not disappear as quickly as the ideal carpet because the random shuffling of the holes results in some of the smaller holes falling into the existing holes of the carpet. This is demonstrated by the Sierpinski fractal dimension deduced from the data of Figure 5.23(c), based on an experimental assessment of the rate of disappearance of the statistical self-similar version of the carpet created in Figure 5.23(a). Figure 5.24 shows the data from an experimental study of the structure of the carbonblack dispersion of Figure 5.21, treated as if it were a Sierpinski carpet, with the black pigment becoming the equivalent of holes in the mathematical treatment of the data.

It can be seen from the display of data in Figure 5.24 that there are two linear relationships in the carpet disappearance data for the carbonblack dispersion data. One interpretation of this dual set of datalines is that the coarse resolution data represents the actual dispersion of the carbonblack and that the secondary

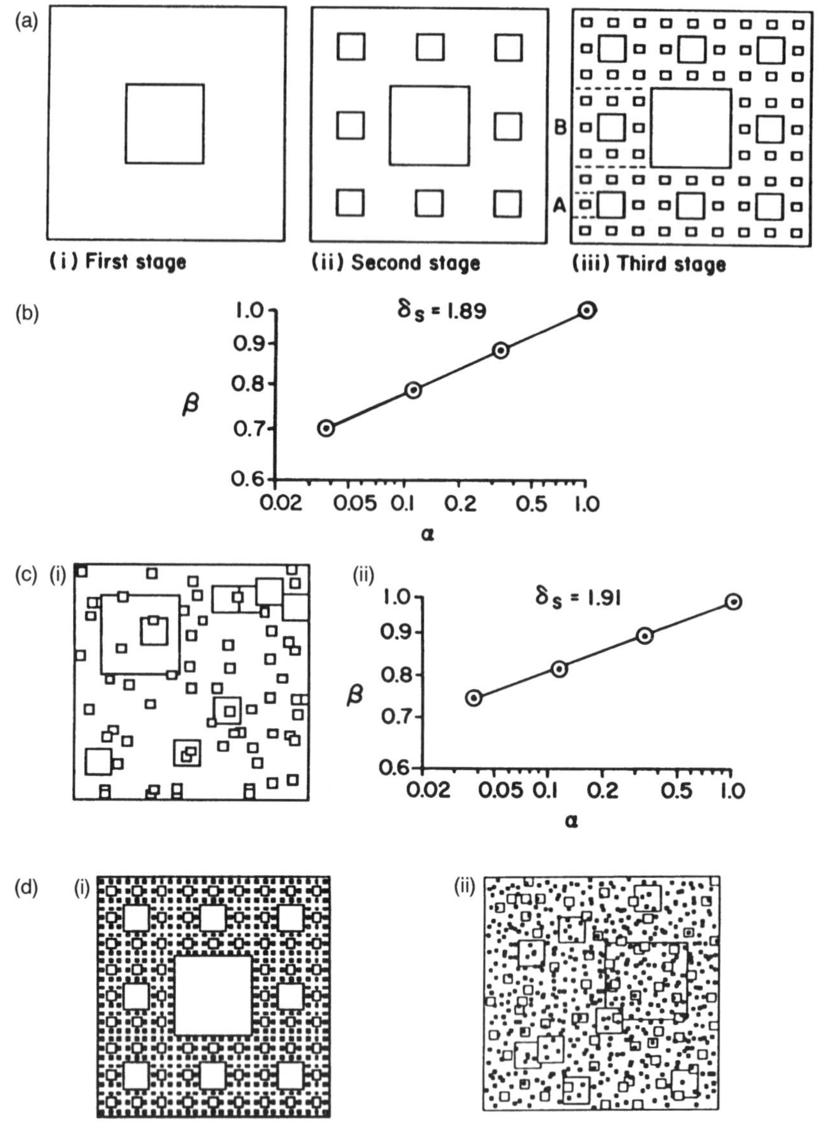

Figure 5.23 The Sierpinski carpet is a mathematical construction which, at the limit, has no area and an infinite number of holes. (a) Construction of a Sierpinski carpet. (b) Richardson plot obtained by plotting the residual area of the carpet, β, against the normalized resolved hole size, α. (c) When the holes of (a)(iii) are randomized, some overlap results and the residual area of the carpet does not disappear as quickly. This carpet can be described as a 'statistically self-similar Sierpinski carpet'. (d) The Sierpinski carpet rapidly becomes more complex as we progress to further stages of construction; shown here is the fourth stage of construction (i) and its randomized counterpart (ii).

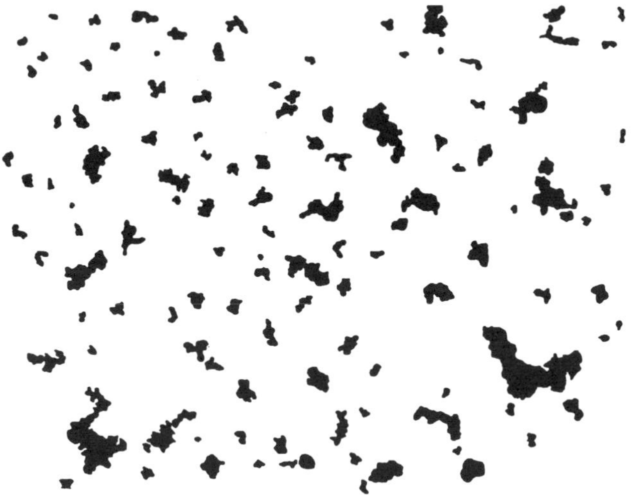

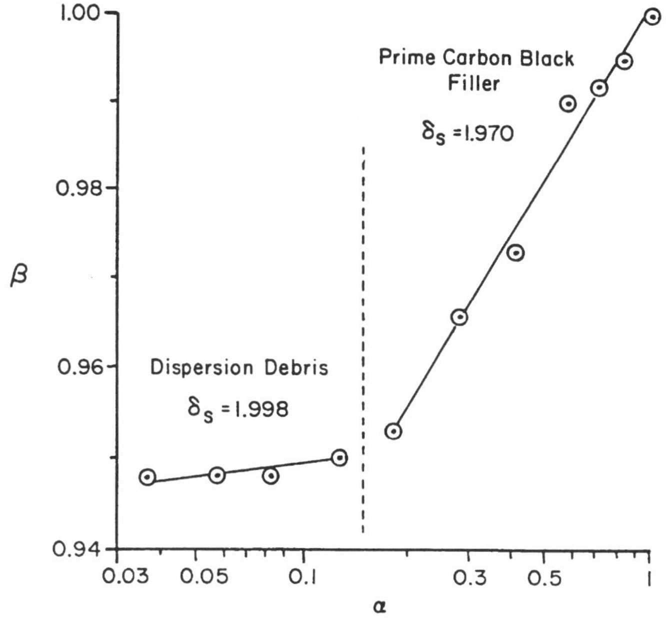

Figure 5.24 Richardson plot of the data generated when the section through a toner bead of Fig. 5.21 is treated as a Sierpinski carpet. The two regions of the graph represent two populations of fineparticles present in the dispersion. The steeper region represents the original carbonblack fineparticles, while the shallow, second region, represents fragments of carbonblack agglomerates created by high shear dispersion of the carbonblack into the plastic matrix material in the creation of the toner bead [39].

dataline represents fragments of the primary carbonblack pigments created during the dispersion shearing stress in preparing the dispersion [39]. It should be noted that the Sierpinski fractal dimension of a dispersion does not contain any information which is not also present in the size distribution of the dispersed items. It does, however, form a convenient summary which is characteristic of the dispersion. It is probable that the physical properties of a dispersion can be related empirically to the magnitude of the Sierpinski fractal. This possibility is one of the many new ideas and concepts being generated by the applications of fractal geometry to materials science, and only further experimentation will indicate whether or not such fractal dimensions are a useful concept when studying dispersions.

Alder and colleagues, in a series of publications, have applied what is known as a distance transform function to the characterization of the quality of a powder mixture [40–43]. To understand what is meant by a distance transform function in the computer-aided processing of images, consider the boundary shown as a set of pixels in Figure 5.25(a). These pixels could represent a river on a map and the stars could denote villages. In their planning of the development of such things as water supply by pipeline, geographers invented what they called the map into a pixelated terrain, with each pixel labeled with a number giving the distance of a specific pixel (in pixel units) to the nearest point on the boundary being considered (in the case of Figure 5.25 a river), as shown in Figure 5.25(b). The scientific processing of images such as Figure 5.25(a) can now use what is known as thresholding operation to select areas within a chosen distance of a boundary. To illustrate the appearance of a pair of operations of this kind to a boundary, Adler and co-workers used the boundary of Ireland, as shown in Figure 5.26. What Adler and colleagues call the 'swollen' outlines at distances from the actual coastlines are called the 'dilated boundary' and the pixel distance from the coastline is the dilation radius. (In their publications Adler and co-workers have described how a study of the dilated boundaries can be used to measure the fractal dimensions of the rugged boundary being studied [40].) When the distance transform function is applied to an image of a mixture, the interparticle distance is dramatically displayed, as shown in Figure 5.27, and this information can describe in a quantitative manner the structure of the mixture.

In the words of Adler and Melia,

> The method in essence measures the shapes of the spaces between the particles of the mixture by calculating the distance from every empty pixel to the nearest particle. In well mixed systems there are more pixels close to particles while in poorly mixed systems greater distances are more common. Note that to implement Adler and Melia's logic it may be necessary to first transform a dispersion of fineparticles into an array of dots representing the profile. Also note that if one is using image analysis systems with color imaging capacity one could use a particular color to

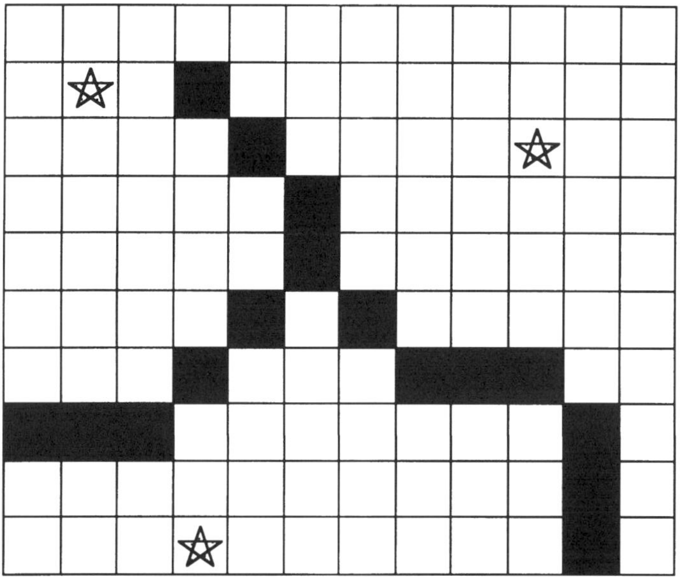

(a)

3.2	2.2	1.4	1.0	1.4	2.2	2.8	3.6	4.2	5	5.8	6.3
3	2	1		1	1.4	2.2	2.8	3.6	4.5	5.1	5.4
3.2	2.2	1.4	1		1	1.4	2.2	3.2	4	4.1	4.5
3.6	2.8	2.2	1.4	1		1	2	2.8	3	3.2	3.6
3	2.8	2.2	1.4	1		1	1.4	2	2	2.2	2.8
2	2	1.4	1		1		1	1	1	1.4	2.2
1	1	1		1	1.4	1				1	1.4
			1	1.4	2.2	1.4	1	1	1		1
1	1	1	1.4	2.2	2.8	2.2	2	2	1		1
1.4	2	2	2.2	2.8	3.6	3.2	3	2	1		1

(b)

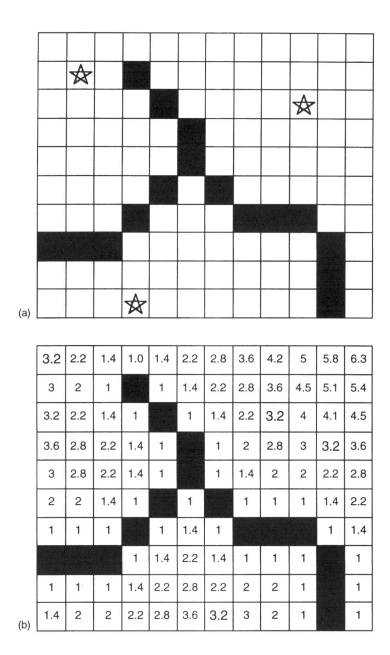

Figure 5.25 Geographers developed what is known as a 'distance transform function' to help them examine such things as water supply to towns from a river. (a) A simple, pixelated map of a river basin and three town sites. (b) Distances to the nearest water supply in terms of the pixel side length.

isolate a set of profiles of one ingredient and then carry out the analysis of the mixture repeating the procedure for other ingredients [42].

Adler and Melia describe two other procedures for studying sections through a mixture by image analysis. In the first, one still uses the distance transform function, but in this case one in fact calculates the watersheds between pixel distances to an adjacent profile to generate the type of pictures shown in Figure 5.28. The particle size distribution of these 'watershed delineated areas' now becomes a quantitative assessment of the quality of the mixture. This is again demonstrated by the data shown in Figure 5.28.

In another publication, Adler and Melia [43] describe what they call a serial dilation for measuring powder mixture homogeneity. In image processing one can cause an individual image to grow by adding pixels to the boundary of an

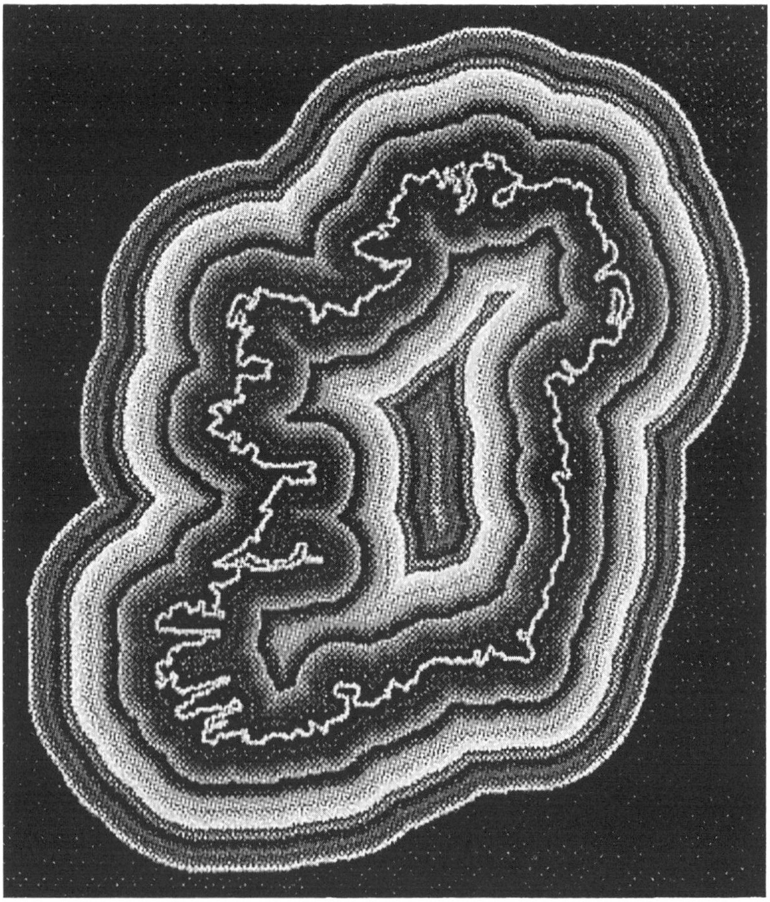

Figure 5.26 The distance transform function can be used to construct contours of common distance from a specified boundary.

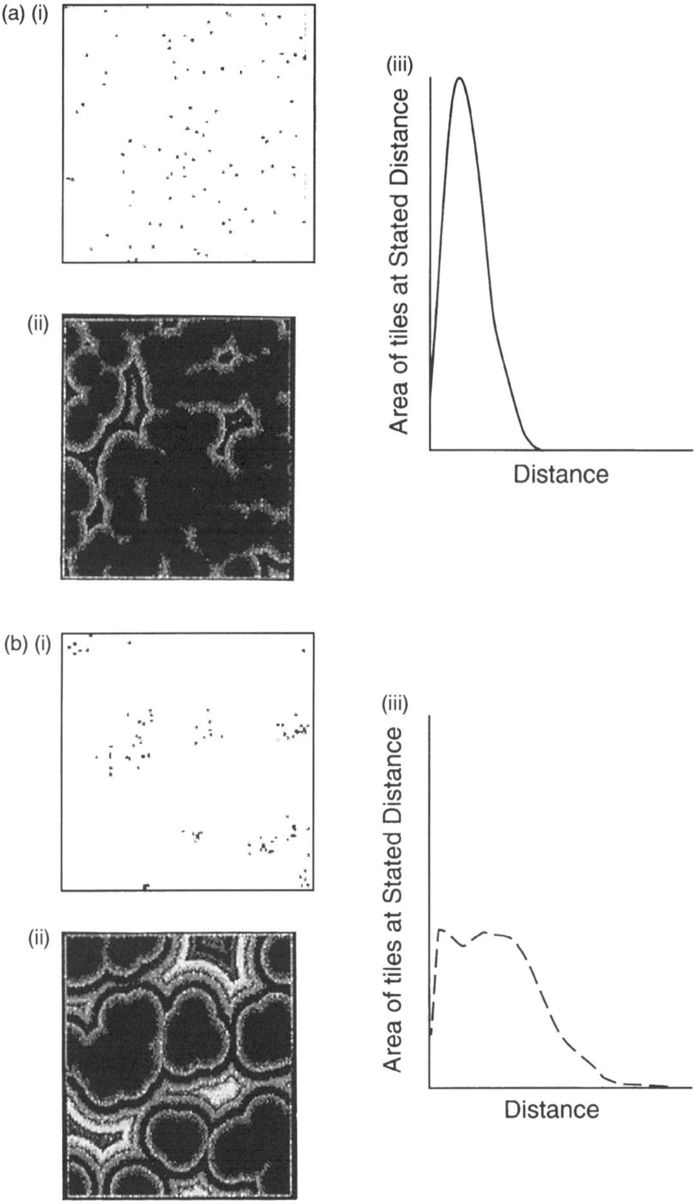

Figure 5.27 The use of the distance transform function to process images of mixtures is a graphic way to show the differences in structure of mixtures. (a) Application of the distance transform function to a random mixture. (b) Application of the distance transform function to a clustered, nonstochastic mixture.

object when moving around the boundary. This process is known as dilation. The complementary method in which one strips off pixels from the boundary of an object viewed in the image analysis is called erosion [44].

As Adler and Melia point out, if one takes a point image of an ingredient in a mixture and dilates that point for a perfect mixture there would be a maximum rapid growth of the area covered by the dilated mixtures, whereas in poorly mixed materials the dilated images of particles in a cluster of ingredient would start to merge with each other, reducing the overall growth of the density of the image under the dilation process. Thus in Figure 5.29 the point image transforms of four types of dispersions of ingredients in a mixture are shown. The rate of growth of the image density as the four different images are serially dilated is shown in Figure 5.29(b). Adler and Melia point out that these curves can be processed in various ways to make quantitative statements about the quality of the mixture. They suggest that a simple mixing index could be based on the studies of this kind [45].

When one looks at a relatively sparse distribution of an ingredient, such as that shown in the first block of Figure 1.15, one can intuitively anticipate that

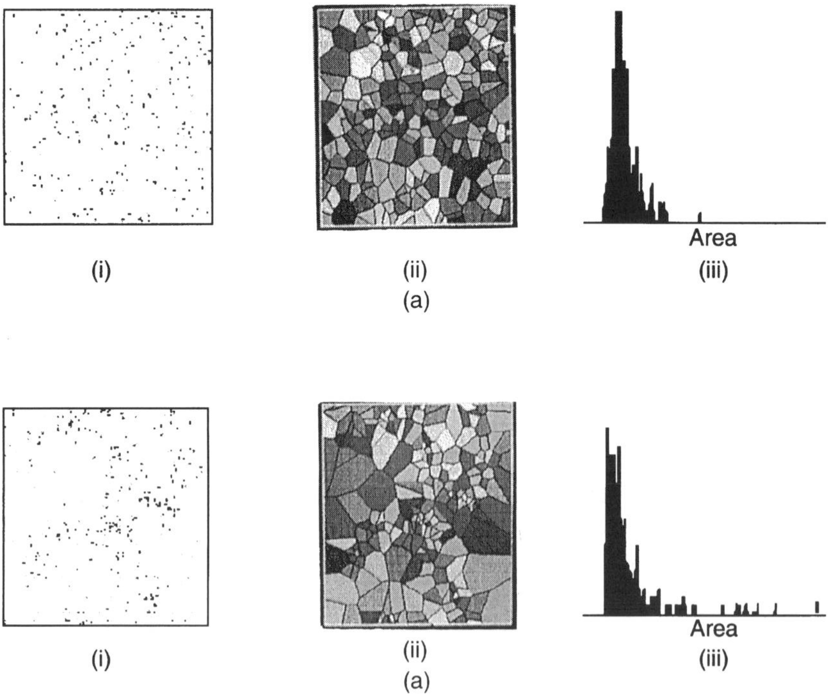

Figure 5.28 Delineating the watersheds between 'pixel zones of influence', transforms a mixture image into a display not unlike the crystals in a rock section. The size distribution of the watersheds is a measure of the mixture quality as shown by the graphs. (a) Results for a random mixture. (b) Results for a clustered mixture.

drawing tracks between the ingredient locations should contain information on the dispersion structure. The dispersion of Figure 1.15 is a simulated 10% dispersion of a black ingredient in a white matrix created by changing every 9 in a random number table into a black pixel. Thus by definition it is a truly random dispersion. To test the hypothesis that one could obtain some measure of the quality of a dispersion by making track to track measurements, Kaye and Clark measured the point-to-point track length as they moved up and down the random number table in vertical and reverse vertical directions. Obviously such an investigation is starting to have problems with clusters of ingredients and also the answer is digitized in units equivalent to the size of the fineparticle. However, as shown in Figure 5.30 the track length between the simulated ingredient fineparticles is distributed according to a log–normal distribution.

Kaye and Clarke extended their simulation experiments to show how the median track length shifted with pigment concentration. Their data are summarized in Figure 5.31. Since the whole array is a stochastic dispersion, one would expect the direct measurement of the track lengths to vary from specimen to

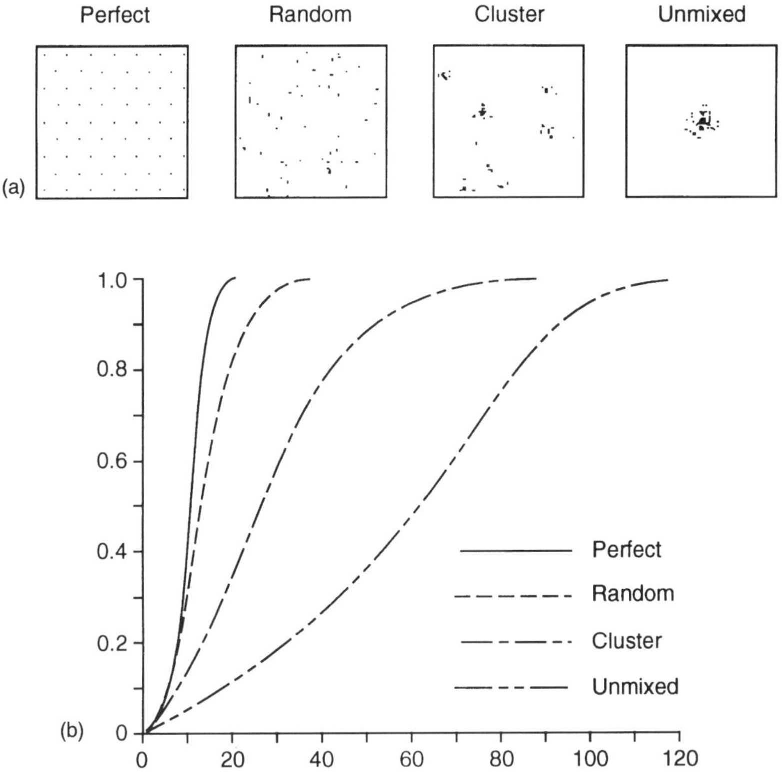

Figure 5.29 The growth in the area occupied by an ingredient when the locations of the ingredient particles are subjected to serial dilation can be used to assess the quality of a mixture [45].

specimen of the simulated mixture. To see how such a structure variation from sample to sample affects the estimates of the richness of the mixture as well as the characterization of the dispersion, measurements were carried out on several simulated ingredient dispersions. In Figure 5.31 the variations in the median track length for a set of investigations are shown, indicating that if one is prepared to take a large number of measurements, the technique could not only generate information on the richness of the mixture, but also on the randomness or lack of randomness of the dispersion. Kaye and Clark pointed out that in a real mixture the shape of dispersed fineparticles influenced the answer obtained by track measurements. To avoid distortion of the measured track lengths by the shape and size of the dispersed ingredients, they discussed in detail how to convert a real dispersion into a point dispersion suitable for carrying out track length evaluation of the structure of the dispersion. Thus in Figure 5.32, various ways of drawing tracks to explore the structure of a real dispersion of carbon-black fineparticles in a white matrix are shown. In Figure 5.32(a) the idea would be to use random tracking in which each time your line touched a profile (or the

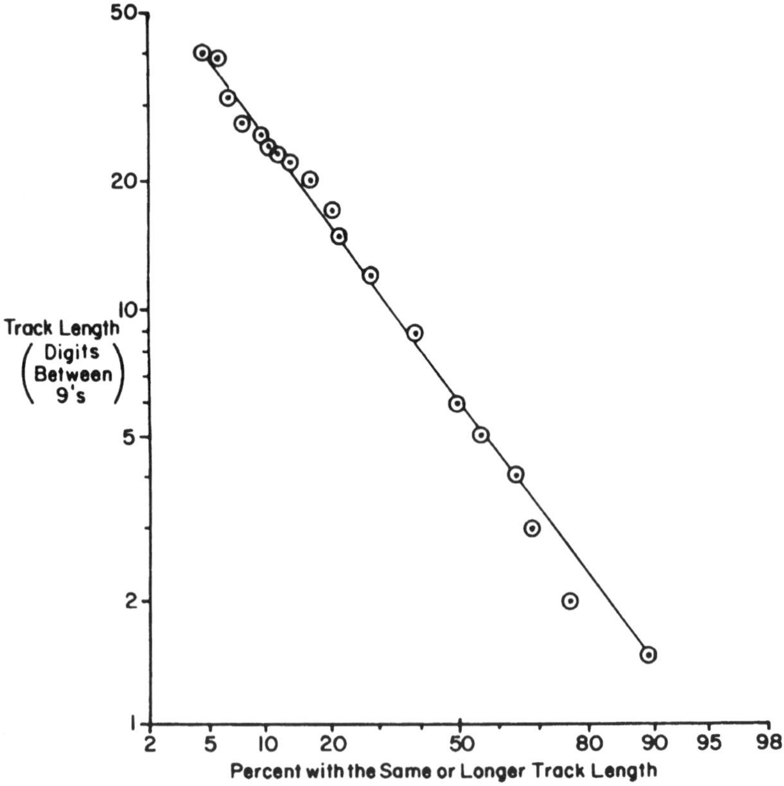

Figure 5.30 Point-to-point tracking of a dispersion should contain information on the quality of the dispersion. (From Kaye and Clark [46].)

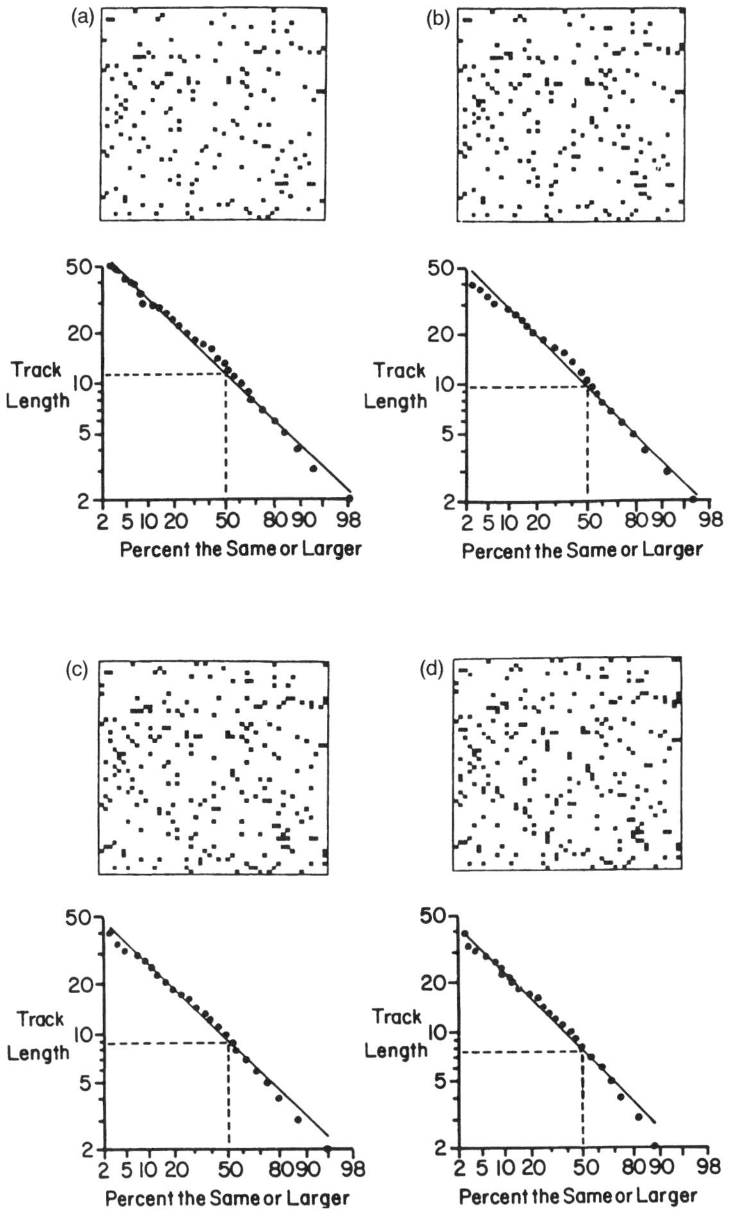

Figure 5.31 Kaye and Clark showed that the range of track lengths between ingredient particles in a truly random mixture changes in a manner related to the richness of the mixture. (a) Appearance of and data for a 7% mixture of black ingredient. (b) Appearance of and data for a 8% mixture of black ingredient. (c) Appearance of and data for a 9% mixture of black ingredient. (d) Appearance of and data for a 10% mixture of black ingredient. (From Kaye and Clark [46].)

border) you would choose a new direction at random to construct the next track. In Figure 5.32(b) the construction of what is known as Aster scanning is shown. Again, although informative, information generated would be subject to shadow problems of the larger profiles, particularly if one unfortunately chose a reference point close to one of the larger fineparticles as illustrated in Figure 5.32(b)(ii). After considering very different methods of scanning the array Kaye and Clark decided to use a parallel line scan (Figure 5.32(c)). They investigated

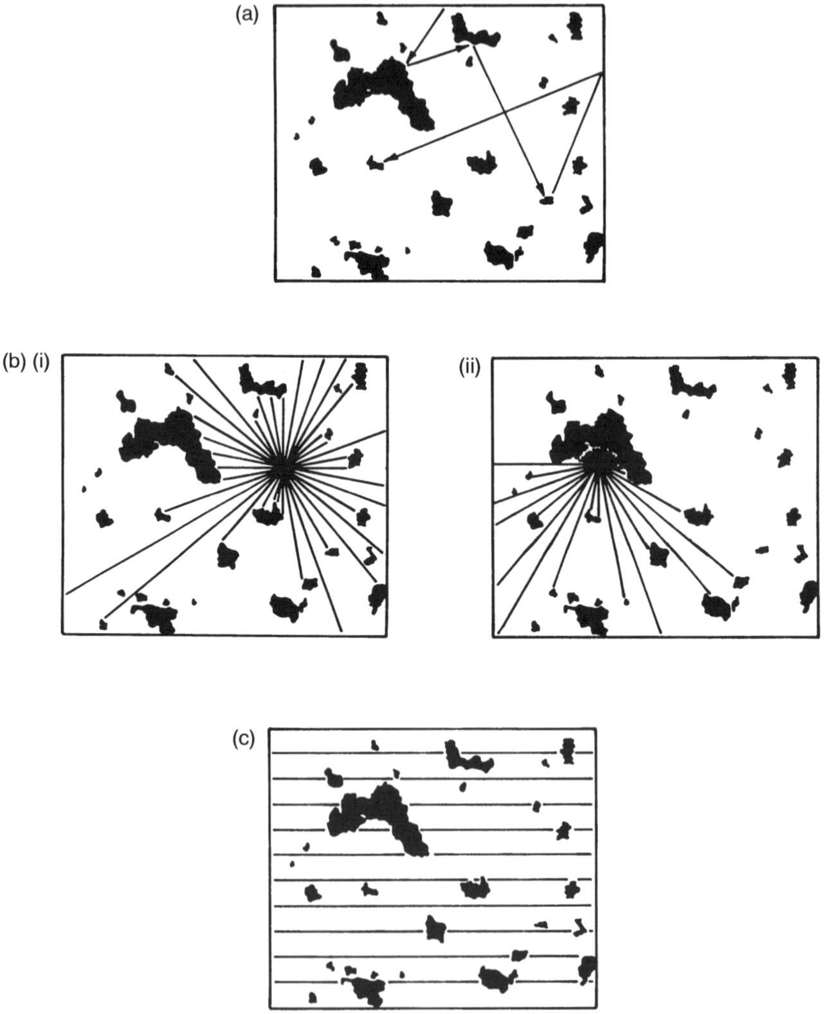

Figure 5.32 Evaluation of the track length between ingredient fineparticles can be carried out using several techniques. (a) Randomly redirected search lines. (b) Aster scan. Note that the starting point chosen can result in a severe 'shadow effect' as shown in (ii). (c) Parallel line scans. (From Kaye and Clark [46].)

the effect of the shadows of the larger fineparticles, and their data is summarized in Figure 5.33.

Thus in Figure 5.33(a) the finite width of the profile was ignored and the next track began at the opposite side of the profile. In (b) the track generated by ignoring the size of the profile instead of starting the next track immediately

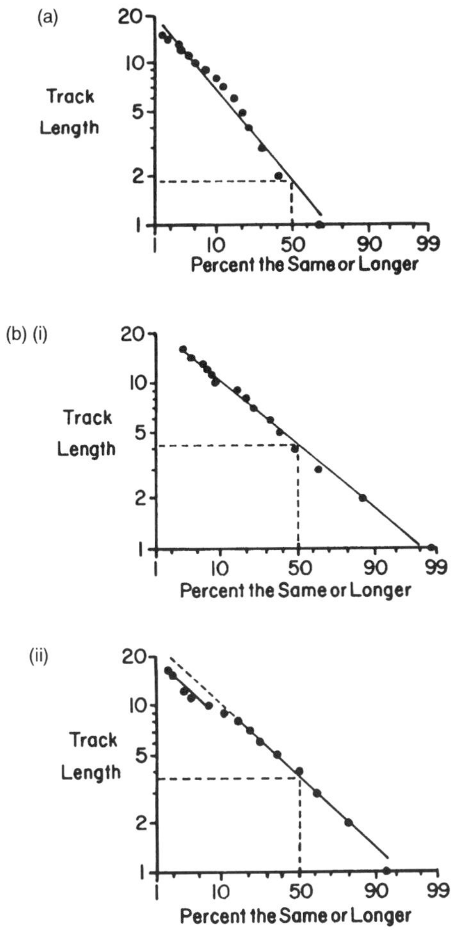

Figure 5.33 The size of the ingedient fineparticles can cause disturbances in the track lengths measured using parallel line scans. Here we demonstrate variations on the technique applied to the image of Figure 5.21. (a) Track lengths generated by line scans spaced at the average size of the fineparticles, when the length is measured by 'stepping over' a profile, that is the track length does not include the distance across the profile. (b) Data generated when the size of the profile is ignored and the next track length starts from the point of first contact for parallel lines spaced at (i) the average size of the fineparticles and (ii) half the average size of the fineparticles. (From Kaye and Clark [46].)

after meeting the profile is shown. As can be seen, the dispersion of the carbonblack in the mixture appears to be completely random, since the search track data is log normally distributed. Kaye and Clark pointed out that the separation of the search lines could influence the answer and in Figure 5.33(b)(ii) the effect of increasing the density of scan lines is shown.

It should be recalled that any real field of view is a limited sample of a larger ensemble and that, because of the stochastic nature of the experiments, there is inherent variation in any one particular sample. Thus in Figure 5.34, track lengths from a sample of simulated 10% ingredient, which was the lowest one encountered in a series of 20 measurements, actually have an ingredient richness of 8.7% with respect to the nominal coverage. The other set of tracks came from a specimen which, although nominally 10% ingredient, had an actual ingredient richness of 11.9%. In Figure 5.34(b) the variations in the mean track length for several simulated 10% ingredient richness mixtures are shown and it can be seen that the variations within a set of measurements are Gaussian. Thus probably the best estimate of the track-to-track length for a monosized ingredient dispersed at random is 7.4 particle diameters (that is the mean of Figure 5.35). In Figure 5.36 the track length between ingredient particles when all of them are reduced to circles of the same size is shown. It should be noted that one encounters a philosophical dilemma when working with an array transformed into a point system. If one uses infinitely thin scan lines and very small dots, there are theoretically no encounters; the scan lines would move between the tiny points representing the fineparticles without any encounters. A way of dealing with the finite sizes of the ingredient particles without transforming them into a point set is to place an array of the profiles in a memory system, and as the scan line proceeds, each profile is discarded from the memory at first encounter. This is a quick way of dealing with finite size, but as shown by the data of Figure 5.35 if the field of view contains many fines they will be missed between the scan lines. To deal with this problem one would have to move the scan line system down half a width after the first scan to look for the fines.

Kaye and Clarke have also simulated agglomerated mixtures and demonstrated that for such systems the track length between the ingredient fineparticles is bimodal on the log-Gaussian scales as illustrated by the data of Figure 5.36. Bimodal track distribution can also be generated if one had selected a field of view which actually varied in ingredient richness from one part to the other, as demonstrated by the data of Figure 5.37. Kaye and Clark suggest that if one is using track length to characterize ingredient structure, one could measure track length for a satisfactory mixture and for an unsatisfactory mixture, with the two sets of data being stored in the memory of the computer. When a new sample was examined the computer could compare the track lengths with the data in memory and describe the structure as being adequate or inadequate. Obviously such an intelligent machine could have many refinements in its data to be able to look for specific problems in any specific mixture [46].

5.7 AUTO- AND CROSS-CORRELATION OF MIXTURE STRUCTURE

Using the mathematical procedures of auto- and cross-correlation is still working with images of cross-sections through the system; however, the mathematics are unfamiliar to many people who must work with powder technology

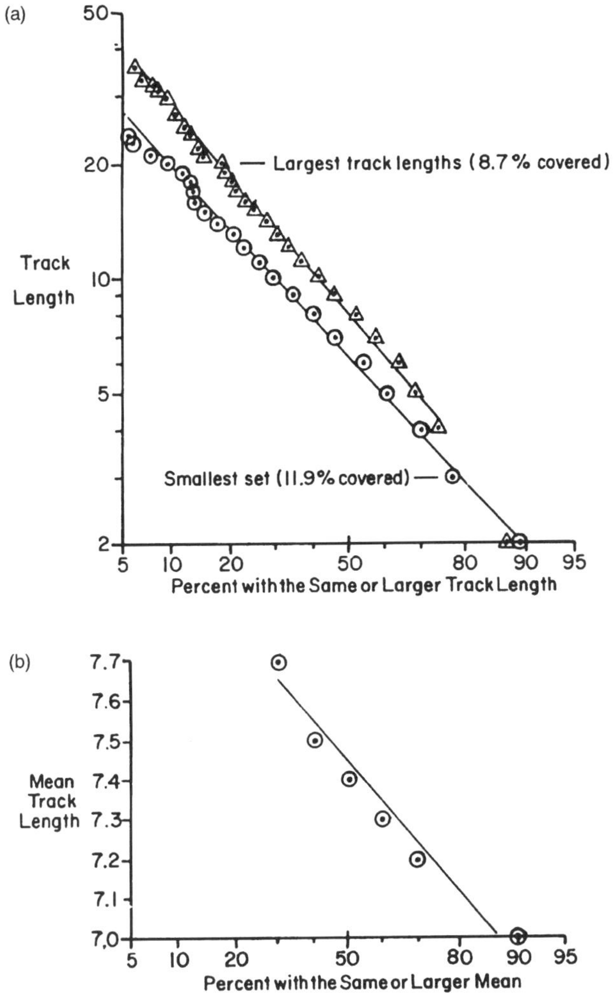

Figure 5.34 Track length distributions for simulated mixtures fluctuates because of inherent variations in the structure of a chaotically assembled mixture. (a) Range of variation in track length distribution found from a series of simulated 10% mixtures. (b) Distribution of the mean track length taken from the track length distributions for the series of 10% simulated mixtures. (From Kaye and Clark [46].)

systems and so the topic is treated separately in this section. Auto- and cross-correlation methods are basically techniques developed by electrical engineers to process noisy signals [47, 48]. To be able to understand how the system can be used with sections through a powder mixture, we will discuss first the use of such mathematical procedures to recognize the basic shapes of profiles from geometric signature waveforms of rugged fineparticles which generate what can

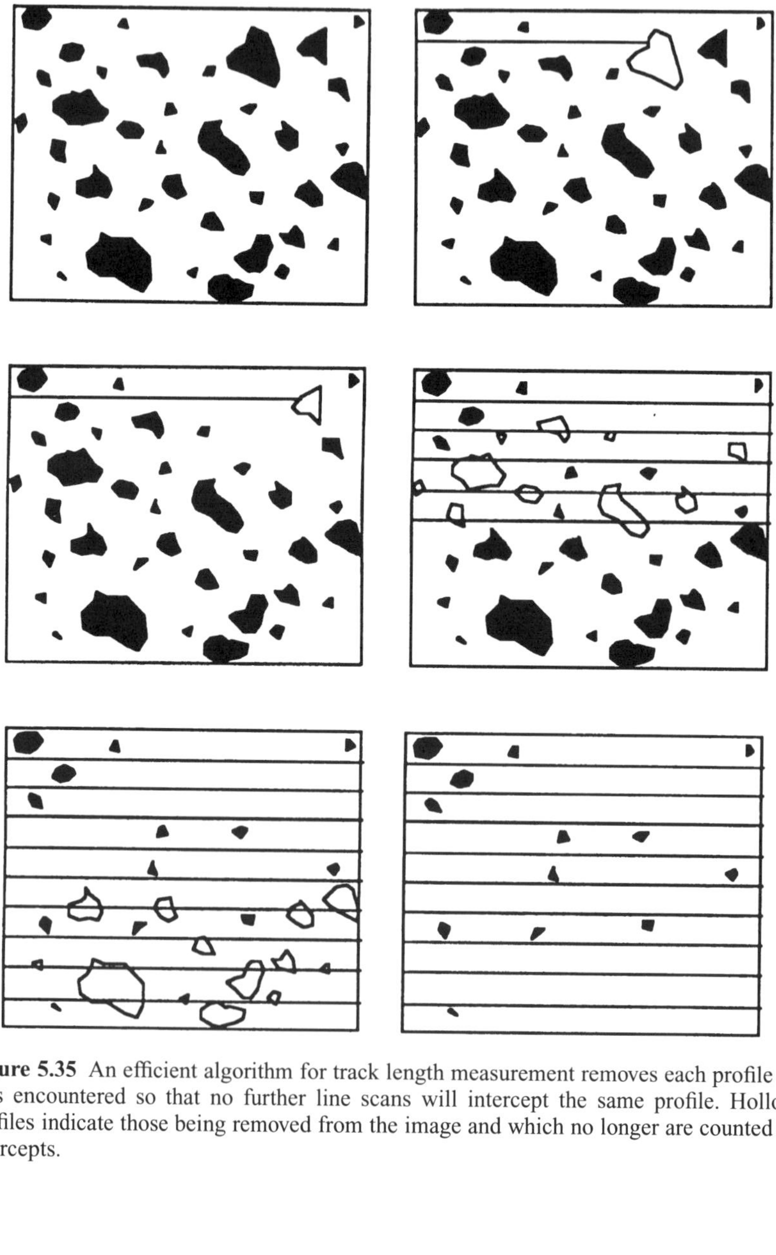

Figure 5.35 An efficient algorithm for track length measurement removes each profile as it is encountered so that no further line scans will intercept the same profile. Hollow profiles indicate those being removed from the image and which no longer are counted as intercepts.

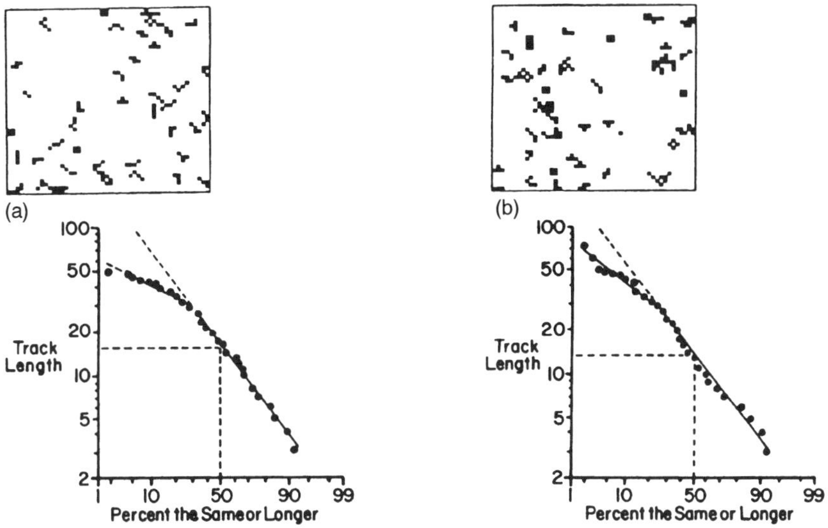

Figure 5.36 Kaye and Clark simulated agglomerated mixtures, representing a mixture that was not completely dispersed. Such mixtures produced track lengths having a bimodal distribution when plotted on log-Gaussian probability paper. (a) Results for a 7% agglomerated mixture; (b) 8%.

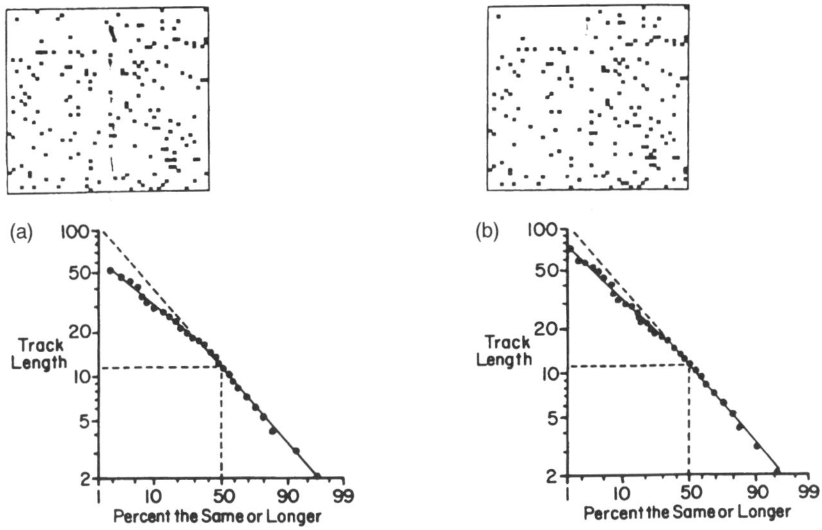

Figure 5.37 Kaye and Clark showed that a bimodal track length distribution could also be indicative of uneven mixing such that one area of a mixture is richer than another. (a) Results for a mixture in which the left half of the field of view is 6% and the right half is 8%. (b) Results for a mixture in which the left half of the field of view is 5% and the right half in 9%.

be considered by the electrical engineer as noisy waveforms. Thus in Figure 5.38 the geometric signature waveforms generated by working with four different ellipses which have different ruggedness of texture are shown [49]. The basic principles of the autocorrelation technique of a noisy waveform can be appreciated from the graphs of Figure 5.39. In Figure 5.39(a) a noisy waveform like those of Figure 5.38 is shown with the known underlying waveform of the basic elliptical shape superimposed. To calculate the autocorrelation function, one

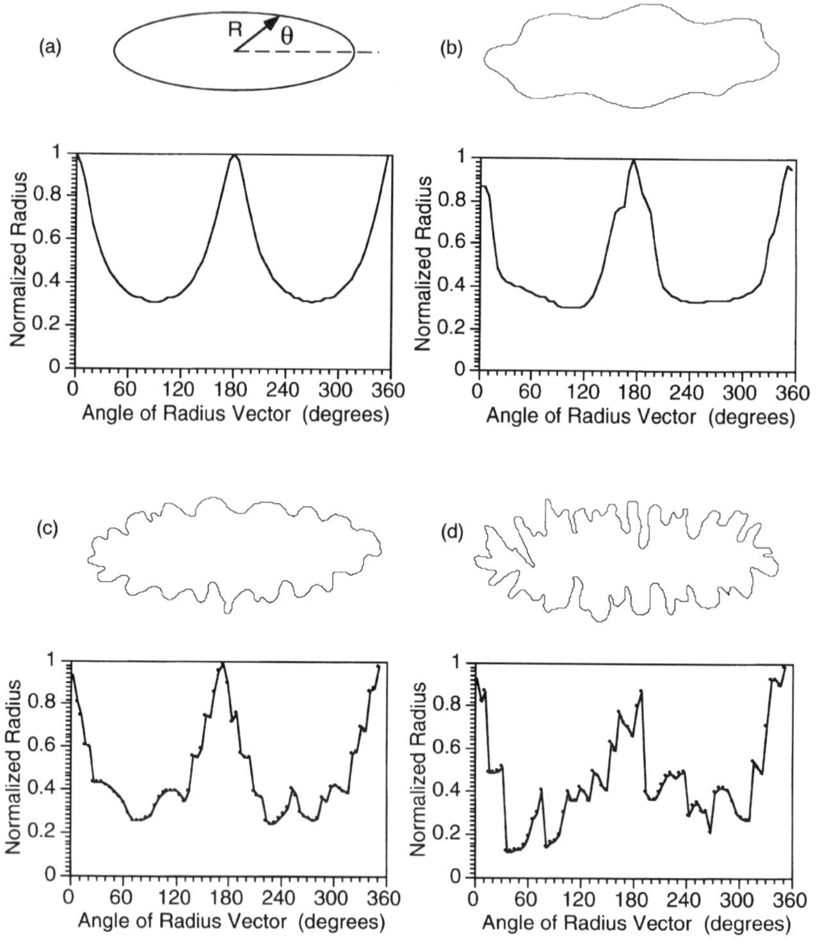

Figure 5.38 The human eye can recognize that all four profiles above are basically elliptical. Using a technique known as autocorrelation of the signature waveforms of the profiles, a computer can learn to ignore the noise on the waveform and recognize the basic shape. Below the profiles are their signature waveforms. Where the radius vector could have taken more than one value, the shortest was chosen, resulting in the sharp changes in the waveforms of the more rugged profiles. (From Kaye [49]. Copyright © 1986 by International Scientific Communications Inc.)

starts at a point 0 as shown, and then moves a distance τ from this reference point. The value at the function at time T is now multiplied by the value of the function at time $T + \tau$. The expression for this mathematical operation is

$$C_{1,1}(\tau) = \lim_{T \to \infty} \int_{-T}^{+T} f_1(t) f_1(t + \tau)\, \mathrm{d}t$$

$$C_{1,1}(t_1) = f_1(t_1) f_1(t_1 + \tau)$$

$$\vdots$$

$$C_{1,1}(t_n) = f_1(t_n) f_1(t_n + \tau)$$

The fact that we are carrying out an autocorrelation is shown by the fact that the function C has the subscripts $1,1$. This is then carried out for a whole series of T and τ as shown. It is obvious that this is a very labor intensive calculation and is only possible with the power of modern computers. The result of

(a)

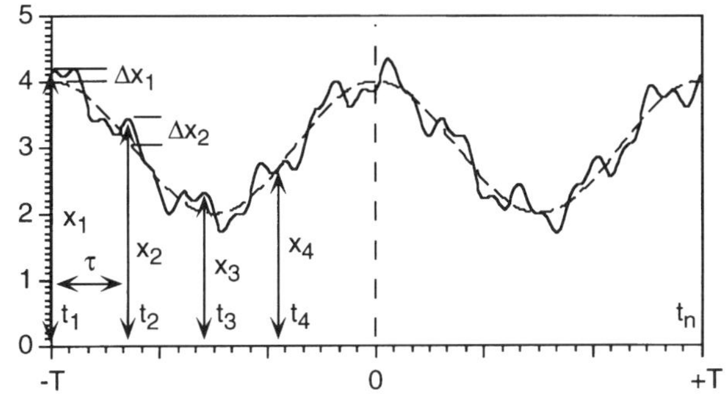

Autocorrelation function:

$$C_{1,1}(\tau) = \lim_{T \to \infty} \int_{-T}^{+T} f_1(t) f_1(t + \tau) dt$$

$$C_{1,1}(t_1) = f_1(t_1) f_1(t_1 + \tau)$$

$$C_{1,1}(t_2) = f_1(t_2) f_1(t_2 + \tau)$$

$$C_{1,1}(t_3) = f_1(t_3) f_1(t_3 + \tau)$$

Figure 5.39 Carrying out the calculation of the autocorrelation on a noisy waveform extracts the basic waveform underlying the noise. (a) Graphical depiction of the concepts involved in the calculation of the autocorrelation function of a noisy waveform. (b) Results of perfoming an autocorrelation on each of the waveforms of Figure 5.38. From Kaye [49]. Copyright © 1986 by International Scientific Communications Inc.)

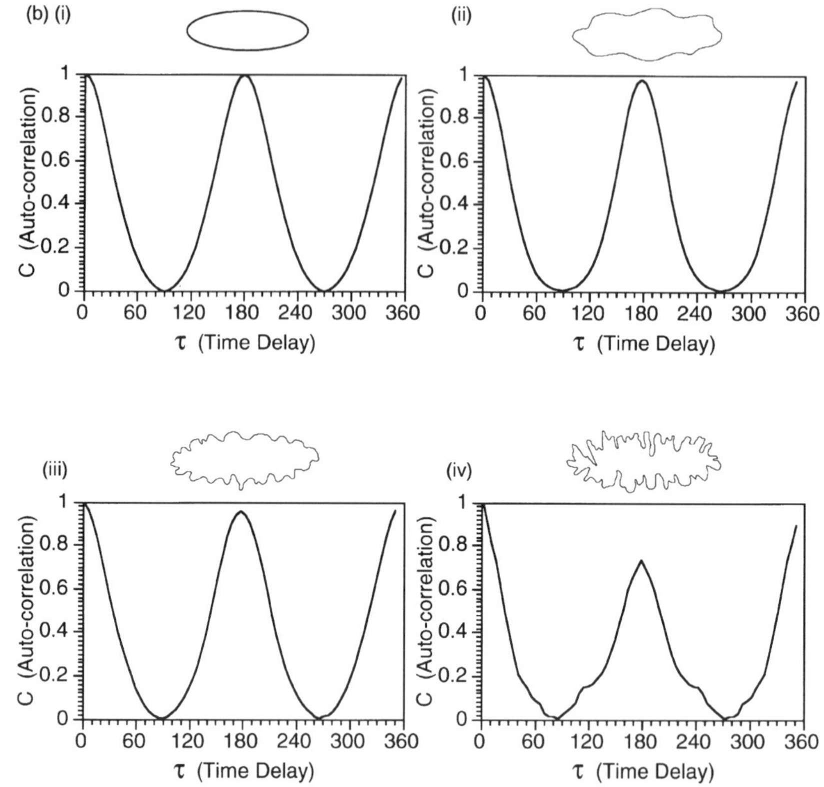

Figure 5.39 *Continued*

calculating the autocorrelation functions of the four waveforms of Figure 5.38 is shown in 5.39(b). It can be seen that the data processing recovers the basic elliptical form from the noisy geometric signature waveforms. (In this case the noise is real in that it denotes texture, not spurious distortion of the signal information.)

In cross-correlation one compares a noisy waveform with different waveforms which might contain the basic signal structure of the noisy signature waveform. Thus in Figure 5.40 the noisy signature waveform generated from a real profile is combined with information drawn from the comparative waveforms to generate what is known as the cross-correlation function, which enables the specialist to see which was the most probable structure buried in the noisy waveform.

If one calculates the autocorrelation function for a completely random noise signal over a long period of time, the magnitude of the autocorrelation is 0 for all values of τ, as shown in Figure 5.41(a). If one is dealing with a wave function which is something like a noisy sine wave, there is an initial decay in the autocorrelation function until one then encounters the recovered signal stripped of its noise (Figure 5.41(b)). To the specialist the early part of this autocorrelation function in Figure 5.41(b) is information on the type of signal distortion present in the waveform being studied and so to the powder technologist it would be information on the texture of the profile for the waveforms such as those shown in Figure 5.38. As set out the information summarized in Figures 5.38 to 5.41 the auto- and cross-correlation functions look at the variation in signal strength along the function.

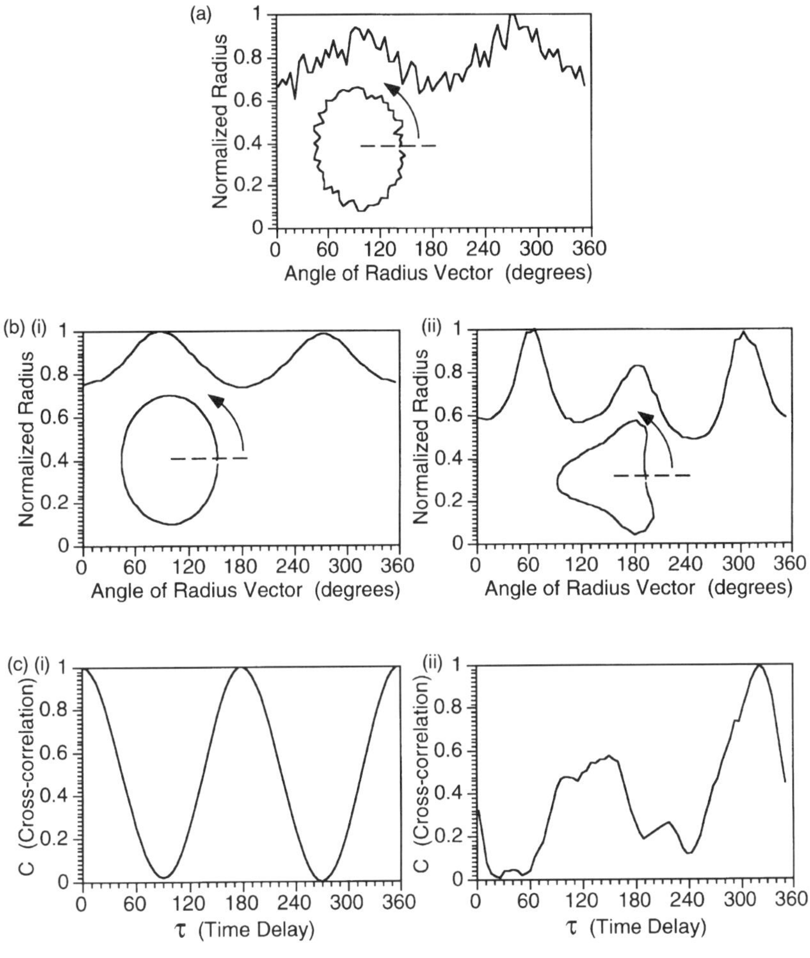

Figure 5.40 Cross-correlation of two different waveforms can help to determine if any similarities exist betweeen a given standard waveform and a noisy waveform containing uncertain structural elements. (a) A noisy signature waveform generated from the rugged profile shown with the graph. (b) Signature waveforms of two different smooth profiles. (c) Result of the cross-correlation of the noisy waveform of (a) with those of the smooth profiles of (b) shows that the rugged profile resembles the profile (b)(i) more closely than that of (b)(ii).

When one comes to look at mixture structure one can scan a section through the powder mixture to generate a waveform. It is interesting to note that C. Schofield explored the use of autocorrelation methods to study powder mixture

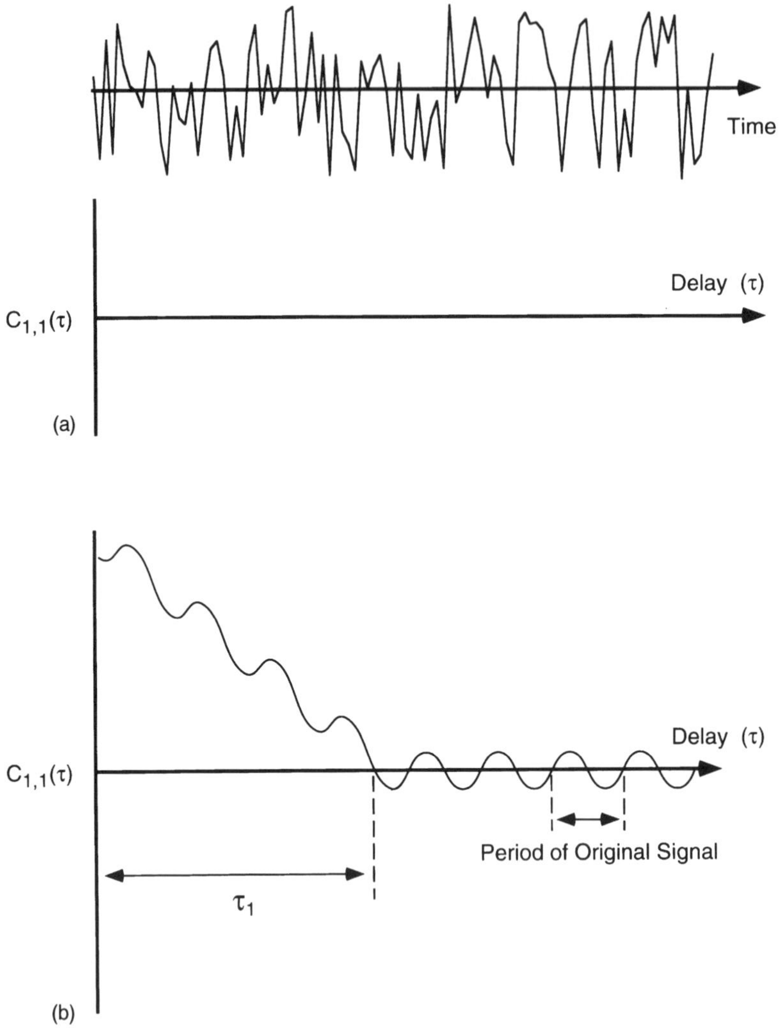

Figure 5.41 In general, carrying out the calculation of the autocorrelation function not only alerts the technologist to the presence of any basic structure in the waveform, but also yields information on the type of noise obscuring the signal. (a) If one calculates the autocorrelation function for a random noise signal, the magnitude of the function will be zero for all values of τ. (b) In a general autocorrelation function the time, τ_1, before the function begins to oscillate about the retrieved signal, is a measure of the magnitude and structure of the noise obscuring the signal. (From Kaye [49]. Copyright © 1986 by International Scientific Communications Inc.)

structures but concluded that, at the time (1970), the methods were too labor intensive for widespread use [50]. Modern computer technology has changed that conclusion. Strip signatures, generated by inspecting good mixtures and bad, can then be used as reference strip signature waveforms in the auto- and cross-correlation process. In fact the same calculation can be performed optically. A direct autocorrelation function of a dispersion of fineparticles in a section through consolidated mixture can be generated. Two negatives of a good mixture could be joined together to give a continuous strip for inspecting the sample negative. When the strip is moved past the sample negative in the field of a photodetector system the cross-correlation functions can be generated directly by following the change in the light intensity. The possibility of using this type of calculation is so new that there is yet to be any published literature on the study of powder mixtures, although work is currently proceeding at Laurentian University. However, Thompson *et al.* have used the basic technique to study the pore structure of sandstone [51]. The holes in a sandstone specimen are obviously analogous to the pieces of ingredients dispersed in a powder mixture. This optical computing procedure could be a very powerful method for looking at large areas of powder mixtures via their negative photographic image.

5.8 INFRARED FINGERPRINTING OF POWDER MIXTURES

In a pioneering study of the possibility of measuring richness of a mixture by inspecting the mixture with a fiberoptic interface to an infrared spectrometer, Ciurczac has shown that the development of a good mixture can be followed over a period of time [52, 53]. As pointed out by Ciurczac, a pharmaceutical mixture may contain up to nine different ingredients. The technique used in the work carried out at the infrared systems laboratory measures the spectrum of each of the ingredients and then the properties of the ultimate mixture are derived by overlaying the curves for all of the ingredient. When this overlay curve matches the curve of a direct examination of the mixture, it can be deduced that the system represents a well-mixed system. The reader should consult the original publications for detailed procedures [52–54].

NOTES

1. Kaye, B.H. and Clark, G.G. (1989) Evaluating the performance of chaotic powder and aerosol sampling devices using tracker fine particles, in *Proceedings Nurenberg Conference on Particle Characterization (Partec)*, May.
2. Kaye, B.H. (1993) *Chaos and Complexity: Discovering the Surprising Patterns of Science and Technology*, VCH Publishers, Weinheim, Germany, Chapter 5.
3. Moroney, M.J. (1953) *Facts from Figures*, 2nd revised edn, Pelican, Harmondsworth, UK.
4. See trade literature of Microtracers Inc., 1370 Van Dyke Avenue, San Francisco, CA 94124, USA.

5. Eisenberg, D. (1992) Microtracers^{TM} F and their uses in assuring the quality of mixed formula feeds. *Advances in Feed Technology*, Spring, 78–85.
6. The Nauta mixer is available from Micron Powder Systems, 10 Chatham Road, Summit, NJ 07901, USA.
7. Hixon, L. and Ruschmann, J. (1992) Using a conical screw mixer for more than mixing. *Powder and Bulk Engineering*, **6**(1), 24–8.
8. van den Bergh, W.J.B., Scarlett, B. and Kollar, Z.I. (1993) Computer simulation model of a Nauta mixer. *Powder Technology*, **77**, 19–30.
9. Moselmian, D., Chen, M.M. and Chao, B.T. (1989) Experimental and neumerical investigations of solid mixing in a gas fluidized bed. *Particulate Science and Technology*, **7**, 335–55.
10. Moselmian, D. (1987) Study of solids motion, mixing and heat transfer in gas fluidized beds. Ph.D. Thesis, University of Illinois in Urbana, Illinois.
11. (a) Plessis, P. and Kaye, B.H. (1991) Powder sampling from mixing chambers, in *Proceedings of the Powder and Bulk Solids Conference*, Rosemont, 6–9 May, Cahners Exposition Group, Cahners Plaza, 1350 E. Touhy Ave, Des Plaines IL 60019–9593, USA. (b) Kaye, B.H. (1991) Optical methods for measuring the performance of powder mixing equipment, in *Proceedings of Powder and Bulk Solids Conference,* Rosemont, Illinois, 6–9 May, Cahners Exposition Group, Cahners Plaza, 1350 E. Touhy Ave, Des Plaines, IL 60019–9593, USA.
12. Gray, J.B. (1957) Dry solids mixing equipment. *Chemical Engineering Progress*, **53**(1), 25–32.
13. Ashton, M.D., Schofield, C. and Valentin, F.H.H. (1966) The use of a light probe for assessing the homogeneity of powder mixtures. *Chemical Engineering Science*, **21**, 843–49.
14. Kaye, B.H., Brushenko, A., Ohlhaber, R.S. and Pontarelli, D.A. (1968–69) A fibre optic probe for investigating the internal structure of powder mixtures. *Powder Technology*, **2**, 243–47.
15. Harwood, C.F., Davies, R., Jackson, M. and Freeman, E. (1972) An optic probe for measuring the mixture composition of powders. *Powder Technology*, 2 (Jan), 77–80.
16. Saito, F. and Kamiwano (1985) An optical method for measuring flow velocity and composition of a mixture of powders continuously, in *Fineparticle Society Meeting*, Miami, August.
17. Alonso, M., Satoh, M. and Miyanami, K. (1989) Recent works on powder mixing and powder coating using an optical measuring method. KONA, No. 7, 97–105.
18. Weinekötter, R. and Reh, L. (1994) On-line measurement of concentration for characterization of powder mixers. *Particle and Particle Systems Characterization* **11**(4), 284–91.
19. For a discussion of the Kubelka–Munk theory see Judd, D.B. and Wyszecki, G. (1963) *Color in Business, Science and Industry*, J. Wiley & Sons, New York.
20. Gratton-Limmatainen, J. (1995) characterizing the structure of powder mixtures by optical methods. M.Sc. Thesis, Laurentian University, Sudbury, Ontario.
21. Calcium carbonate powders supplied by the White Pigment Corporation, Florence, Vermont 05744, USA.
22. A section through concrete and other composite bodies is shown in Chapter 6 of Kaye, B.H. (1994) *A Randomwalk Through Fractal Dimensions*, VCH Publishers, Weinheim, Germany.
23. Many of the appropriate statistical relationships involved in the assessment of rock structure by techniques such as the Rosiwal intercept method are to be found in Chayes, F. (1956) *Petrographic Model Analysis*, J. Wiley & Sons, New York.
24. See discussion of legal variation in richness of simulated fields of view in Chapter 1

and also in Kaye, B.H. and Clark, G.G. (1992) Computer aided image analysis procedures for characterizing the stochastic structure of chaotically assembled pigmented coatings. *Particle and Particle Systems Charaterization,* **9**, 157–70.

25. Kendall, M.J. and Moran, P.A.P. (1963) *Geometrical Probability. Griffin's Statistical Monographs*, Number 10, Hafner, New York.
26. Underwood, E.E. (1970) *Quantitative Stereography*, Addison-Wesley, Reading, MA.
27. For a readily accessible discussion of the Rosiwal intercept method, see Herdan, G. (1960) *Small Particle Statistics*, 2nd edn. Butterworths, London.
28. For an introduction to the basic theory of tomography see Swindell, W. and Barrett, H.H. (1977) Computerized tomography taking sectional X rays. *Physics Today* (Dec), 32–41.
29. Henderson, H. (1979) A sideways look at scanners. *New Scientist* (6 Dec), 782–85.
30. Kruger, R.P. (1980) Industrial applications of computed tomography at Los Almos Scientific laboratories. *Optical Engineering*, **19**(3), 273–82.
31. Hall, N. (1977) X-rays slice into the heart of matter. *New Scientist* (15 Oct), 54–56.
32. Somasundran, P., Huang, Y.B. and Gryte, C.C. (1987) CAT scan characterization of sedimentation and floccs. *Powder Technology*, **53**, 73–77.
33. Lin, C.L., Miller, J.D. and Cortes, A. (1992) Applications of X-ray computed tomography in particle systems. *KONA*, **10**, 88–95.
34. Dickin, F.J., Dyakowski, T., McKee, S.L., Williams, R.A., Waterfall, R.C., Xie, C.G. and Beck, M.S. (1992) Tomography for improving the design and control of particulate processing systems. *KONA*, **10**, 4–14.
35. Huange, S.M., Xie, C.G., Salkfeld, J.A., Plaskowki, A., Thorn, R., Williams, R.A., Hunt, A. and Beck, M.S. (1992). Process tomography for identification, design and measurement in industrial systems. *Powder Technology*, **69**, 85–92.
36. Hosseini-Ashrafi, M.E. and Tuzun, U. A tomographic study of voidage profiles in axially-symetric granular flows. *Chemical Engineering Science*, **48** (Nov), 53–67.
37. Broadbent, C.J., Bridgwater, J., Parker, D.J., Keningley, S.T. and Knight, P. (1993) A phenomenological study of a Batch mixer using a positron camera. *Powder Technology*, **76,** 317–29.
38. Kaye, B.H. (1993) *Chaos and Complexity: Discovering the Surprising Patterns of Science and Technology*, VCH Publishers, Weinheim, Germany, Chapter 9.
39. Kaye, B.H. (1994) A *Randomwalk Through Fractal Dimensions*, 2nd edn, VCH Publishers, Weinheim, Germany, p. 271.
40. Adler, J. and Hancock, D. (1994) The advantages of using a distance transform function in obtaining boundary fractal dimensions. *Powder Technology*, **78**(3), 191–96.
41. Danielson, P.E., Euclidean distance mapping. *Computer Graphics and Image Processing*, **14**, 227–48.
42. Adler, J. and Melia, C.D. (May 1994) Using a distance transform function in image processing, microscopy and analysis. Preprint supplied by authors, Department of Pharmaceutical Sciences, University of Nottingham, Nottingham NG7 2RD, UK.
43. Adler, J. and Melia, C.D. (1992) An image analysis technique based on serial dilation for measuring mixing homogeneity. *Journal of Pharmacy and Pharmacology*, **44**, 1064.
44. For a discussion of the basic use of erosion and dilation in powder technology processing see Kaye, B.H. (1993) *Chaos and Complexity: Discovering the Surprising Patterns of Science and Technology*, VCH Press, Weinheim, Germany.
45. Adler, J. and Melia, C.D. (1994) An image analysis technique based on serial dilation for measuring mixing homogeneity. Preprint provided by the authors, Department of Pharmaceutical Sciences, University of Nottigham, Nottingham NG7 2RD, UK.

46. Kaye, B.H. and Clark, G.G. (1992) Computer aided image analysis procedures for characterizing the stocastic structure of chaotically assembled pigment coatings. *Particle and Particle Systems Characterization*, **9**, 157–70.
47. Hieftze, G.M., Horlick, G. (1981) Correlation methods in chemistry laboratories. *American Laboratories* (Mar), pp. 76–78.
48. Horlick, G. and Hieflje, G.M. (1978) Correlation methods in chemical data measurements, in *Contemporary Topics in Analytical and Clinical Chemistry*, Vol. 3 (eds Hercules, D.M., Hieflje, G.M., Snyder, L.R. and Evenson, M.A.), Plenum Press, New York, Chapter 4.
49. Kaye, B.H. (1986) Fractal dimension and signature waveform characterization of fineparticle shape. *American Laboratories*, **18** (Apr), 55.
50. Schofield, C. (1970) Assessing mixtures by autocorrelation. *Transactions of the Institution of Chemical Engineers*, **48**, T28–T34.
51. Thompson, A.H., Katz, A.J. and Krohn, G.E. (1987) The microgeometry and transport properties of sedimentary rock. *Advances in Physics*, **36**(5), 625–94.
52. Ciurczac, E.W. (1991) Pharmaceutical mixing studies using near infrared spectroscopy. *Pharmaceutical Technology*, **15**(9), 140–45.
53. See applications bulletins from NIRS Systems Inc., 12101 Tech Road, Silver Spring, Maryland 20904, USA, entitled 'Measurement of particle size of raw materials and mixtures with near infrared reflectants spectroscopy' and 'Following the process of pharmaceutical mixing using near infrared spectroscopy'.
54. Near IR analysis benefits Europe's pharmaceutical industry. *R & D Magazine* (1994), 54–58.

6
The impact of chaos theory and experimental mathematics on powder mixing theory and practice

6.1 INTRODUCTION

This chapter is a short one but contains very important ideas. Possible trends in research in two important areas of powder technology are reviewed.

There is no doubt that what is known as chaos theory has important implications for the study of liquid mixing, as demonstrated by the publications of Ottino [1, 2]. It is also possible that the theorems put forth by Ottino will have many applications when looking at the mixing of pastes as one incorporates powders into mixtures. This topic lies outside the scope of this book [1, 3].

In Chapter 1 we outlined briefly the growth of the new subject known as deterministic chaos or simply chaos. Chaos theory deals with systems which are essentially deterministic, but which are so complex, because of the multivariate causes interacting, that they seem to have chaotic outcomes. Such systems have to be studied empirically to see if one can discover patterns of behavior to enable one to understand the complex system [4].

In the late 1960s and early 1970s, when statisticians realized that they could test their statistical theories by studying powder mixing systems, the powder technologist was flooded with information on statistical theorems and indices. As pointed out in Chapter 1, these statistical studies did not prove to be too useful to the working powder technologist. In the same way, now that mathematicians and physicists have started to study what they call chaotic systems and critically self organized systems, they are turning their attention to the avalanching behavior of sand piles and also to the functioning of powder mixers because a powder mixer to them is

> a confined universe of large quantum in which chaos is deliberately introduced to achieve stocastic mixing of the ingredients [6].

In the 1990s we face the prospect that we will see many studies of the chaotic behavior of powder mixers, not because people are interested in powder mixing, but because powder mixers are examples of chaotic universes [5, 6].

At the start of this discussion it should be pointed out again that the situation with regard to fluid mixing is very different compared to powder mixing. Again, as we discussed in Chapter 1, it is sometimes a handicap if one approaches powder mixing from the perspective of fluid mixing, because in powder mixing what is known as Brownian motion in classical science is a much more dominant dispersing mixing mechanism for powder, compared to the situation in liquids. This primary difference will be discussed in more detail later in this chapter.

In the title of this chapter we make reference to a new subject termed **experimental mathematics**. Increasingly this term is being used in the scientific community to describe the use of computer simulations to study the physical behavior of systems by generating various models of behavior [6]. In these studies the computer is not really used in a computational mode; rather one is experimenting with physical systems in an abstract manner. In recent years several groups have started to apply the massive modeling capacity of modern computers to the dynamics of powders in general and also to powder mixing. Thus at the First International Conference on Computational Physics held in Buenos Aires, October 1993, H.J. Herrmann presented a video generated by studying simulated powder segregation which had been created using six hours of computational power on a Cray computer. In this video he modeled the movement of three large spheres in a thousand small spheres, following their behavior as all of the members of the ensemble were given small random motions in sequence. For example, he simulated the diffusional behavior of large and small fineparticles when the bed was dilated and subjected to turbulence. He showed in a graphic manner the type of segregation we discussed in Chapter 1 for which we gave the example of Brazil nuts moving up through the peanuts in a mixed nut can, when subjected to dilation and vibration. Dr Herrmann and his colleagues are continuing their studies on the simulation of powder mixing and segregation [7–9]. Extensive computer simulation of powder mechanics in food mixtures has been carried out by G.C. Barker and colleagues [10].

Similar experimental mathematical modeling on computers of the movement of granular solids is underway at Cambridge University in Great Britain. In Figure 6.1, computer modeling of the flow of powders through an orifice, both for chunky grains and for elongated grains, is shown. These pictures were generated by Hogue and Newland [11]. Other workers who have been active in this area are Mehta (see Barker in [10]) and Tsuji *et al.* [12]. In Figure 6.2, some of the simulations carried out by Tsuji are shown. The reader should be cautioned that although these simulations are fascinating, technology has a long way to go before such simulations can be applied to the myriad odd-shaped fineparticles of a multi-ingredient real powder mixture.

6.2 RANDOMWALK MODELS OF POWDER MIXING

Many of the Monte Carlo routines used in experimental mathematical investigations of powder technology use randomwalk strategies. From to time to time in our discussion of powder mixing mechanisms we have referred to Brownian motion as a dispersion mechanism in liquid mixing. In a study of Brownian

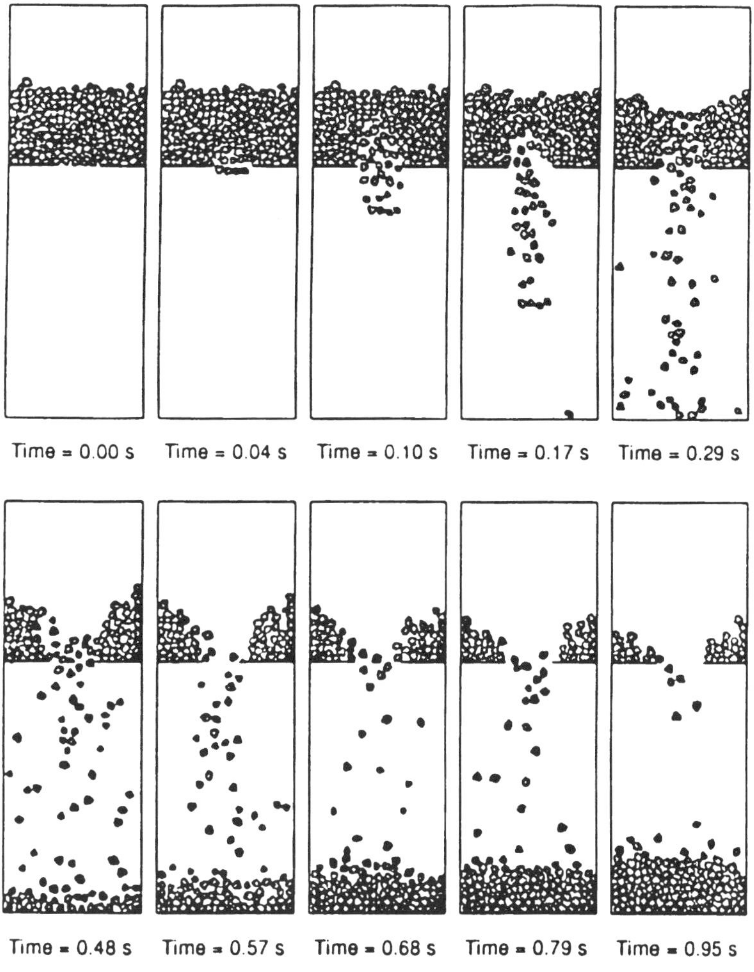

Figure 6.1 Hogue and Newland have carried out extensive computer modeling of the flow of granular solids. Flow out of a mixer is a potential source of powder segregation, and one can anticipate that future modeling may elucidate powder segregation phenomena [11]. (a) Bin flow simulation using particles of random shape and size. (b) Bin flow simulatiuon using only elongated grains.

motion of a dispersion of colloidal fineparticles, one can demonstrate that the average rate of diffusion as predicted by Fick's law can be modeled with a randomwalk [13, 14]. Such a modeling is often introduced to undergraduate students by considering how drunks undergoing a randomwalk are dispersed about a lamppost [15].

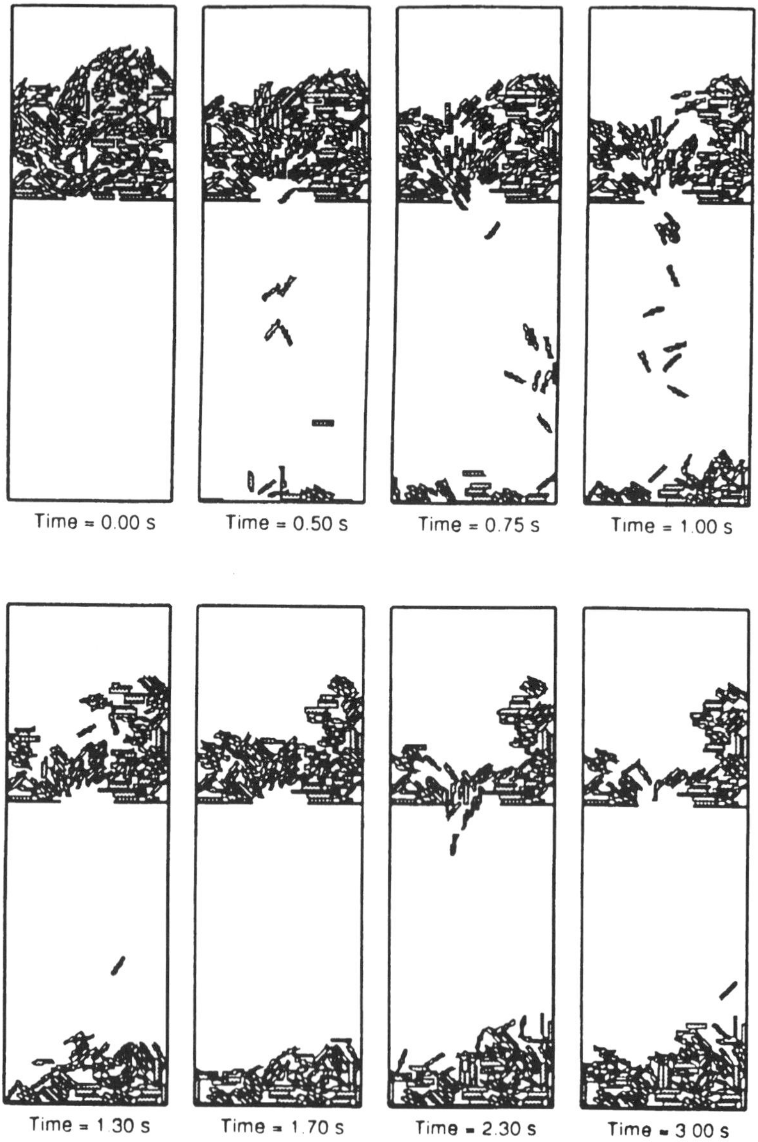

Figure 6.1 *Continued*

The basic steps in simple modified randomwalk modeling of diffusion are illustrated in Figure 6.3. It has been discovered that if the drunks are taking N equal steps then on average they arrive at a distance of square root of N times the step size from the lamppost. It can be shown that the clustering of the actual distances reached by an ensemble of drunks around the expected distance is a Gaussian distribution [14].

In Figure 6.4, the use of random numbers to select step size and direction in a simulated Brownian motion walk is shown. The relevance of the concept of a randomwalk of powder mixing can be appreciated from the data summarized in Figures 6.5 and 6.6. When I was looking through Dr Mandelbrot's book on fractal geometry in 1977, I thought, when I saw the picture reproduced in Figure 6.5, that Mandelbrot was going to discuss the dispersing of powders. However, when I looked closely at the text I found that Mandelbrot was describing what he called a Levy dust model for the dispersion and clustering of galaxies in outer space! [16]. Further exploration of Mandelbrot's text gave the information that a

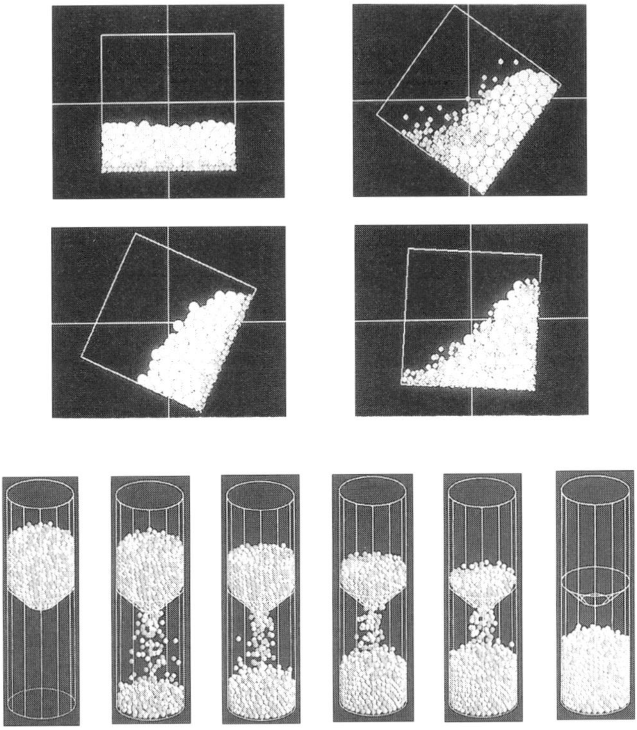

Figure 6.2 Yutaka Tsuji of Osaka University has applied discrete particle dynamic simulation on large digital computers to (a) powder mixing and (b) powder flow. (From Tsuji [12], reproduced by permission of the Council of Powder Technology and the Hosokawa Powder Technology Foundation, Osaka, Japan © 1995 KONA. All rights reserved.)

Levy dust is a Levy flight without connecting lines being drawn. (The lines drawn between successive positions in Brownian motion have no physical significance and are only drawn to show sequential step information.) In Figure 6.6, one of the Levy flight randomwalks described by Mandelbrot is shown. Mandelbrot associates a fractal dimension with the spatial density of the Levy dust, recording that the progress of the Levy flight of Figure 6.6 has a fractal dimension of 1.5. (This fractal dimension is a measure of how efficiently the flight occupies space.) To the uninitiated, the appearance of Figure 6.6 is that of

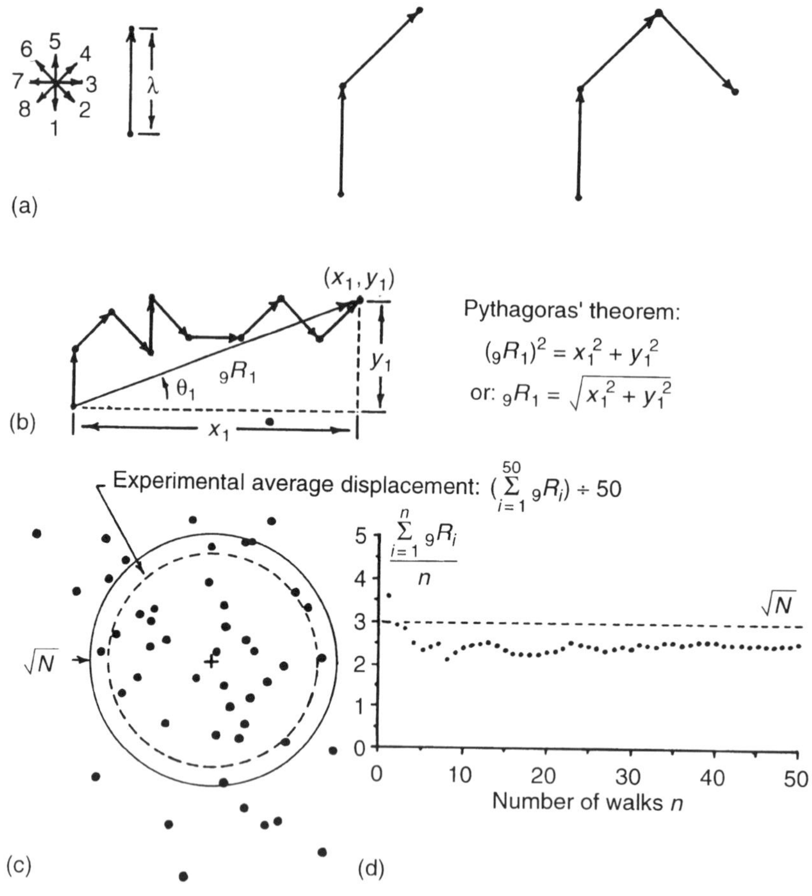

Figure 6.3 The dispersal of ingredients in a powder mixture can be modeled using randomwalks. (a) A random number is selected to choose the direction a step will take. In this case all the steps will be the same length. (b) The distance from the starting point, the displacement vector, can be calculated using Pythagoras' theorem. (c) Final positions of 50 particles from a single starting point after each has undergone nine steps. (d) Calculation of the average displacement for a group of particles approaches the theoretical value of $R_{av} = LN^{1/2}$ as more and more walks are attempted. R_{av} = average displacement vector; L = length of an individual step; N = number of steps of length L.

the track of an ingredient in a powder mixture being dispersed by the random actions occurring in the mixer, and indeed it would appear to be a record of a mixture in which there is localized randomization with the occasional large convective movement. In fact a **Levy flight** is a generalized diffusion theory involving not only localized movement but the occasional large leap. The process is named after the French mathematician Paul Levy (1886–1971). Dr Levy was one of Mandelbrot's teachers. A Levy flight can be used to describe

Step N	Length R	Angle θ	Step N	Length R	Angle θ
1	15	111	26	7	348
2	2	42	27	2	267
3	5	9	28	20	100
4	4	140	29	16	26
5	6	324	30	17	253
6	17	10	31	9	278
7	18	86	32	4	123
8	17	31	33	16	293
9	9	302	34	1	258
10	1	279	35	13	249
11	10	139	36	10	295
12	14	95	37	16	215
13	5	106	38	13	85
14	9	132	39	19	11
15	9	301	40	6	200
16	3	79	41	5	201
17	9	240	42	14	132
18	13	259	43	14	267
19	17	75	44	7	298
20	20	239	45	8	67
21	14	290	46	14	315
22	14	84	47	5	35
23	20	67	48	16	252
24	1	335	49	3	132
25	9	21			

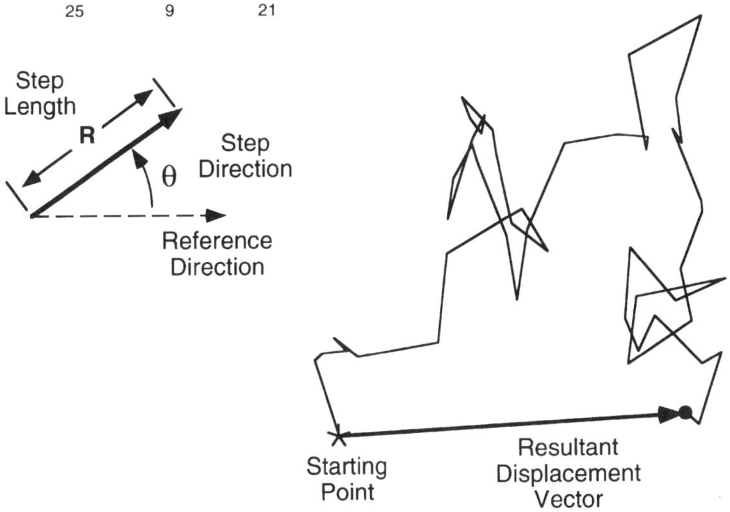

Figure 6.4 To model Brownian motion in two-dimensional space, one must allow both the step size and direction vary at random.

very different systems such as the spread of an epidemic or the metastasis of a cancer cell. When modeling a randomwalk involving a Levy type flight the probability of movement includes a small number of random number selections that will generate large steps in the randomwalk. Thus if we had a hundred

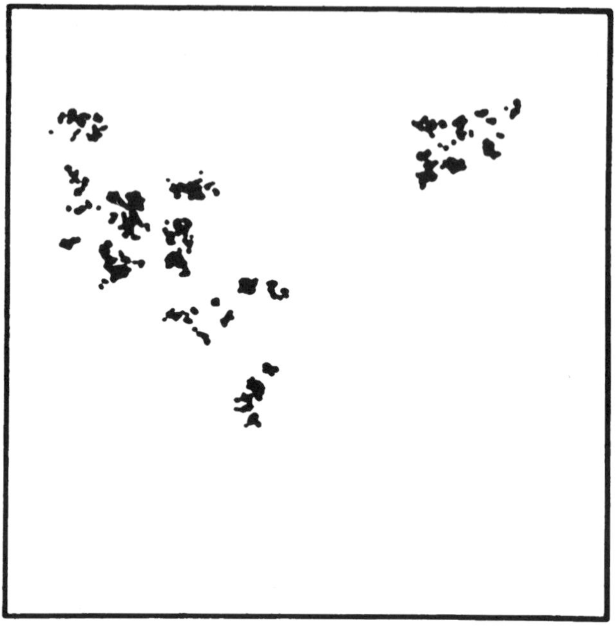

Figure 6.5 Describing the fractal structure of galaxies requires the same mathematical methods as those required to describe the dispersal of a black powder in a white matrix. (From Mandelbrot [16], reproduced with permssion of the author.)

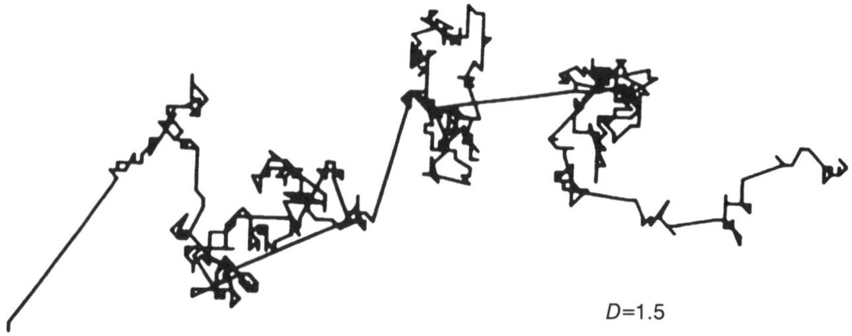

Figure 6.6 A Levy flight is a type of random walk in which, at any time, there is a small probability of a large step. The result is a series of small steps interspersed with an occasional large leap. (From Mandelbrot [16], reproduced with permssion of the author.)

random numbers representing various steps possible in a Levy flight, then numbers from 1 to 90 could be relatively small steps, with 91 being a four-step, 92 being a six-step and so on, with the directions of the steps also being randomized in space using an ancillary random number table. For a randomwalk to be an exact Levy flight according to the definition given by Mandelbrot the probability of a set of different size steps must be determined by a given mathematical function. A Levy flight following a selection procedure outlined in this paragraph would be more properly called a **pseudo-Levy flight**. Until recently, the idea that a Levy flight could model the eccentric randomwalk taking place within a mixer appeared to be an intellectual novelty; however, the recent work on radioactive tracing of random pathways in a mixer carried out by Bridgwater and colleagues makes it possible to identify probable Levy flights in a powder mixer and then to use such information to predict the overall intermingling of ingredients within a powder mixer [17].

In the early 1960s I developed a mixing device which in fact mixed the ingredients of a free flowing powder by allowing the fineparticles to undertake a randomwalk with large Levy flights [18]. A prototype model of the system developed is shown in Figure 6.7. The ingredients to be mixed are added to the top of the tower. In the body of the tower are angled plates with holes in them. The typical plate is shown in Figure 6.7(b). As the powder cascades over the sloping distributor plate, local tumbling creates a randomizing effect corresponding to a random diffusion in a dilated powder system. This type of random motion in a cascading powder has been shown to be the dominating mechanism in creating powder mixtures in a horizontal mixer rotating about an axis down through the center of the mixer [18]. Movement down through the random holes in the sloping distributor plate creates the equivalent of convective disturbances in an active powder mixer. This leap from plate to plate is a Levy type element in the randomwalk being followed by the fineparticles of the ingredients. The varying distance between the holes of one plate and the cascading powder in the next distributor plate constitutes the variable length Levy type flight leaps. The secret in using this randomwalk in three dimensions to achieve powder mixing is that the grains of the powder walking down through the plates and through the holes should never move so fast that the bigger components of the mixture gain a momentum advantage to create segregation [18]. In the original patent application describing the randomizing tower was envisaged that air jets placed at random around the wall of the tower might be a useful adjunct to create randomization within the tower [18]. From a randomwalk perspective, it can be stated that the grains of the mixture moving down through the tower take part in a three-dimensional randomwalk, and that the stack of material in the container at the base of the tower constitutes a three-dimensional dispersion assembled chaotically. The original intent behind the design of the randomizing tower was to deliver premixed powder ingredients to a process over a larger area without the mixture being able to segregate *en route* to the final processing of the powder mixture in a furnace. (The original work was carried out with the free flowing

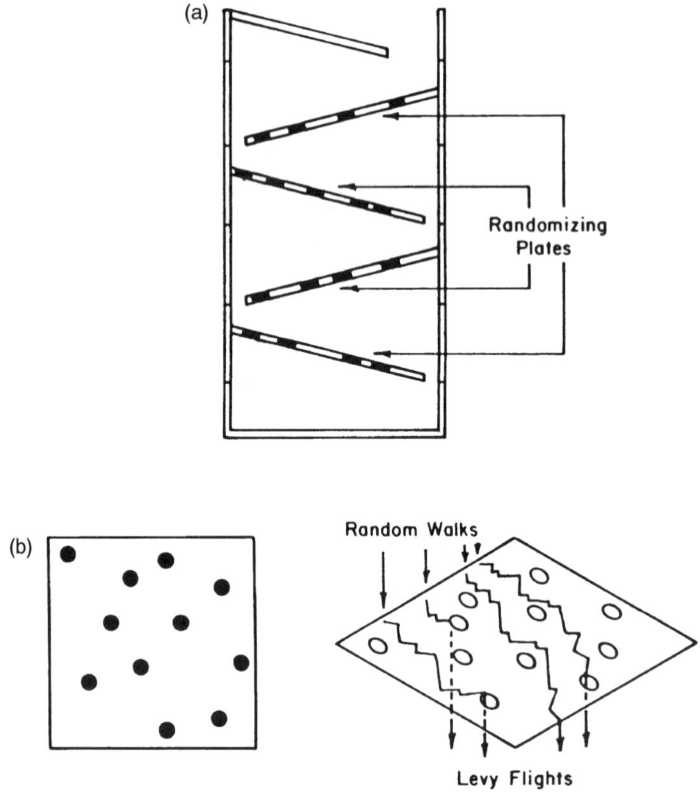

Figure 6.7 A randomizing tower causes the powder flowing through it to undergo what is essentially a Levy flight in three-dimensional space. (a) A randomizing tower containing plates with randomly dispersed holes. (b) Typical motion of powder grains across a randomizing plate.

ingredients feeding a glass furnace.) In fact from one perspective the randomizing tower is a passive mixer, and we will describe in Chapter 8 how passive mixers could find more employment as intermediate connectors between batch mixers and an industrial process.

NOTES

1. Ottino, J.M. (1989) The mixing of fluids. *Scientific American* **258**(1), 56–67.
2. Ottino, J.M., Leong, C.W., Rising, H. and Swanson, P.D. (1988). Morphological structures produced by mixing in chaotic flows. *Nature*, **333** (6172), 419–25.
3. Ottino, J.M. (1990) *The Kinematics of Mixing: Stretching Chaos and Transport*, Cambridge University Press, Cambridge.
4. For an introduction to the basic concepts of chaos theory see Gelick, J. (1987) *Chaos*, Viking Press, New York.
5. Samdani, G. (1991) The simple rules of complexity. *Chemical Engineering*, (Jul), 30–35.

6. Studt, T. (1994) Finding order in chaotic mixing, *R&D Magazine*, (Feb), 82–84.
7. Gallas, J.A., Herrmann, H.J. and Sokolowski, S. (1992) Two dimensional powder transport on a vibrating belt. *Journal de Physique I.* (Paris), **2**, 1389–1400.
8. Gallas, J.A.C., Herrmann, H.J. and Sokolowski, S. (1992) Molecular dynamics simulation of powder fluidization in two dimensions. *Physica A* **189**, 437–46.
9. Lee, J. and Herrmann, H.J. (1993) Angle of repose and angle of marginal stability; molecular dynamics of granular material. *Journal of Physics and Mathematics* (26 Jan), 37–3 to 38–3.
10. Barker, G.C. (1994) Computer simulations of granular materials, in *Granular Matter, an Interdisciplinary Approach* (ed. A. Mehta), Springer-Verlag, New York, pp. 35–83. In this communication, segregation in a powder mixture is simulated on a computer.
11. Hogue, C. and Newland, D. (1994) Efficient computer simulation of moving granular particles. *Powder Technology*, **78**, 51–56.
12. Tsuji, Y. (1993) Discrete particle simulation of gas-solid flows (from dilute to dense flows). *KONA*, No. 11, pp. 57–68.
13. Kaye, B.H. (1989) *A Randomwalk Through Fractal Dimensions*, VCH Publishers, Weinheim, Germany, Chapter 4.
14. Kaye, B.H. (1993) *Chaos and Complexity: Discovering the Surprising Patterns of Science and Technology*, VCH Publishers, Weinheim, Germany.
15. Gamow, G. (1947) *123 Infinity*, Viking Press, New York.
16. Mandelbrot, B.B. (1977) *Fractals: Form, Chance and Dimension*, Freeman Press, San Fancisco.
17. Broadbent, C.J., Bridgwater, J., Parker, D.J., Keningley, S.T. and Knight, P. (1993) A phenomenological study of a batch mixer using a positron camera. *Powder Technology*, **76**, 317–29.
18. Kaye, B.H. (1993) Improvements in/or relations to a method and apparatus for handling particles. *British Patent Applications*, No. 38871–63, October.

7
Active Mixing Machines

7.1 RIBBON MIXERS

As indicated in Chapter 1, in this book we refer an active mixer as one in which there are either internal moving parts to randomize the positions of the ingredients, or air jets to create convection currents and turbulence in the powder container. The first type of industrial mixer that we will consider is the **ribbon mixer**. In a ribbon mixer, a long complicated single paddle is mounted axially and used to disperse the ingredients of the mixture. The ribbon is usually so constructed that the powder near the outside of the container is moved in one direction, whereas in the middle it is moved in the opposite direction. Ribbon mixers are available from several manufacturers and they all have their own type of ribbon [1–5]. The reader should note that many of the companies mentioned with respect to a particluar type of mixer in the discussion often make other type of mixers, as set out in their catalogs.

Ribbon mixers are widely used in the food industry and the pharmaceutical industry. This type of mixer can have dead pockets, especially near the end of the mixer close to the axis of rotation [6]. Loading and cleaning of a ribbon mixer is usually done from the top of the trough. The discharge valve can be located at either end or in the middle of the mixture. (The manufacturers usually recommend emptying the system at the end of the mixer.) Van den Bergh points out that attention must be paid to the design of the mixers, bearings and glands to avoid lubrication contamination. He also states that ribbon mixers are not easy to clean and not recommended for sticky materials. Because the mixing device turns over the whole charge of powder, the power demands of this type of mixer are relatively high, thus van den Bergh gives an estimate of 12 kW per 1000 kg of charge. (Many of the figures on power demands and of available capacity in industrial mixers are taken from an excellent review article by van den Bergh [7].) One does not normally fill this type of mixer to more than 50% of the available volume. Normally the capacity of this type of mixer is restricted to an upper limit of 15 m^3.

Sometimes, to improve the dispersion of ingredients in ribbon mixers, the ribbon blades are built to be close to the cylindrical wall to give high shearing. Sometimes the ribbons are also fitted with rubber wipers to ensure complete intermingling of the ingredients by lifting any packed material near the wall into the middle of the ribbon ensemble. It is difficult to predict the scale-up performance if one attempts to increase mixing capacity by going to a larger mixer of the same type [8].

In a ribbon mixer the paddles move relatively slowly, as distinct from the movement in another class of mixer which is variously referred to as a paddle or a plowshare mixer. In the Forberg mixer, the twin paddle system shown in Figure 7.1(a) rotates very rapidly to generate a fluidized chaotic array of the ingredients of the mixture. The Forberg mixer consists of two horizontal drums that are joined on one side. The mixing elements consist of two shafts that rotate anticlockwise along the axis of each of the two drums. Scaling up the capacity of the Forberg mixer is expensive because of its stainless steel construction. (This type of equipment has been referred to as the tossed salad approach to mixing.) One of the advantages of the Forberg mixer is the very short mixing times; usually, 15 seconds are sufficient to achieve full mixing. There are virtually no shear forces in the mixer, so it is not suitable for use with cohesive materials. On the other hand, the way in which the powder is dumped through a bottom door often prevents segregation in the initial post-mix handling of the powder. If there are any fine components in the mixer, the way that the mixture is dumped from the Forberg mixer can give rise to fugitive emissions. Forberg mixers are available in sizes ranging from 6 dm^3 up to 5 m^3. The operational power requirement can be up to 50 kW, but this is offset by the very short time of mixing [9, 10]. For some purposes the Forberg mixer can be equipped with ancillary transverse bars, as shown in Figure 7.1(a). When the paddles are operating, the ingredients impact on the lump breaking bar to deagglomerate any lumps in the feed material. The 'flow distortion bar' in this diagram is also known as an intensifier, since it increases the efficiency of the mixing system. The system can incorporate a device for adding liquid to the powder ingredients. The addition of liquid may be an integral part of the ultimate mixture or it may be a technique for stabilizing the mixture via agglomeration to prevent segregation when the system is emptied and moved away from the mixer.

Another type of mixer in which the rotation of the dispersing paddles is so intense that the ingredients are suspended in the moving air to create a turbulent mixture is the Littleford mixer shown in Figure 7.2(b). This mixture is known in Europe by the term Lodige mixer [11–13]. Because the rotating paddles are shaped like plowshares to enable the paddles to keep the wall clean as the mixer operates, this type of mixer is sometimes referred to as a plowshare mixer. Normally the equipment can be filled with a range of occupied volumes from 20% to 70% of the mixing drum. Van den Bergh states that the entire material mass moves so that power consumption is relatively high. The mixer is available in a range of sizes from 0.13 to 30 m^3. Mixing times are relatively short, and mixing is often complete by the time the paddles are up to top speed, so that the limiting time for the mixer is the time required for the paddles to reach top speed and then to become stationary. To increase the rate of mixing with difficult

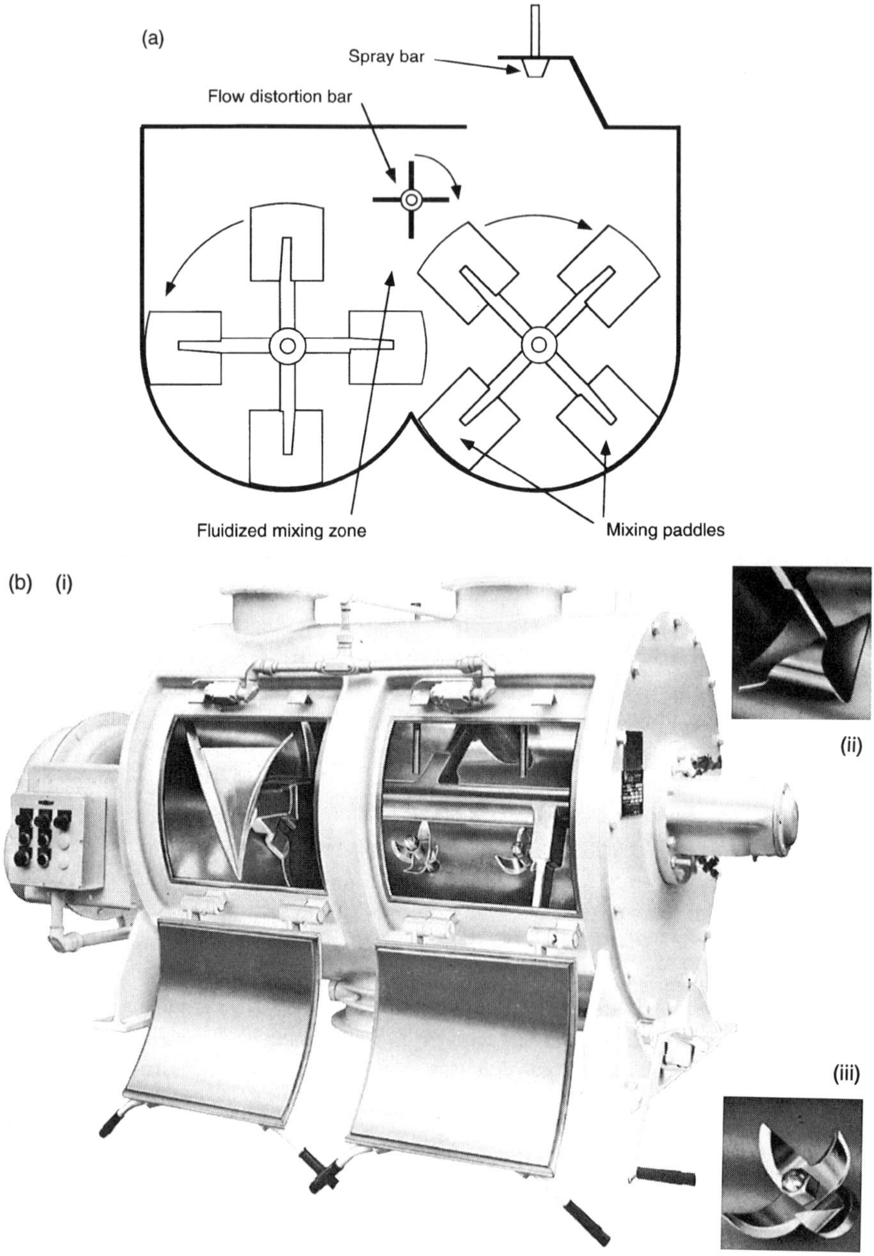

Figure 7.1 In some mixing equipment, high-speed paddles, of various shapes, rotate rapidly to create a fluidized, turbulent zone in which mixing takes place. (a) The Forberg mixer. (b) (i) The Littleford–Lodige mixer with (ii) plow paddles and (iii) intensifier choppers. (a) From A.B. Flower, Forberg Mixing Ness, 4–89. (b) Used by the permssion of Robert E. Blank, Littleford Day Inc.

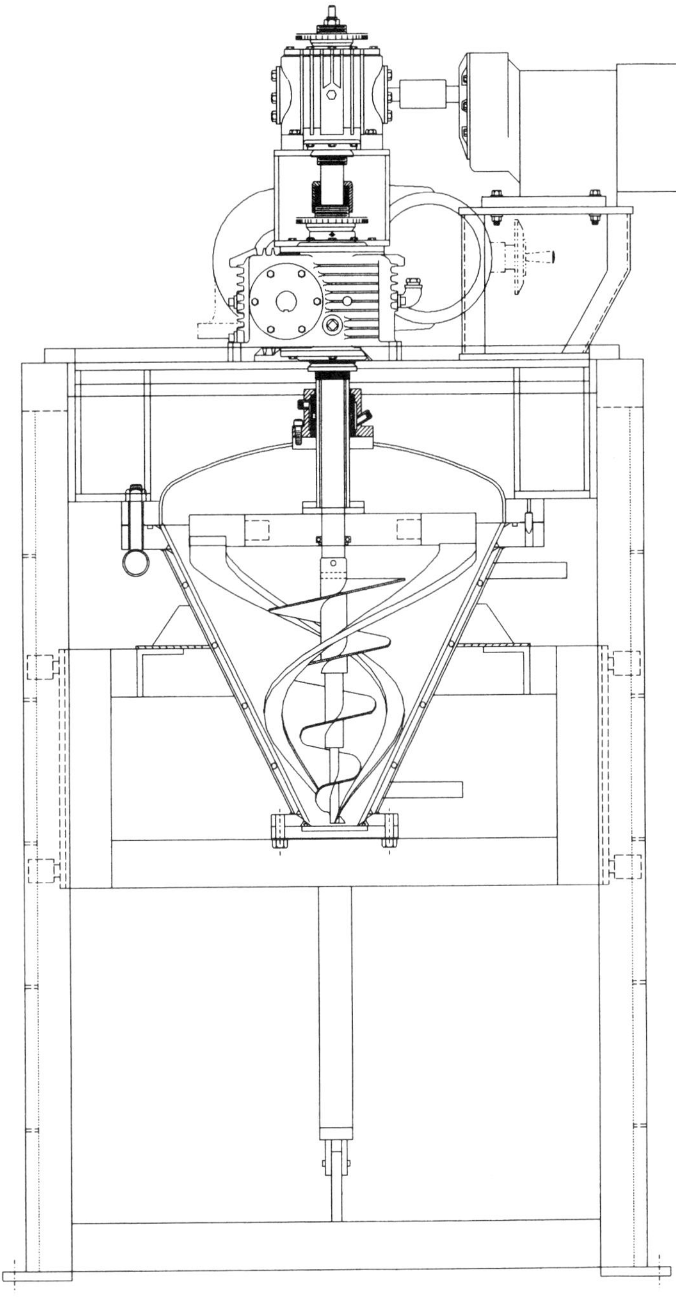

Figure 7.2 The Helicone™ mixer creates turbulent convective mixing currents by using counter-rotating and lift-screw elements. (Copyright © 1995 Design Integrated Technology, Inc. All Rights Reserved. Used by permission only.)

powders, and to disperse agglomerates which can exist in the ingredients, so-called intensifier choppers are mounted in the side of the mixer and driven independently of the movement of the high-speed paddles. If cohesive powders are mixed in this equipment, it may be necessary to intimize the final mixture by passing it through a pinmill. The basic system of the pinmill (or pegmill as it is sometimes called) was described in Chapter 1. Equipment of this kind is also available from the Alpine Company, which is now owned by the Hosokawa Company. (See trade literature on the pegmill from Hosokawa [14].)

A vertical ribbon mixer is manufactured by National Bulk Equipment Incorporated [15]. Closely related to the vertical ribbon mixer is the Helicone™ mixer [16] shown in Figure 7.2. The Nauta® mixer* is similar to the Helicone™ mixer. It employs a rotating lift screw to create convective mixing inside a conical mixing chamber [17–19]. The Nauta® mixer is available in several configurations as shown in Figure 7.3. Van den Bergh has studied this type of mixer extensively [20, 21] and makes the following comments:

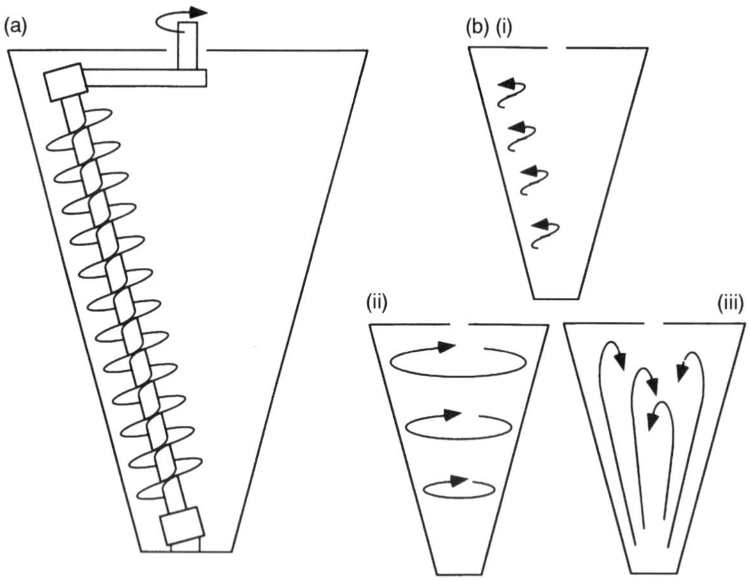

Figure 7.3 The Nauta® mixer uses a single convective lift screw which also rotates around the conical blending chamber. (a) Internal stucture of the Nauta® mixer. (b) Three mixing actions which take place in the Nauta mixer: (i) around the lift screw; (ii) around the mixing chamber; (iii) convection currents. (This material is reproduced by the permission of Hosokawa Micron Powder Systems, a division of Misokawa Micron International Inc., 10 Chatham Road, Summit, New Jersey 07901. © HMPS. All rights reserved.)

*Registered trademark of Hosokawa Micron International Inc. and affiliated entities.

> The power consumed and heat generated are relatively low on average. The unit requires 3.5 kilowatts per thousand kilograms of charge. The mixers can be constructed in sizes ranging up to 60 cubic meters. Mixing time can be halved with a turnscrew mixer and by 30 to 40% with a tapered screw unit. In a tapered screw design the flights are wider at the top of the screw than those at the bottom thus a tapered screw is useful if one component of a material tends to float i.e. if one of the components have a very low internal friction or when there is a difference between the bulk densities of the materials being mixed. That is because the wider screw flights at the top of the screw are closer to the center of the mixer and so floating particles are drawn into the mixing action [21].

See also the work of Hixon and Ruschmann [22].

As mixers become more complicated in their internal structure they become more difficult to clean, especially in the avoidance of trace contamination from one mixing product to another, and therefore this is an important consideration in the pharmaceutical industry. It is sometimes recommended that mixers as complicated as the Helicone™ mixer and Nauta® mixer should be dedicated to a given product line. Another type of mixer making use of lift screws is manufactured by Prater Industries Incorporated [23]. In the equipment known as a Vertical Batch Mixer manufactured by the Koppers Division of Sprout Waldron [1] a central lift screw creates mixing convection currents.

In mixers such as the Airmix® equipment, mechanical paddles are dispensed with and the powder is made to move around the mixer under the influence of air jets. Not only do the air jets create convective currents within the mixer, but they also dilate the powder bed to permit randomization of local ingredients in the turbulence between the boundaries of the moving jets and the passive powder surfaces over which they move. The Airmix® equipment was originally developed in Germany and is available in North America from Sprout Waldron [1]. Another air jet type mixer is termed the Blendicon® and is available from Dynamic Air [24, 25]. The operation of the Sprout Waldron Airmix® as described in the manufacturer's brochure is as follows:

> The mixing head located at the bottom of the unit contains a manifold with a series of nozzles and a cone valve. Compressed air enters the tank through the manifold and especially designed round nozzles. A device is installed to prevent fine material from passing through the nozzles and into the manifold when air is not flowing. The cone valve is merely a discharge gate to be opened at the completion of the mixing cycle. It has been designed to prevent the segregation of particles of varying size and density during discharge. The mixing cycle usually consists of a series of intermittent air blasts. The air enters the mixer through the nozzles in the mixing head at a velocity near the speed of sound. The nozzles are adjusted at an angle to cause the material to spiral upward along the side of the tank, materials in the center of the mixer flow down toward the

> mixing head to complete the circulation of material. A typical cycle might require a 2 to 6 second blast of air with a two second interval between blasts. The total mixing cycle is of the order of 24 seconds.

The manufacturers estimate that to fill, operate and empty, the mixer requires no more than 20 minutes. Fauver has described the retrofitting of a pulsed air type mixing device to a silo [26]. Note that the air jet type mixers differ from fluidized bed mixers in that in equipment such as the Airmix® the aim is not to fluidize the powder to be mixed but to move it around turbulently using air jets instead of paddles. Harnby [27] has discussed the use of fluidization to mix powders and makes the following comments:

> In a fluidized bed system air is made to levitate the particle. In such a system the bulk density of the powder is reduced and the individual particle mobility is increased. If the gas flow is sufficiently enlarged there will be considerable turbulence within the bed and the combination of turbulence in particle mobility can produce excellent mixing. A constant danger in the fluidized bed is that if the turbulence is not complete then the constituent particles can readily segregate because of variable settling rates.

Harnby also comments that

> Very few commercially available fluidized mixing units exist. Because of the diversity of its application the fluidized bed is usually designed for a specific process and is not available as a standard line product.

Finally he states that

> Several fluidized beds have been purposely built for the pharmaceutical industry.

The Airmerge system manufactured by the Fuller-Kovako company is a hybrid air-jet–fluidized-bed mixer. The basic system of the Fuller-Kovako Airmerge equipment is shown in Figure 7.4 [28]. The various quadrants of the base of the mixing chamber are ultimately the source of fluidizing air. The variations in these air currents from zone to zone of the base of the mixer create turbulent currents and dilated powder beds in which the fineparticles can move freely, with the overall result of rapid mixing. A similar piece of equipment used on a larger scale to homogenize large supplies of materials such as cement and flour has been described by Harnby and is shown in Figure 7.4(b).

7.2 TUMBLER MIXERS

The first group of mixers in this category can be described as tumblers with captive mixing chambers. This group includes the familiar V, Y and double cone type mixers. This type of equipment is available from several manufacturers [29–32].

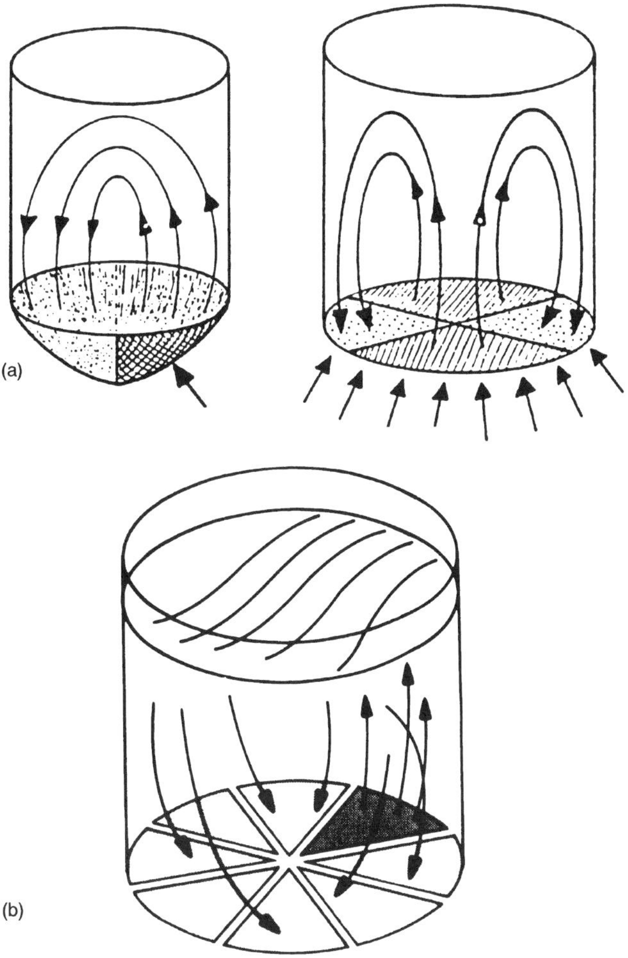

(a)

(b)

Figure 7.4 The Airmerge blender and homogenizing silo both employ air fluidization to achieve mixing of powders. (a) The Airmerge system manufactured by Fuller-Kovako Corp., has a completely fabric-covered fluidized bottom divided into four quadrants that are fludized in sequence to achieve strong, varying convection currents. (b) The homogenizing silo systems uses aeration pads on the silo floor divided into eight segmented areas. This silo also fluidizes the segments in sequence resulting in turbulent convection currents. (Copyright by the Fuller-Kovako Corporation, Bethlehem, PA 18016–0805, U.S.A.)

The reader should note that many of the companies mentioned with respect to a particular mixing device also manufacture several other types of mixers. Many of these manufacturers have helpful application brochures available on request. (Full addresses of these companies are given in the notes.)

Figure 7.5 shows one of the most familiar examples of this type of mixer, the

Patterson-Kelley V-mixer. This system should not be filled to more than half capacity in order to allow freedom of movement of the powder as the mixer operates. As the mixer is inverted, the powder falls and splits into two streams, with each stream being turbulently mixed by the upward moving air. Better mixing is likely to be achieved in this type of mixer if it is inverted quickly, with a pause to allow the powder to move down the system. This is not always possible in many industrial situations, however, where the system moves relatively slowly and, as a consequence, rather long times are required for mixing. Often a so-called intensifier bar is placed across the mixer to increase the turbulence of the falling powder stream. It is probable that the use of an

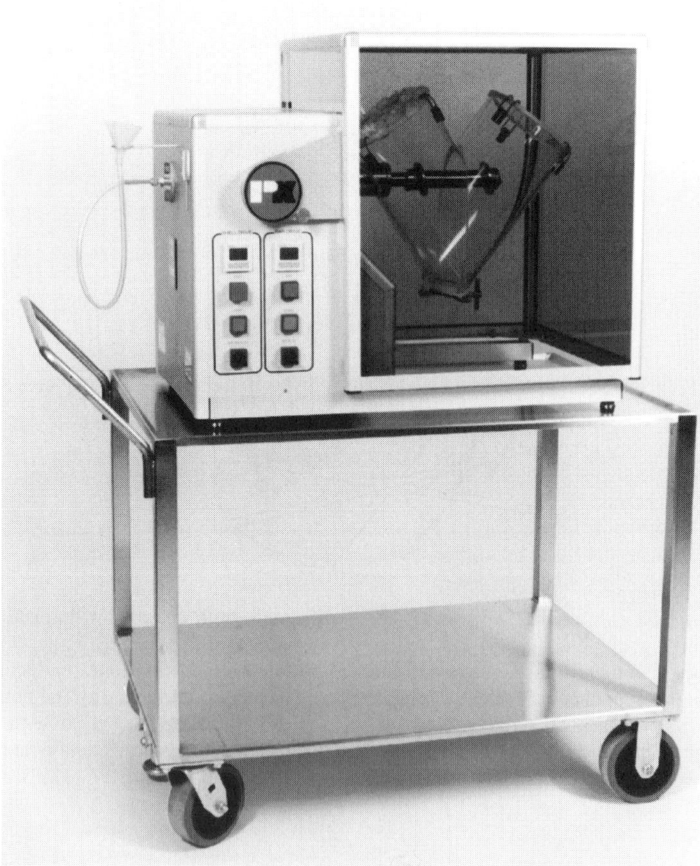

Figure 7.5 The Patterson-Kelley Porta-Shell V-mixer; indicating some of its main features. Mixing occurs when the powder divides and flows turbulently as the mixer is turned. Note the intensifier bar, which can also be used to add liquid to the mixture. Copyright permission given by the Patterson-Kelley Co., Division of HARSCO Corp., East Stroudsburg, PA 18301, USA.

intensifier bar was empirically discovered when hollow tubes were placed in the position of the intensifier bar of Figure 7.5 so that the liquid could be sprayed into the falling powders to achieve granulation. As discussed in Chapter 1, one can intensify the mixing with several inserts, but one must always balance the cost of cleaning the system for the overall mixing process. With regard to the industrial operation of this type of mixer, van den Bergh makes the following comments:

> For larger capacities the cost becomes prohibitive because the mixers require massive supports to handle the large out of balance load. In general however tumbling mixers are relatively inexpensive because of their simple design. Since the entire mixer is being moved power consumption is of the order of 5 kilowatts per thousand kilograms of charge [7].

He also points out that this type of mixer requires a rather larger floor area when operating in an industrial situation [7]. A slightly different type of captive mixing chamber mixer is manufactured by the Mixing Equipment Company [33]. The basic design of their equipment known as a Lightnin® Mixer is described in the manufacturer's brochure:

> Containers are separated from the mixing drive in order to optimize the work flow. A variable speed mixing bar inside the container optimizes the mixing action. Materials module containers can be changed in five minutes to avoid cross contamination between batches of different materials. Containers are charged, discharged and cleaned away from the central mixing site [33].

Several companies manufacture equipment which can take varied types of containers to achieve mixing by complicated motion of the mounted mixing chamber. The equipment manufactured by Heimbeck is called the Sto-mixer [34] (in North America it is sold under the name of Dynatherm). Once the mixing chamber is mounted in the equipment, in the words of the manufacturers,

> The tumbler's inclination angle is altered at random i.e. irregularly and stochastically. Simultaneously the speed and the direction of rotation of the mixer axis is controlled stochastically.

The manufacturers call this movement a double stochasym and claim that 'it is efficient for mixing and that it mostly avoids segregation' [34]. Another mixing device in which one can mount various mixing chambers is the Turbula® mixing system shown in Figure 7.6 [35]. This system is available either at laboratory scale or at manufacturing level thus shown by the various parts of Figure 7.6(a). The mounted mixing chamber is moved through the complicated mixing pattern shown in Figure 7.6(b). Other companies which manufacture equipment to convert powder containers into powder mixing devices include Custom Metalcraft [36], Morse Manufacturing [37] and Bulk Unit Load Systems [38]. The only active mixer in which the powder container is tumbled with the tumbling

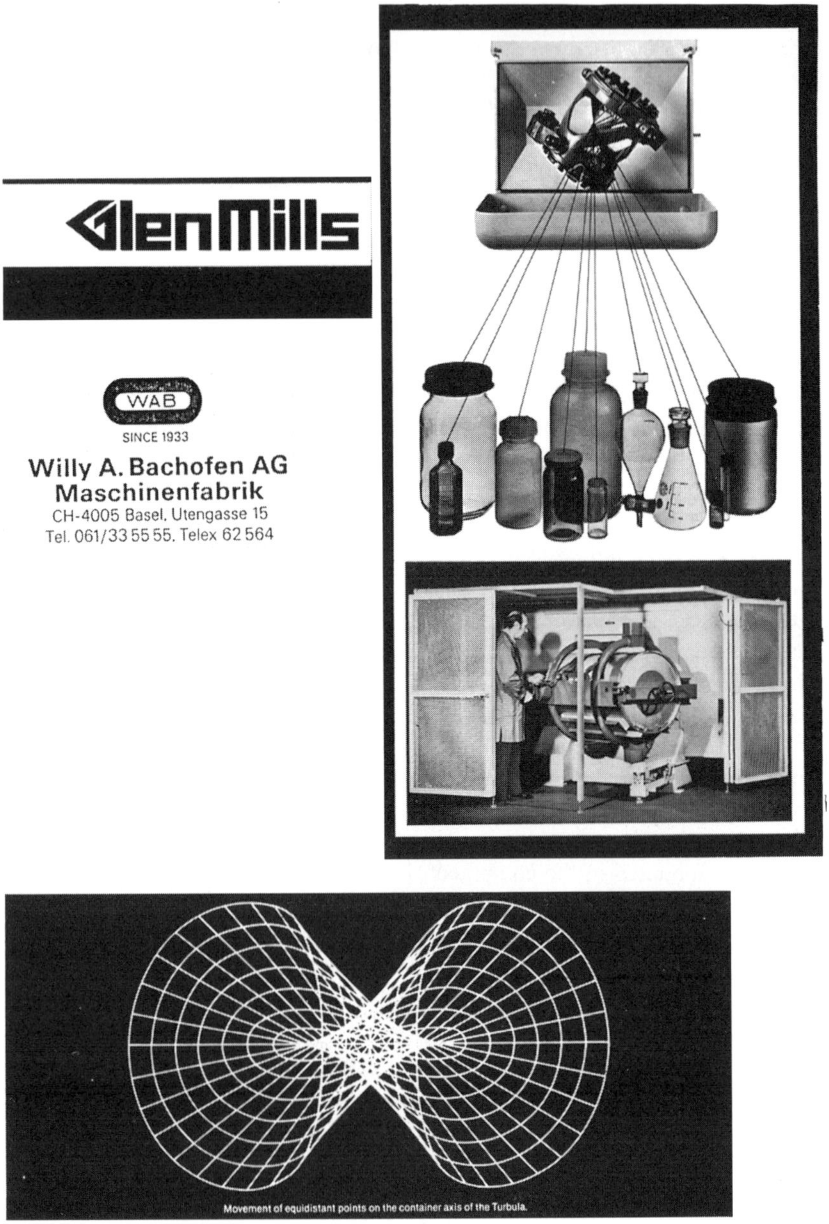

Figure 7.6 The Turbula mixer.

chamber being free of any connection to the tumbling device is the AeroKaye® sampler/mixer device referred to in Chapter 5.

7.3 HIGH SHEAR MIXING AND MULTIMECHANISM MIXERS

In Chapter 1 we discussed how mulling using large wheels moving around a pan can be used to achieve high shear mixing. Equipment of this kind is manufactured by the National Engineering Company [39].

In the piece of equipment known as a Turbulizer, high-speed rotating paddles are used to mix powders together and liquid can be added to the material. The way in which the equipment operates is illustrated in Figure 7.7 [40]. It should be noted that the Turbulizer equipment can be used to achieve continuous mixing. The subject of continuous mixing arises in many discussions about powder mixing. The problem with continuous mixing is the cost of equipment required to accurately feed the mixer with small quantities of powder on a continuous basis. Developments in the area of powder feed weigh feeders are required before the cost of continuous mixing becomes acceptable in many industrial situations. The Munson Machinery Company manufactures powder mixing equipment in which zigzag blending elements mounted in a cylinder are rotated to mix the ingredients. This equipment can also be used for continuous blending and has been used in plastics and paint mixing equipment [41].

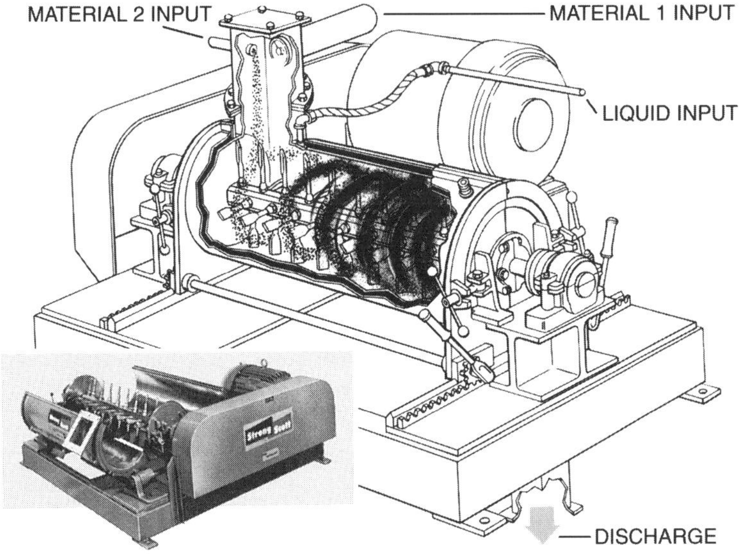

Figure 7.7 In the Bepex Turbulizer a combination of high-speed paddle and feed mechanism can be used to achieve continuous mixing of powder [41].

NOTES

1. Ribbon mixers are available from Sprout Waldron & Company Incorporated, Muncy, Pennsylvania, 17756, USA. This equipment is manufactured in the United States under license from Gebrüder Grün, KG, Lissberg, Germany.
2. V-mixers and ribbon mixers are available from O'Hara Manufacturing Ltd, 65 Skagway Avenue, Toronto, Canada, M1M 3T9.
3. Ribbon mixers are manufactured by SCOH Equipment Company, 605 Fourth Avenue N.W., New Prague, MN 56071, USA.
4. Ribbon and other mixers are available from Teledyne-Redco, 901 South Richland Avenue, PO Box M-552, York, PA 17405, USA.
5. Ribbon mixers are also manufactured by several companies including Beardsley and Piper Process, Equipment Division, 5501 W. Grand Avenue, Chicago, IL 60639; USA. The controlled circulation magazine *Powder and Bulk Engineering* has its May issue each year dedicated to powder mixing, and this issue has a comprehensive listing of the manufacturers of powder mixing equipment.
6. See the illustration in Editorial (1993) Tips for improving mixture sampling. *Powder and Bulk Solids* (Jan), 41–45.
7. van den Bergh, W.J.B. (1994) Removing the uncertainty in solids mixer selection. *Chemical Engineering*, (Dec), 70–77.
8. Brennan, A.K. Jr (1990) Selecting the right mixer: batch or continuous. *Powder and Bulk Engineering*, CSC Publishing, Inc., 1300 East 66th Street, Minneapolis, MN 55423, USA, pp. 38–42.
9. Forberg Mixers, Halvor, Forberg A/S, Hegdal, 3250 Larvik, Norway.
10. In Canada, agents for Forberg Mixers are A & J Mixing Specialists, 8-2345, Wyecroft Road, Oakville, Ontario, Canada.
11. Littleford Bros. Inc., 15 Empire Drive, Florence, KY 41042, USA. Manufactured under license from Geruder Lodige, GmbH.
12. Lodige mixers are available from Geruder Lodige, GmbH, Elsenser Strasse, P-4790, Paderborn 1, Germany.
13. Lodige Mixers are also available from Processall Inc., Cincinnati, OH, USA.
14. Hosokawa Micron Systems, Hosokawa Mikropul Environmental Systems, 20 Chatham Road, Summit, NJ 07901, USA.
15. National Bulk Equipment Inc., 12727 Riley Street, Holland, MI 49424, USA.
16. The Helicone mixer is available from Design Integrated Technology Inc., 100 East Franklin Street, Warrenton, VA 22186, USA.
17. The Nauta mixer was developed by N.V. Nautamid, Haarlem, Holland, and is manufactured under license by several companies.
18. Nauta mixers are available from Micron Powder Systems, 10 Chatham Road, Summit, NJ 07901, USA.
19. Nauta mixers are also avaialble from the J.H. Day Company, 4932 Beech Street, Cincinnati, OH 45212, USA.
20. van den Bergh, W.J.B. (1994) Part 1: Simulation of the dynamic mixing performance of an orbiting screw mixer. Part 2: Influence of particle breakage on the wall friction coefficient of brittle particulate solids. Ph.D. Thesis, Delft University of Technology, Delft, The Netherlands.
21. van den Bergh, W.J.B., Scarlett, B. and Kollar, Z.I. (1993) Computer simulation model of a Nauta mixer. *Powder Technology*, **77**, 19–30.
22. Hixon, L. and Ruschmann, J. (1992) Using a conical screw mixer for more than mixing. *Powder and Bulk Engineering*, **6**(1).
23. Prater Industries Inc., 1515 South 55 Court, Chicago, IL 60650, USA.

24. Dynamic Air Conveying Systems, Ward Ironworks Ltd, 123 Victoria Street, PO Box 511, Welland, Ontario, Canada.
25. Dynamic Air Inc., 1125 Wothers Blvd, St Paul, MN 55110, USA.
26. Fauver, V.A. (1994) Pulsed air mixing solving a blending problem in *Proceedings of the Powder Show*, Rosemont, Illinois, Reeds Exhibition Group, 1350 E. Touhy Avenue, P.O. Box 5060, Des Plaines, IL, USA.
27. Harnby, N., Edwards, M.R., Nienow, A.W. (1992). *Mixing in the Process Industries*, 2nd edn Butterworth & Co. Ltd., London.
28. Fuller-Kovako Corporation, 3225 Schoenersville Road, PO Box 805, Bethlehem, PA 18016-0805, USA.
29. V-mixers are available from Patterson-Kelley Co., Division of Harsco Corp., East Stroudsberg, PA 18301, USA.
30. Ribbon and other mixer systems are available from O'Hara Manufacturing Ltd, 65 Skagway Avenue, Toronto, Canada, M1M 3T9.
31. Double cone mixers are available from Sybron-Pfaudler, Division of Sybron Corporation, Rochester, NY 14603, USA.
32. Tower Iron Works Inc., Division Valley Industries, Plant & Home Office, 10 Tower Road, Seekoup, MA 02771, USA.
33. Mixing Equipment Co. Inc., 128 MI Road Boulevard, Rochester, NY 14603, USA.
34. The Sto-mixer is manufactured by Werner Heimbeck GmbH, Hochfield 15, D-8201, Schechen, Germany.
35. The Turbula® mixer was developed by Willy A. Bachofen, AG of Switzerland, CH-4005, Basel Utengasse 15. Exclusive US agents are Glen Mills Inc., 395 Allwood Road, Clifton, NJ, 07012, USA.
36. Custom Metalcraft Incorporated, 217 East Main, PO Box 67, Plymouth, Nebraska 68424, USA.
37. Morse Manufacturing Company Incorporated, West Manliuss Street, East Syracuse, NY 13057, USA.
38. Bulk Unit Load Systems Limited, Stratford Works, Shipston-on-Stour, Warwickshire, UK.
39. National Engineering Company, 1716 West Habbard Street, PO Box 66369, Chicago, IL 60680, USA.
40. Hosokawa Bepex Corporation, 333 N.E. Taft Street, Minneapolis, MN 55413, USA.
41. Munson Machinery Company Incorporated, PO Box 855, 210 Seward Avenue, Utica, NY 13503-0855, USA.

8
Passive powder mixing systems

8.1 BAFFLED PASSIVE MIXERS

The term passive mixer is used in this book to describe any mixing equipment in which there are no moving parts and in which the mixing of ingredients is achieved by the use of baffles or by bypass pipes moving parts of a mixture from one portion of a storage container to another. The passive mixer systems described in this chapter are essentially limited to the mixing of free flowing powders. There has been little development work on the use of possible baffled in-line systems to mix cohesive powders. Various manufacturers use different names for their passive mixers, some of them being proprietary registered names.

Fan and co-workers describe their experiments with a baffled powder mixer as studies with a motionless mixer [1]. The use of passive, i.e. motionless, mixers with baffles has not been widely applied to the mixing of powders, although Fan and colleagues have also applied it to the mixing of objects such as grain seeds [2, 3]. The motionless mixer used in their study carried out in 1970, was fabricated from a Pyrex glass tube 1.5 inches (38 mm) in internal diameter. They constructed baffles out of brass plates cut to a length of 3.25 inches (83 mm) to give helices with a 180° twist. This passive mixer is similar to one available from the Kenics Corp. known by the trademark Static Mixer® [4].

The mixers, such as the Static Mixer®, have been successfully used to mix fluids and have been widely used as turbulence creators and heat exchangers. Kenics static mixers are available in very different sizes ranging from one with baffles as large as a meter down to the small in line mixer shown in Figure 8.1. It is probable that the small, in-line, mixers of the type shown in Figure 8.1 could be made to function efficiently with powders, particularly if bypass holes were drilled in the baffles to simulate the type of motion that was described with respect to the randomizing tower described in Chapter 6 and similar to those described in the Japanese mixer described in section 7.3. In other work Fan and colleagues examined the use of another passive mixer patented by Sulzer Brothers Limited of Switzerland and available in North America from Koch Engineering Company [5–7]. This mixer is known by the trade name Motionless Mixer.

The basic type of baffle used in the Koch static mixing unit, which is available in several different sizes, is shown in Figure 8.2. As in the case of other mixers discussed in this book, the design engineer faces the dilemma that the more complex the baffles in the passive mixer, the better the mixing but the harder the cleaning if one changes from product to product. In studies underway at Laurentian University, a passive mixer similar to the ones discussed in this section has been made by crumpling wire mesh, such as the type used to protect chicken runs, into a complex structure, and placing it in a tube leading from the supply of the ingredients to the position where the mixture is to be assembled. The advantage of such low-cost internal baffling is that, at the end of an industrial mixing process, one can afford to discard the wire mixing complex and start afresh with a new one [8].

Figure 8.3 shows three different types of mixers manufactured by Charles Ross & Son Company [9]. As far as I know these systems have not been tested with powder, but they probably have potential for powder mixing, although again they may be difficult to clean if the powders are at all sticky.

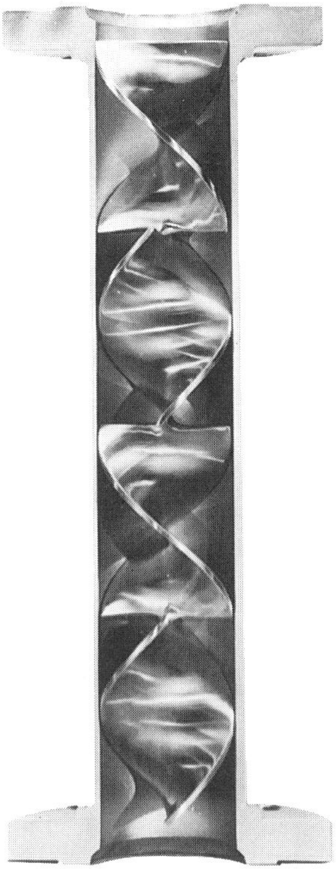

Figure 8.1 Cheminer Inc. manufactures a passive mixer known by the trade name Static Mixer®. The mixer is available in a wide range of sizes [4].

Passive mixers are manufactured by Torrey Industries [10], Luwa Corporation [11] and the Mixing Equipment Company [12]. Another passive mixer is made by EMI Mixing Technology Group, the Statitec Motionless Mixer [13].

8.2 GRAVITY IN-BIN MIXING DEVICES

A different type of passive mixer which is known by various names such as 'gravity blender' is used by industry to homogenize powdered material flowing out of a bin or to mix ingredients as they flow through a bin or silo [14–20]. Although we have avoided the use of the word 'blending' throughout this book, preferring to use the word mixing, several of the devices discussed in this section have the term blender in their commercial names, and so the word will be used when describing such equipment. Stein, in a review of such equipment, described the systems shown in Figure 8.4, which illustrates their mode of operation. De Silva and colleagues have made extensive use of gravity blenders and have published a report available from Postec Research, Norway [17, 21]. the performance of these types of device have been discussed by Johanson [18]

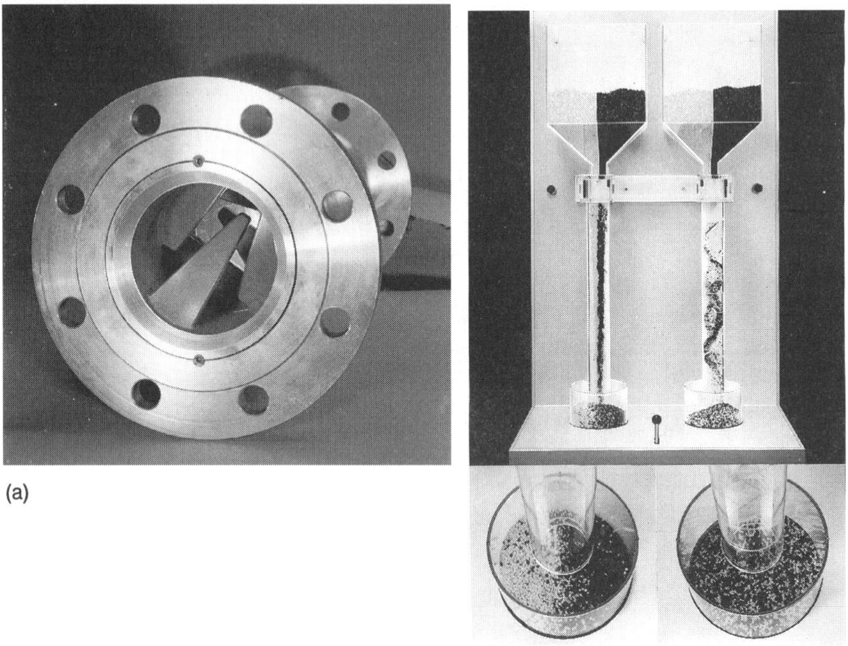

(a)

(b)

Figure 8.2 In the static mixing unit manufactured by Koch Engineering under license from Sulzer of Switzerland, the ingredients of a mixture are randomized by a complicated baffle system. (a) In-line passive mixer assembled using a series of baffle units. (b) A single baffle unit. (a) Comparison of the action of the Sulzer SMF mixer with the results obtained from a straight pipe. (b) Sulzer static mixing unit type SMF NPS 100 mm.

and by Roberts [20]. Other studies have been carried out by Cassidy and co-workers [22].

It is interesting to note that, in their introduction to their study of the performance of gravity blenders, Cassidy and colleagues make the following comment:

> Because the flow of granular materials, in all but the simplest flow domains, is too complex to be described in an analytical or numerical way industry normally uses experimental results and past experience for the design of blenders in silos [22].

De Silva and colleagues, when discussing the performance of gravity blenders, describe the two systems shown in Figure 8.5(a, b). Commenting on the design of these systems, de Silva says that

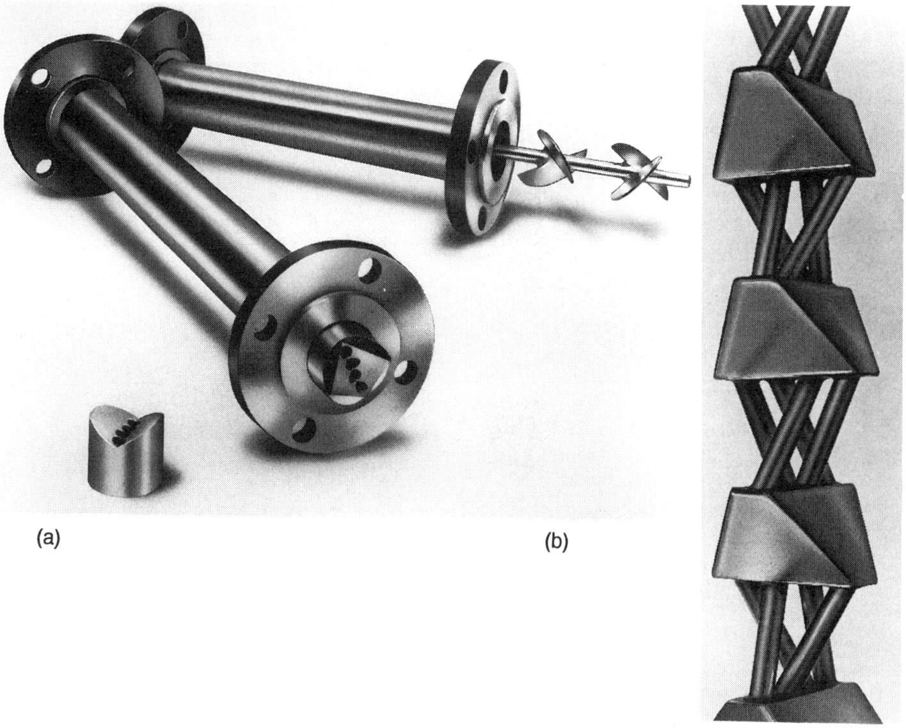

(a) (b) (c)

Figure 8.3 Charles Ross makes three different kinds of passive mixer [9]. (a) The Interfacial Surface Generator (ISG). (b) The Low Pressure Drop (LPD) mixer. (c) Blendex system. (Copyright permssion from Charles Ross & Son Company, Hauppauge, New York.)

The Young blender uses the principle of simultaneously drawing down materials from different levels in the silo and further promotes homogenization (mixing) by providing two mixing stages [21, 23].

(a)

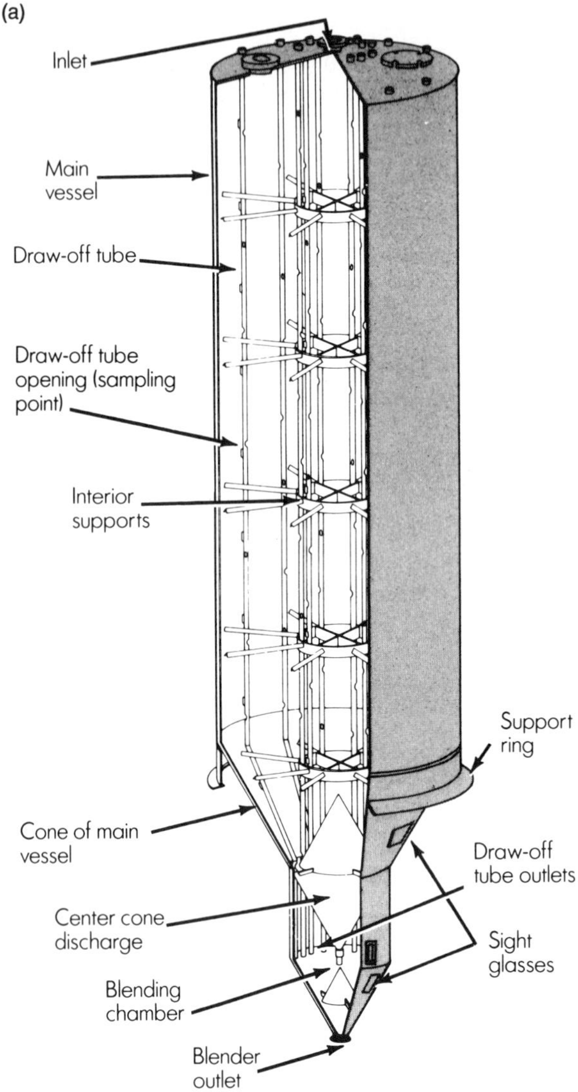

Figure 8.4 Stein, in a review of gravity blenders, described three systems used in industry. (a) Multiple sampling tube gravity blender. (b) Internally recirculating pneumatic gravity blender. (c) Partition gravity blender. (From Stein [14]. Reprinted with permssion, © 1990 by CSC Publishing, Inc., 1300 E. 66th St., Minneapolis, MN 55423, USA.)

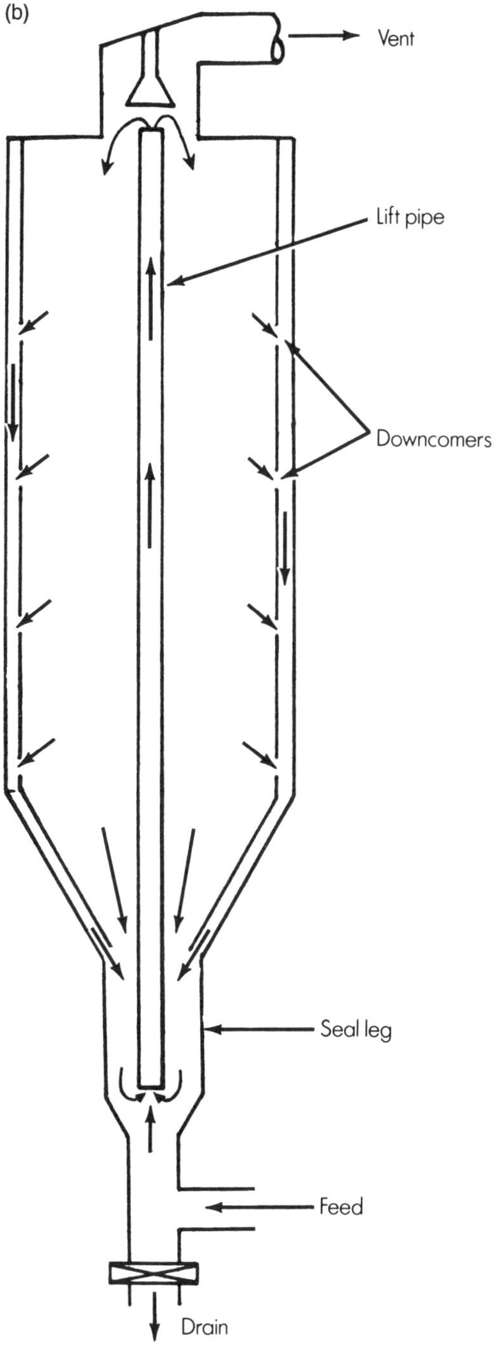

Figure 8.4 *Continued*

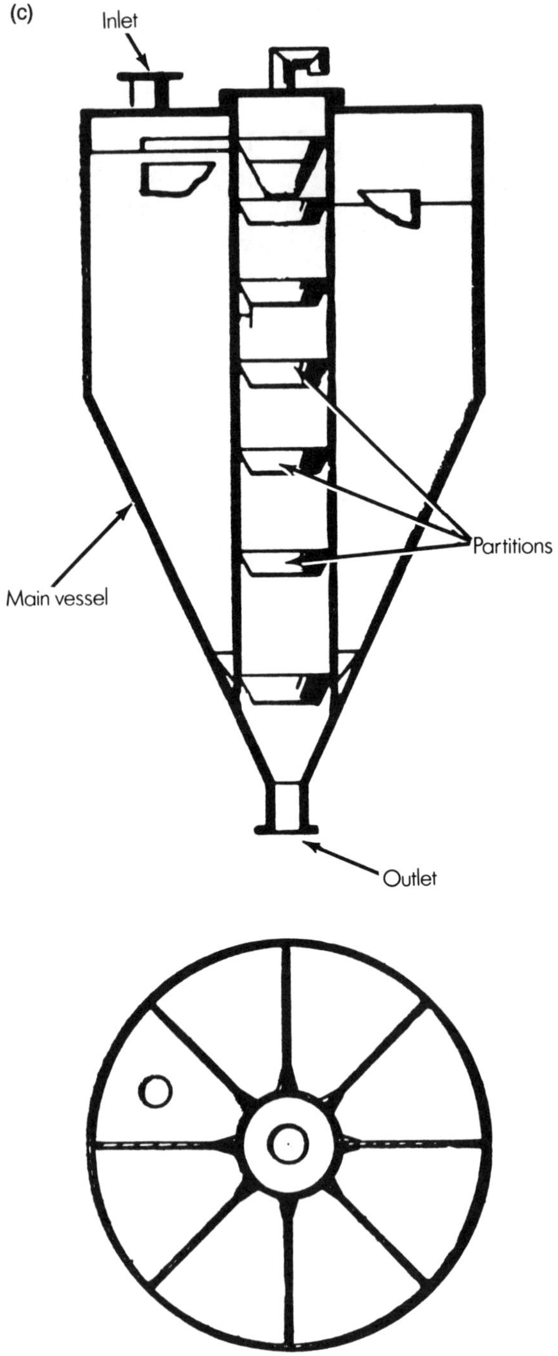

Figure 8.4 *Continued*

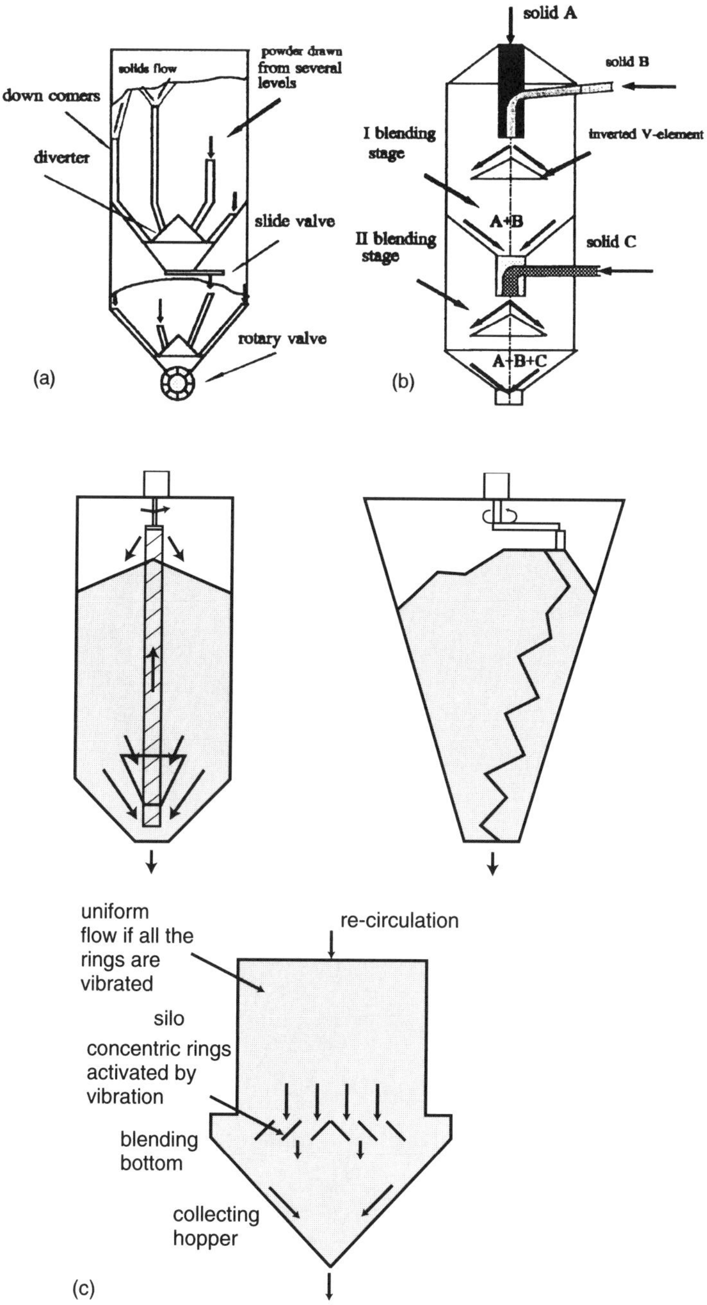

Figure 8.5 Various designs of industrial gravity blenders have been patented by several workers. (a) Young's blender [23]. (b) Roth's blender [24]. (c) Peschl's universal blender [26]. (Used by permission of S. de Silva, Norway.)

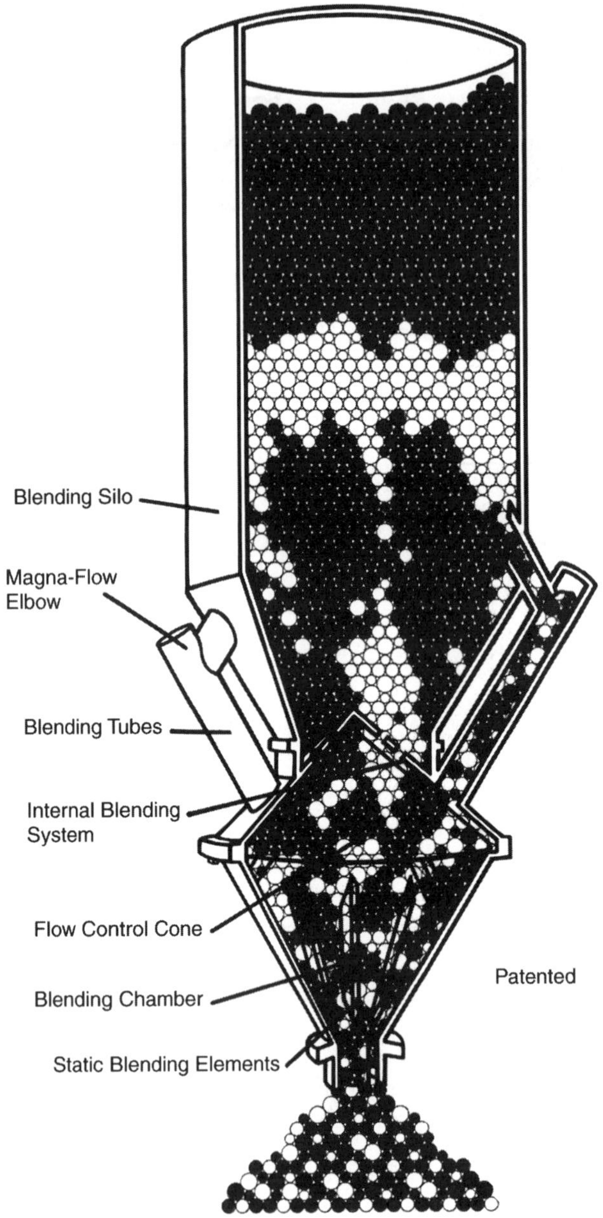

Figure 8.6 A multiport gravity blender system (Used by permission of W. Mahoney, The Young Industries Inc., Muncy, PA, USA.)

De Silva comments that Roth's blender has 'Three product streams'. It will be noted that in the Roth blender there are interests which distribute the powder at different stages of the movement through the equipment [24]. The use of special inserts for gravity blenders has been studied extensively by Johanson [19, 25] and de Silva [21]. The gravity blender of Figure 8.5(c) was developed by Peschl [26]. This system is also discussed in detail by de Silva [21]. The Fuller-Kovako Corporation manufactures gravity flow blenders [27].

Van den Bergh makes the following comments on the use of gravity blenders in industry:

> Mixing is achieved by simultaneously withdrawing material from two or more zones in a silo and then mixing them in the discharge section. The flow velocities and the last flow ratios of the material that are withdrawn from different regions in the silo differ from the velocities in the rest of the silo. This results in different resident times and intensifies the mixing in the discharge section. Mixing can be improved by recirculating the whole load two or more times using separate internal or external recirculation lines. The recirculating material is spread over the free material surface at the time of the silo, the mixing quality increases with the manner of draw up points and the number or recirculations. To avoid stagnant zones, gravity flow blenders have to be designed as mass flow silos based on the individual flow properties of the material [28].

Young Industries manufactures a gravity blending system shown in Figure 8.6 [29].

NOTES

1. Fan, L.T., Chen, S.J., Eckoff, N.D and Watson, C.A. (1971) Evaluation of a motionless mixer using a radioactive tracer technique. *Powder Technology*, **4**(6), 345–50.
2. Walawendeer, W.P., Gelves-Arocha, H.H., Fan, L.T. and Watson, C.A. (1973). A study of the axial mixing and de mixing of grain particles in a motionless mixer, in *Proceedings of the First International Conference on Particle Technology*, 21–25 August, Research Institute, Chicago, IL 60616, USA, pp. 166–71.
3. Gelves-Arocha, H.H., (1972) Mixing and segregation for particulate solids in a motionless mixer. M.Sc. Thesis in Chemical Engineering, Kansas State University.
4. The Static Mixer® is available from Chemineer Incorporated, PO Box 1123, USA.
5. Wang, R.H. and Fan, L.T. (1976) Axial mixing of grains in a motionless Sulzer (Koch) mixer. *Industrial Engineering and Chemistry Process Design and Development*, **15**(3), 381–87.
6. Koch Engineering Company Incorporated, 161 East 42nd Street, New York, NY 10017, USA.
7. Sulzer Brothers Limited, CH-401, Winterthur, Switzerland.
8. The work with the 'chicken wire' randomizing tower is part of graduate studies being undertaken by N. Faddis and it is hoped that a detailed publication of the work will be available in the not-too-distant future.
9. Charles Ross & Son Company, 710 Old Willets Path, Happauge, NY 11787, USA.

10. The passive mixer manufactured by Torrey Industries is known as the Hi-Mixer™, see trade literature of Torrey Industries Incorporated, 3-3 Nakanoshima, Kita, Ku, Osaka, 530 Japan.
11. The equipment made by Luwa is also known as Blendex®. In North America one should direct inquiries to LCI corporation (formerly Luwa) Process Division, 4433 Chesapeake Drive, Charlotte, NC 28297, USA.
12. The Lightnin® Inliner™ Mixer is made by the Mixing Equipment Co. Inc., 128 MI Road Boulevard, Rochester, NY 14603, USA.
13. Information on the Statitec mixing system is available from EMI Incorporated, P.O. Box 912, Clinton, CT 06413, USA.
14. Stein, M.R. (1990) Gravity blenders storing and blending in one step. *Powder and Bulk Engineering*, **4**(Jan).
15. Carson, J.W. and Royal, T.A. (1991) Techniques of in-bin blending, in *Proceedings of the Institution of Mechanical Engineers, Bulk Materials Handling – Towards the Year 2000*, London.
16. Wilms, H. (1992) Blending silos, an overview. *Powder Handling and Processing*, **4**(3)
17. Manjunath, K.S., de Silva, S.R., Roberts, A.W. and Ballestad, S. (1992) Determination of the performance of gravity blenders with emphasis on plane symmetric designs. POSTEC-Research, 9211600-2. Avaialble from POSTEC-Research A/S, Kjolnes Ring, 3914, Porsgrunn, Norway.
18. Johanson, J.R. (1970) In-bin-blending. *Chemical Engineering Progress*, **66**(6)
19. Johanson, J.R. (1982) Controlling flow patterns in bins by use of an insert. *Bulk Solids Handling*, **2**(3).
20. Roberts, A.W. (1991) Design of bins and feeders for anti-segregation and blending. *Proceedings of the Institution of Mechanical Engineers, Bulk Materials Handling – Towards the Year 2000*, London.
21. de Silva, S.R. and Manjunath, K.S. (1993) Homogenization in silos using inserts to control flows [of powders], in *Proceedings of the Powder and Bulk Solids Conference*, Rosemont, IL, May, Cahners Exhibition Group, P.O. Box 5060, Des Plaines, IL 60017-5060, USA.
22. Cassidy, D.J., Cribens, B.G. and Michaelids, E.E. (1992) An experimental study of the blending of granular materials. *Powder Technology*, **72**(2), 177–82.
23. Young, H.T. (1982) Apparatus for gravity blending of particulate solids. US Patent 4,353,652 (12 October).
24. Roth, C.E. (1982) Blending system for dry solids. US Patent 4,358,207 (9 November).
25. Johanson, J.R., BINSERT®. US Patent 44 286 883.
26. Peschl, I.A.S.Z. (1986) Universal Blender – a blending and mixing for cohesive and free flowing powders. *Bulk Solids Handling*, **6**(3).
27. Paul, K.D. and Romanchik, R.E. Technological advancements in gravity blenders. Preprint available from the Fuller-Kovako Corporation, Bethlehem, PA 18016-0805, USA.
28. van den Bergh, W. (1994) Removing the uncertainty in solids mixer selection. *Chemical Engineering* (Dec), 70–77.
29. Young Industries Incorporated, Box 30, Muncy, PA 17756-0030, USA.

9 Turning powder mixtures into crumbs, pastes and slurries

9.1 FROM POWDER TO PASTE

Although this book is concerned essentially with powder mixing, we have referred several times to the fact that, when a mixture of powders has been assembled, it is sometimes stabilized by spraying liquid into the mixture to achieve granulation. In this chapter we will present a brief review of how liquid bridges hold together the constiuents of a powder mixture and how pastes are created for use as extrudates.

Sometimes the process of adding liquid to the powder mixture proceeds until one achieves a crumbly mass or a cohesive structure, which is known as putty. Further addition of liquid produces pastes and slurries. The engineers should have a basic understanding of the properties of slurries, since in many industries the final stage in the stabilization of a powder mixture is to feed a slurry to a spray drying process to create free flowing granules to be fed into further processing machinery. Sometimes a paste or a slurry is dried and then pulverized to produce the granules. Whole books have been written on the process of agglomeration, and all we hope to do in this chapter is to outline the basic physics and to give references to enable the technologist to track down the information required [1–3]. Sometimes the final granulation of a powder mixture is achieved by extruding and chopping a paste mixture.

My introduction to paste making, in the scientific sense of that word, began in 1962 when I was introduced to the art of making putty. A British dictionary defines **glazier's putty** as a cement of whiting and linseed oil. In British English, **whiting** has two meanings. If after learning the definition of putty given here you checked up on the definition of whiting you would be in for a surprise if you used the same dictionary that I was using back in 1962 [4]. In that dictionary whiting is defined as

> A small seafish allied to cod so called for its white color.

In the late 1950s and early 1960s, the Research Association of the Whiting Manufacturers was a thriving institution located in southern England. I worked there from 1962–1963. The confusion caused by the name is emphasized by the fact that one day an enterprising salesman tried to sell me some fishing nets to capture my whiting until he learned that powdered whiting is also a British term for

> ground chalk free from stoney matter.

At several locations in Great Britain, it is possible to mine naturally occurring **chalk (calcium carbonate)** and pulverize it for direct use as a white pigment. This material, when mixed with the correct proportions of linseed oil, makes a good putty. **Linseed oil** is the oil pressed out of flax seed. **Flax** is a crop grown for its fibrous content from which linen is made. The flax plant was called 'lin' in old English. Linseed oil was widely used as a raw material for making artist's paint and in the manufacture of commercial paints. It has been largely displaced by synthetic oils in the paint industry, but is still used in good quality artist's paints and to finish highly polished wooden surfaces. When exposed to the air, linseed oil dries to form a tough film. When whiting is mixed with linseed oil to make a stiff paste called glazier's putty (or simply **putty**), this paste can be used to fix panes of glass in position. It has the property of being able to give slightly, with changes in temperature. The outside surface dries to a tough film which can be covered with a coat of paint.

An essential step in the manufacture of putty is to calculate how much linseed oil can be added to the whiting powder to create a stiff paste. A quick, *ad hoc* test, to calculate how much oil is needed can be carried out by placing a known weight of the powder in a beaker and then watching the behavior of the powder as water, or some other suitable fluid, is added using a burette, and the mixture is shaken. Surfaces of commercially available powders have sometimes been treated during the manufacturing process and one may need the advice of a surface chemist to know which liquid can be used with a given powder if a test of this kind is to be carried out. Thus, several years ago, I was having trouble dispersing a copper powder into water until I discovered that, during the manufacture of the powder by ballmilling, stearates had been added to the ballmill to act as what is known in engineering circles as a **grinding aid**. During the milling of a powder such as copper where fresh surfaces are created by the fracture of the surface, the freshly created surfaces are very energetic and will spontaneously weld to each other if they chance to meet in the tumbling of the ballmill. By adding a stearate to the ballmill, one arranges for each freshly created surface to be quickly coated with the stearate to prevent the spontaneous welding of the surfaces on contact. Thus the addition of stearates to a ballmilling process can increase the rate of creation of a fine powder by preventing welding of contacting surfaces of freshly broken fragments of the powder. However, this means that powders produced in this way have a tenaciously held film of stearate which can interfere with attempts to disperse the material into a liquid. This example illustrates the need to discuss the history of any powder in the preparation of powder mixtures, pastes, slurries and other dispersed systems.

When I carried out the water to whiting test, the powder was straight from the quarry, but many calcium carbonates intended as use as fillers or pigment in composite materials have been treated with surface active agents to make them either **hydrophilic** or **hydrophobic** (technical terms which mean water loving or water hating respectively).

In the initial stages of the adding of water to the ground whiting in a beaker, the powder remains dry to the external observer as crumbs begin to form. These crumbs are quite strong and resist deformation. They are held together by liquid bridges. This first stage of crumb formation is illustrated in Figure 9.1(a). The liquid bridges hold the grains of the crumb together by means of **surface tension** forces; it is the surface tension of a liquid that holds it together [2]. When a drop of water is placed upon a waxed glass slide, the surface tension creates a spherical drop. For this situation, the forces of the molecules inside the drop pulling the molecules on the surface of the drop have more attraction for these molecules than the waxed molecules on the surface of the slide. The situation for water droplets on a waxed surface is sketched in Figure 9.2(a). An important property of a liquid in such a situation is the **contact angle** made by a large droplet which sags under its own weight. This contact angle determines the structure of the interface between a surface such as the waxed surface and the droplet; it is an important factor in capillary rise or depression, a common phenomenon illustrated in Figure 9.2(b). The surface tension between a liquid and a surface can be dramatically altered by the presence of a **surface active agent** such as a **soap** or a commercial detergent. Thus in Figure 9.2(c), the structure of a common soap known as sodium palmitate (the sodium salt of palmitic acid) is shown. It can be seen that such a compound has a long oil-like chain which sticks out from the sodium salt part of the molecule. When such a molecule is placed in water, the sodium salt end of the compound has an affinity for water (it is said to be hydrophilic) whereas the other end of the chain is described as being hydrophobic or **oleophilic** (meaning oil loving). Thus the 'oil chain' part of the palmitate molecule would have great affinity for the wax surface. Therefore the sodium palmitate would tend to concentrate in the surface of the water droplet and form a strong link with the wax surface. In such a situation, the droplet would no longer maintain its simple outline and would rapidly spread over the surface of the wax, a situation described as the liquid **wetting** the wax. The reader is warned that the use of liquid detergents to wash glassware and everything else in the laboratory is so widespread in modern laboratories that it is almost impossible to have water uncontaminated by commercial detergents, unless one takes extreme precautions and implements stringent cleanliness routines. Thus although many textbooks quote the surface tension of water as 0.07 Nm^{-1}, if you attempt to measure the surface tension of water in the laboratory it will often have a lower value and therefore will behave a little differently when used to make crumbs or granules.

Returning to our discussion of crumbs of the type illustrated in Figure 9.1(b), a physical appreciation of the forces holding the crumb together can be gained by carrying out a simple experiment. If one places a drop of liquid between the contacting surfaces of two disks, this will enable one disk to move over the other very easily because of the lubricating effect of the liquid. However, if one tries to physically separate the two disks by pulling them apart without sliding them one over the other, it soon becomes apparent that a very large force is required. The formula for the force holding the two disks together, which is equal to the area of the disk times the negative pressure created between the disks by the effect of

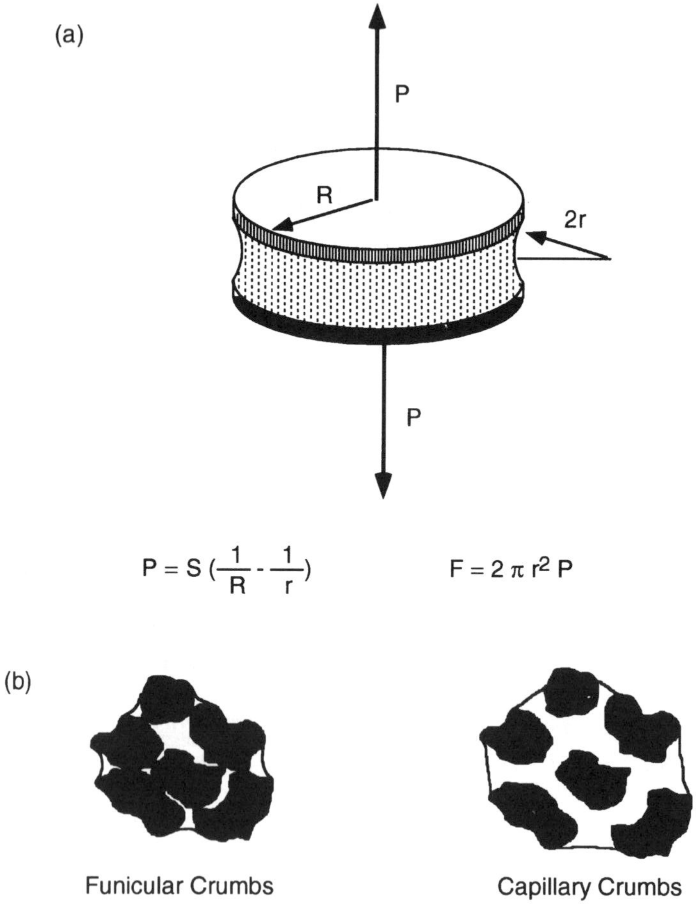

Figure 9.1 In the intial stages of the addition of water to a powder, such as chalk, crumbs held together by liquid bridges between powder grains are formed. (a) Crumbs created by liquid bridges between primary fineparticles are held together by surface tension. S = surface tension; P = pressure in the liquid; R = radius of the disk; r = radius of the exposed surface of the liquid; F = force required to pull disks apart. (b) Crumbs become larger and weaker as more liquid is added to the powder.

intermolecular forces within the liquid between the disks, is given in Figure 9.1(a). A calculation shows that if the disks are 3 cm across, then the force required to pull the discs apart is equal to 1 kg weight when a thin film of water is present between the disks. This is the reason why the crumbs formed in the initial stages of granulation by the addition of water to the powder are very strong. Note that if one later tries to dry out the crumbs in a subsequent process,

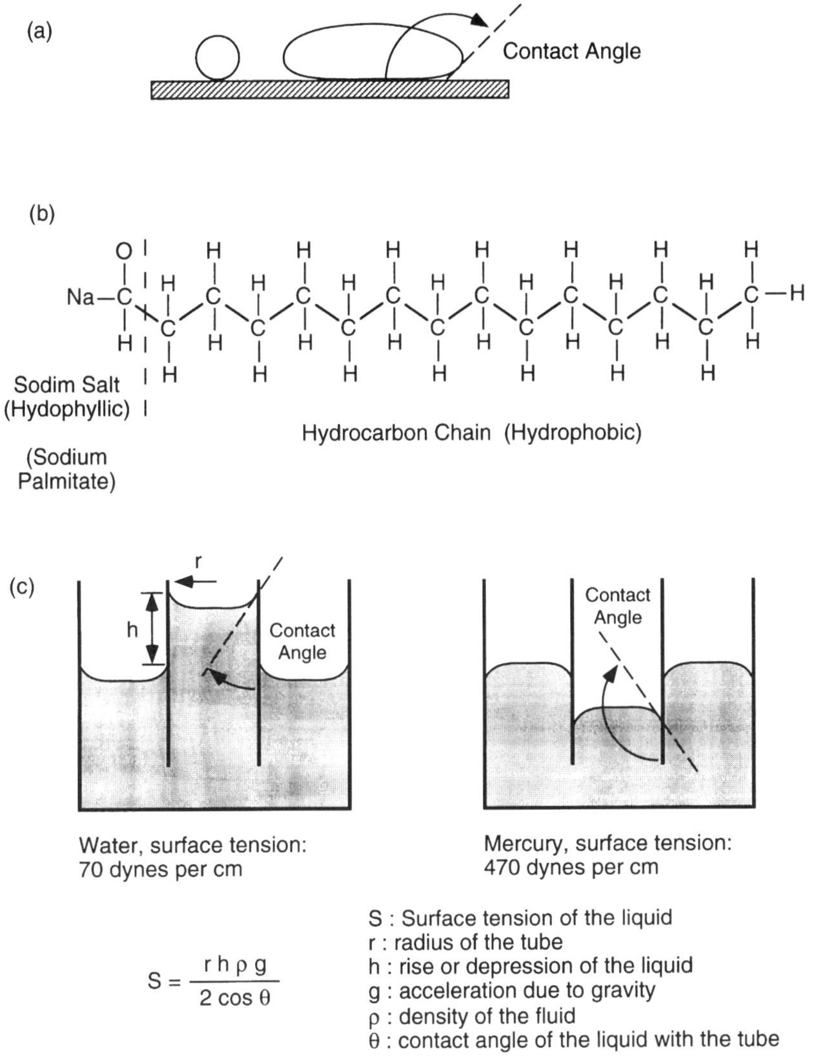

$$S = \frac{r h \rho g}{2 \cos \theta}$$

Figure 9.2 The surface tension of a liquid which is manifest by capillary rise/depression can be modified by surface active agents (surfactants). (a) Drop of water on a waxed surface. (b) Surface tension of water and mercury. (c) Structure of sodium palmitate.

it is very difficult to evaporate these liquid bridges. If the powder is at all soluble in the liquid of the intergrain bridges, permanent crystalline bridges can fuse the grains together as the drying process evolves [3].

As already discussed, sometimes a powder can be deliberately granulated into crumb-type granules by spraying liquid into a moving powder. This can be done for example using a V-mixer equipped with spray injector, so that after a certain mixing time, when the ingredients of the mix will be chaotically assembled, the addition of liquid can permanently granulate the mixture for subsequent handling. Provided the percentage of liquid is kept low, one can often be unaware that any liquid is present in the mixture. Sometimes, where one wishes to add a very small quantity of a powdered active ingredient (e.g. a vitamin) directly to a powder mixture, one should consider dissolving the small amount of additive in a compatible liquid, so that it could then be used as a spray to granulate the other well-mixed ingredients.

The first scientist to describe the progress of granulation by the continued addition of liquid to a powder described the second stage of crumb formation as funicular crumbs. In this situation, the liquid bridges between the grains have increased in quantity and size, but some air is still trapped within the structure of the crumb. (The word funicular comes from a Latin word *Funiculus* meaning a small rope; a **funicular railway** uses a rope or cable to pull carriages up a hill. Presumably the idea is that a **funicular crumb** is one which is a mixture of liquid, air and powder grains which appears to be held together by a small invisible rope.) At this stage of the crumb formation, the crumbs increase in size and can be altered by the application of pressure, in a way which will be discussed later in this chapter, to cause slump in commercial putty.

There is often no clear line between the formation of funicular crumbs and the final stage of crumb formation known as capillary crumbs. In a **capillary crumb**, all of the space between the grains of the powder is occupied by the liquid and the capillary crumb can appear moist to the external observer. Rolling the capillary crumbs around together often results in them forming larger and larger crumbs.

The term **capillary** used to describe the narrow bore tubes of the type used in the experiments to demonstrate surface tension sketched in Figure 9.2(b) comes from a Latin word meaning 'the hair of the head'. Thus a capillary tube has a thin 'hair like' hole down the middle of it. Forces causing the rise of liquid in a tube such as that of Figure 9.2(b) are often described as **capillary forces**, hence the name 'capillary crumb' when the crumb is held together by liquid filling the capillary-like gaps in the internal space of the crumb.

In the experiement in which one adds water to chalk, one quickly reaches a stage where all of the powder forms one large fairly strong ball. However, the further addition of a few drops of liquid causes this large ball to slump down into a wet paste. The volume of liquid required to turn the dry powder into a single solid ball is a good estimate of the amount of linseed oil that will turn the whiting powder into a good putty. However, one has to be cautious, because there is some skill required to carry out the addition of the liquid to the powder in a smooth continuous manner to make the single ball. If one adds the liquid too quickly, one can very often end up with some funicular crumbs with small air pockets inside the crumbs. If the putty made by adding the linseed oil to the chalk contains these small air pockets, they can be eliminated by kneading (to **knead** is defined in the dictionary as to work and press together into a mass, as when making flour into dough) and the subsequent contraction of the crumbs or putty structure will result in surplus oil, which causes the putty to slump into a messy oily mix. Sometimes a whole load of putty being transported will slump due to the vibrations and pressure experienced from the other parts of the load, and what was a dry workable mix at the factory comes out at the purchaser's location as an oily mix. In fact, in modern hardware stores, putty is sold in small sealed plastic bags and is often in the slumped condition when purchased. This product has to be worked manually with the fingers to turn it back into a workable putty. In this case, the manual working of the putty causes dilation of the powdered matrix with subsequent reabsorption of the oil. (This explanation will become more comprehensible when we discuss dilatancy of slurries of irregularly shaped fineparticles in the next section, concerned with the behavior of non-Newtonian fluids.) Unless the powder mixture being added to a liquid is being specially treated, one can often experience great difficulty in the dispersal of crumbs formed by the first contact of the powder with the surface of the liquid. It is usually a better strategy to add small amounts of liquid to a powder and apply high shear forces to make a paste which can then be diluted to a suspension or slurry. Powders which can be added directly to fluids have usually been treated with dispersing agents and/or wetting agents to increase the dispersability of the powder. One of the problems which often prevents efficient use of slurry in a manufacturing process is the fact that powder addition to a liquid often dramatically increases the viscosity of the resultant suspension, and such suspensions can have what are known as non-Newtonian liquid properties. The behavior of non-Newtonian fluids will be discussed briefly in the next section.

9.2 DILATANT AND THIXOTROPIC SUSPENSIONS

A paste is defined technically as a high-solids liquid mixture which can hold an external shape until subjected to pressure. As the liquid content increases a paste changes to a slurry, and a further addition of liquid will make the slurry into a suspension of particles in a liquid. The boundaries between pastes, slurries and suspensions are not always clearly manifest because they depend upon the forces being experienced by the system. This is because many pastes and slurries are what are known as non-Newtonian fluids [5]. To be able to understand what is meant by non-Newtonain fluids one must first of all review the concept to flow under applied forces of liquids such as water and pure mineral oil. Let us imagine that one can apply a force F on the top of an element of fluid at a distance H above the surface of the vessel containing a fluid, as illustrated in Figure 9.3(a). This force produces what is known as a **shearing stress** which causes the various layers of the fluid to slide over each other like a pack of cards. Because of friction between the layers of the fluid, the top layer moves at the highest speed and the layer immediately in contact with the surface over which the fluid is moving is at rest. This static layer of fluid next to a containing vessel such as a pipe is known as the stagnant **boundary layer**. Because of this differential movement of the layers of fluid in the liquid, a **velocity gradient** is established of magnitude v/H, as illustrated in the figure. Isaac Newton, who was the first to study the resistance of fluids to applied forces, found that if one

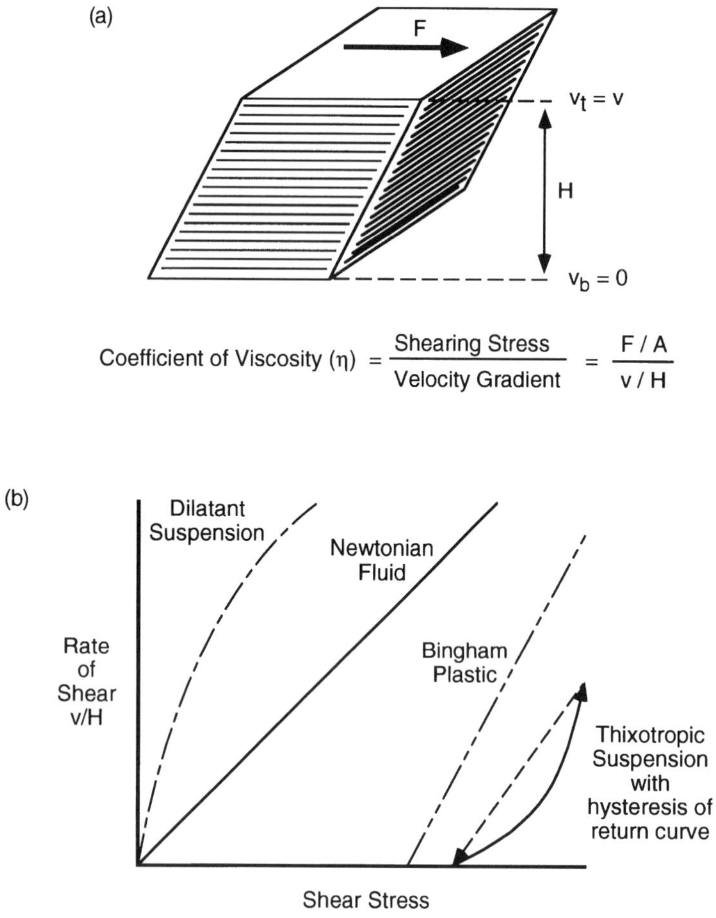

Figure 9.3 There are various types of non-Newtonian fluids that exhibit varying behavior under applied stress. (a) Definition of the coefficient of viscosity. (b) Physical behavior of various types of non-Newtonian fluids.

plotted a graph of the rate of shear of the fluid (the velocity gradient) against the shearing stress applied to the fluid that one obtained a straight line of slope $1/\eta$ for many simple fluids such as water, alcohol, acetone etc. The slope of the line varied from liquid to liquid and was taken to be the characteristic parameter fixing the resistance of the fluid to applied stress. The parameter η became universally known as the **coefficient of viscosity** of the fluid. The type of liquid that gives a straight line on a graph of rate of shear versus shear stress became known as a **Newtonian fluid**. As scientists began to investigate the behavior of mixtures of sand in water, they found two important deviations from Newtonian behavior. Anyone who has walked across a wet beach becomes aware of the properties of slurries known as **dilatancy** [5]. We discussed in an earlier chapter that to dilate means to expand by means of shaking or applied pressure. As one puts a foot down on the wet beach, the sand appears to dry out beneath the footprint. This is because in its drained state, before the foot disturbs it, the sand grains of the beach have aligned to create a minimum voidage structure which has the intergrain spaces filled with fluid. The application of the foot to the wet sand causes the sand to dilate by creating a random packing of the grains which occupies more space than the resulting sand grains on the wet beach. In British English, one describes the rearrangement of the grains as being a higgledy-piggledy structure, illustrated in Figure 9.4(d). This higgledy-piggledy structure requires more intergrain liquid to lubricate it, and in the absence of the required amount of fluid to fill the intergrain space it appears to dry out. In this dried-out state, it is more resistant to movement than when one begins to disturb it. This is shown by the curve of Figure 9.3(b) labeled **dilatant suspension**. As the shear stress is increased, it requires more stress to create the same rate of shear than at the beginning of the disturbance of the suspension. Therefore the rate of shear versus shear stress curve for the dilatant suspension is convex with respect to the shear stress axis. In one's walk across the sand, if one looks back to the footprints which one thought had become islands of dryness on the beach because of foot pressure, one observes that in fact the footprints have become sloppy indentations in the sand. The reason for this is that during the dilated state, which persists as long as one's foot is pressing on the sand, water drains into the dilated structure from adjacent areas of the beach. When one's foot is removed from the sand, the higgledy-piggledy sand settles back into its original drained state and the excess liquid that drained into the dilated structure now becomes excess fluid on top of the footprint.

The other important type of non-Newtonian fluid that the powder technologist is likely to encounter when adding powder to a liquid is what is known as a **thixotropic suspension**. Anyone who has attempted to make mortar for holding bricks together will have encountered a thixotropic suspension. Sometimes a pile of mortar can look dry, and one is tempted to add extra fluid to the mortar to make it easy to apply to a brick. However, before yielding to such temptation, one should move the mortar with a trowel, because under the shear stress created by the trowel, the apparently dry mortar becomes mobile, and if one has yielded to the temptation to add more fluid, one now finds that one has to add more sand and cement to bring it back to a workable consistency. Technically, what has happened is that one has had to apply an initial stress to cause the system to move, but once the system starts to move, very small increases in stress cause big increases in the velocity gradient, i.e. in the mobility of the suspension. This is the reverse of the situation that we have encountered in our discussion of dilatant suspensions. In the resting state, the pile of mortar has sand grains and cement powder grains arranged in a random (higgledy-piggledy) structure, and when the trowel is used to work the mortar, the grains inside the suspension line up to decrease the effective viscosity of the suspension (Figure 9.4(c)). Such mixtures are described as thixotropic suspensions. The word **thixotropic** comes from two Greek words, *thixis*, meaning 'a touch', and *trope*, meaning 'a turn' or 'a turning'. Thus thixotropic materials start to turn easily when touched. The strict definition of thixotropy given in the dictionary is that it is the property of becoming fluid when shaken or disturbed. It should be noted that sometimes thixotropy is called **inverse dilantancy**. From our brief discussion of dilatancy

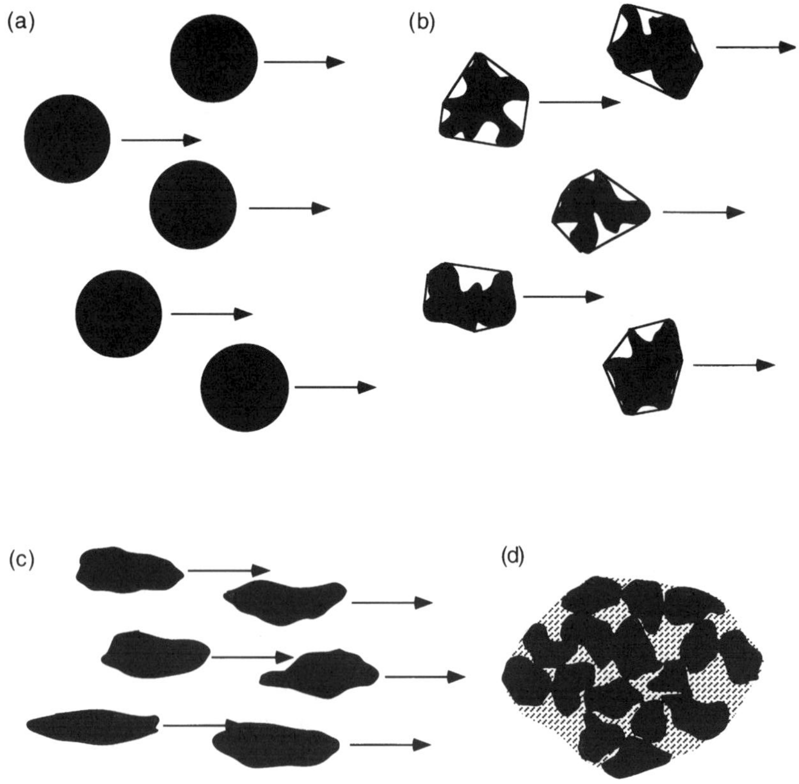

Figure 9.4 The thixotropic nature of some suspensions is due to the shape of the grains of powder in suspension. A = area of the surface upon which F is acting.

and thixotropy, it should become apparent that if one is using a powder, the more that the grain shape differs from spherical (Figure 9.4), the more likely one is to run into difficult problems as one pumps the suspensions through pipes and into spray drying equipment [6–9].

It should be noted that sometimes a thixotropic suspension does not follow the same rate of shear versus shear stress curve as the shear stress is reduced, thus in some situations the reduction of the shear stress results in the behavior of the suspension following the double curve of Figure 9.3(b). This lagging of the return curve behind the increasing shear curve is described as hysteresis. The word hysteresis comes directly from a Greek word meaning 'coming late'. The basic idea is that the return curve lags behind the advance curve. Note that materials which exhibit thixotropic behavior but which do not require a threshold shear stress before the system starts moving are described as **pseudoplastic liquids**. A liquid which behaves as a Newtonian fluid once it starts to move under the initial applied stress is known as a **Bingham plastic**. This type of suspension is named in honor of one of the pioneers of modern rheology [10]. One of the problems that one can encounter when handling thixotropic suspensions is that, if one allows the suspension to come to rest in for example a stirred storage container, the start-up forces required to get the system moving again after it has settled down into a maximum packing situation can break the shaft of the stirring devices. This type of problem was experienced by scientists who were using suspensions of coal in power stations originally designed to use oil. When a storage tank was left unstirred for a period, the shaft of the propeller broke when the system was started up again. A good piece of advice to be followed when handling suspensions of powders in liquid is to make them restless fluids and never give them a chance to settle down!

Some of the effects of fineparticle shape and texture on the viscosity of a suspension are illustrated by the data of Figure 9.5 generated in an MSc study by S. Akhter [11]. In his study Akhter used a simple Couette viscometer of the type shown in Figure 9.5(a). This piece of equipment is to be found in most undergraduate training laboratories and in various industrial situations [12, 13]. More advanced versions are available from commercial vendors of equipment. In the equipment used by Akhter, the viscosity was measured by the turning speed of the internal cylinder as various weights were added to the pans, applying a torque to the central cylinder by the means of the threads shown. The basic fluid used by Akhter in his experiments was a clear mineral oil of sufficient viscosity that virtually no settlement of the suspension occurred during the experimental investigations. All viscosities in his study were expressed as dimensionless numbers obtained by dividing the viscosity of the suspension being investigated by the viscosity of the clear fluid. In a first series of experiments, the viscosity of suspensions of glass spheres was investigated as the volume fraction of the spheres was increased. It can be shown that the data of Figure 9.5(b) for the spheres agrees with formulae developed by earlier workers [7–9]. Measurements made on suspensions of nickel tailings in the same mineral oil are also shown for comparison with the data for glass spheres. It can be seen that up to a value of approximately 10% by volume solids, the glass beads and the nickel ore tailings have essentially the same relative viscosity. However, for spheres suspended in a volume fraction from 0.1 to 0.25, the relative viscosity of the nickel ore tailings suspension increases dramatically. As can be seen from the typical profiles given in Figure 9.5(b), the nickel tailings fineparticles are not of extreme shape and the rapid increase in the viscosity

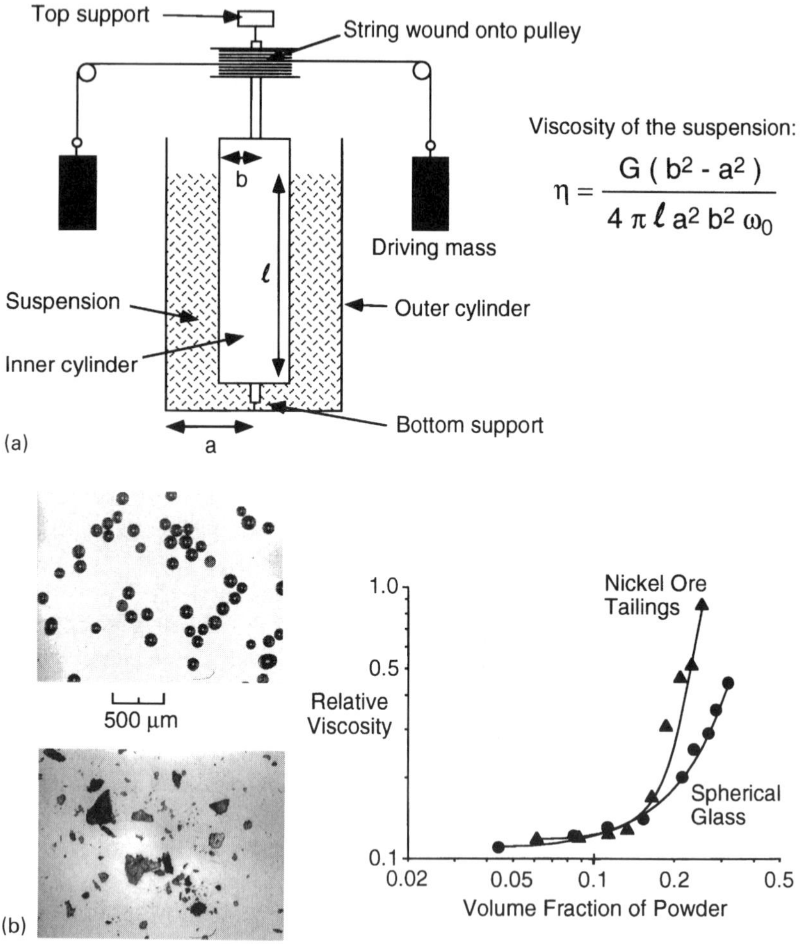

Figure 9.5 Structure and texture can have enormous effect on the viscosity of a suspension as demonstrated by data generated by Akhter. (a) Simple viscometer for measuring the viscosity of a suspension under steady shear. G = gravitiational constant; ω_0 = angular velocity of the inner cylinder. (b) The effect of concentration on the viscosity of the suspension for two powders, glass beads and nickel ore tailings. (c) Viscosity of cubic crystals in suspension as compared to glass beads. (d) Viscosity of two sizes of coal fineparticles compared to glass beads.

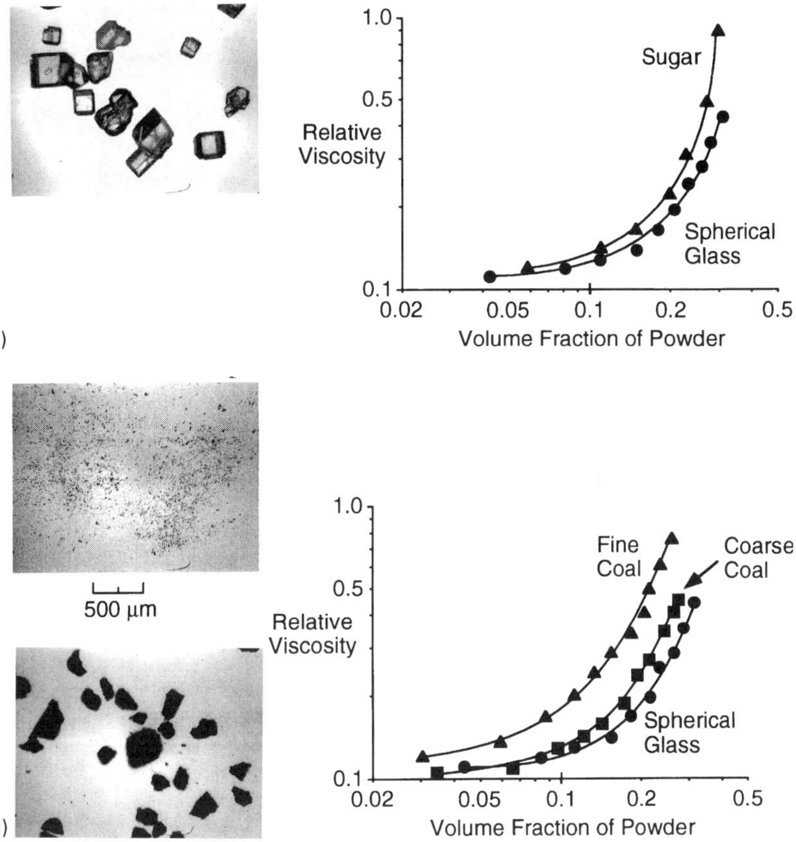

(c)

(d)

Figure 9.5 *Continued.*

must be due to the fact that the rugged exteriors of the fineparticles entrap stagnant fluid in the surface convolutions, as shown in Figure 9.4(b). This has the effect of increasing the effective volume of the tailings in the suspension [14]. It can be seen that by the time one reaches the volume concentration of 0.25, the viscosity of the rock tailings is at least five times higher than that of the glass spheres. It is probable that curves of the types summarized in Figure 9.5 may be very useful in predicting the flow properties and packing properties of a powder, apart from the physical behavior of a mixture of fineparticles incorporated into a liquid. Figure 9.5 also shows relative viscosity measurements made for sugar crystals and powdered coal.

The different sets of fineparticles studied in these experiments summarized in Figure 9.5 are all basically chunky, with or without ruggedness. If one moves to a study of fineparticles which are more fibrous in stucture, then the viscosity of the suspension can increase drastically because of the liquid friction between the high surfaces of the fibers in suspension [15]. Historically, asbestos has been used to make asbestos cement pipes, and one of the problems encountered early in the development of such technology was that **chrysotile asbestos (white asbestos)** of the type mined in Canada has a curly, fibrous structure from which liquid does not drain readily and suspensions of which have a high viscosity. On the other hand, **crocidolite asbestos (blue asbestos)** of the type mined in South Africa and other places has relatively straight thin fibers which are much easier to handle in suspension and from which liquid will drain readily when the suspension is poured into a mold. Unfortunately, the very same properties that make blue asbestos

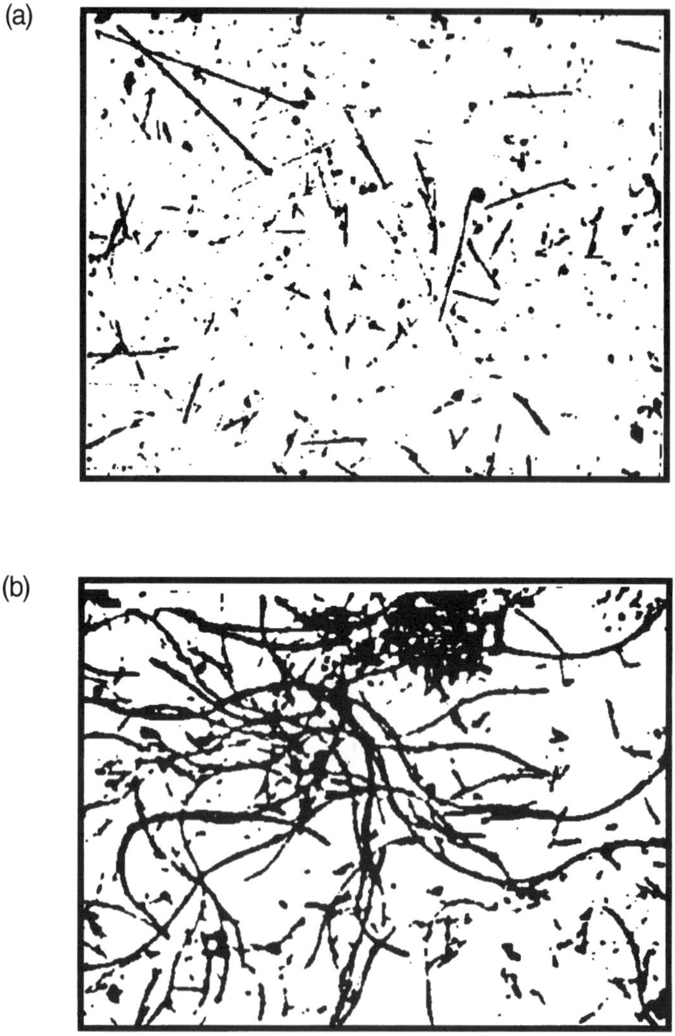

Figure 9.6 Fibers in suspension can generate very high viscosity suspensions. (a) Blue asbestos; (b) White asbestos.

easier to handle when in suspension also create much higher health hazards than the chrysotile asbestos when the dry form of the material is inhaled. Therefore a viscosity–drainage property that was advantageous for manufacturing purposes resulted in a widespread use of a potentially dangerous form of asbestos. The safety of chrysotile asbestos versus crocidolite asbestos is the subject of an ongoing debate, which is too complicated to include in this discussion. The difference between the two froms of asbestos is illustrated in Figure 9.6.

Because of the very high viscosities encountered when making pastes, one has to have mixers in which high shear zones exist, and usually power consumption by such mixing equipment is high because of the work needed to process the paste. This fact means that some mixers for creating viscous pastes require cooling systems. Random diffusion of the components is almost nonexistent, and one can generate good paste mixing by forcing the system through a passive mixer, with the multiple layering produced by successive stages of the mixer leading eventually to a good mixture.

After an extensive study of the mechanics of various paste mixers, Schofield and co-workers developed a special design of rotating mixing elements which created very high shear in the mixer. The design of these special rotors and the way in which their configuration changes as they rotate in opposite directions is

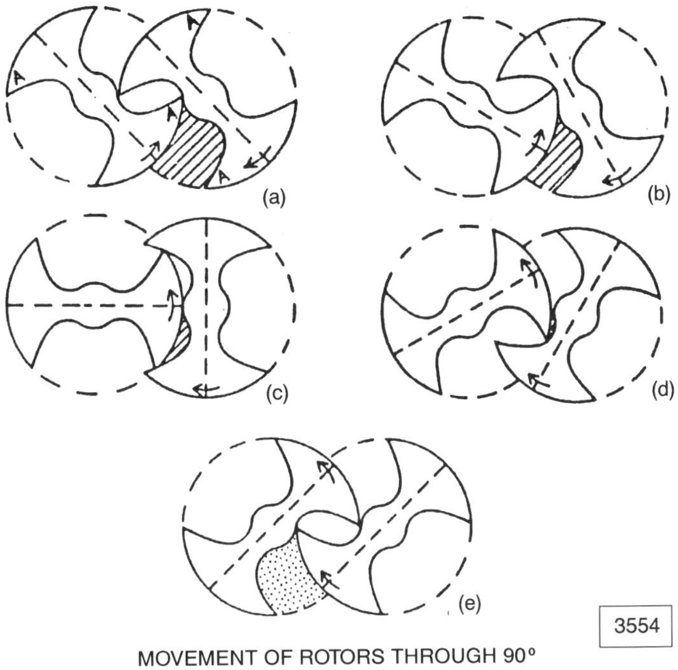

Figure 9.7 Schofield's special paste dispersion rotors are effective for pastes of high viscosity. (a)–(e) shows successive movement of rotors through 90°.

illustrated in Figure 9.7. The tip of either one of the rotors is always in contact with the opposite rotor and with the walls of the mixing chamber. The mixing action is obtained by squeezing the material in the spaces between the rotors and expelling it through the small clearances between them. Details of the processing of a slurry to create granules by spraying is discussed in most chemical engineering textbooks [16–18].

NOTES

1. Kousaka, Y. and Endo, Y. (1994) Liquid bridge adhesion force and dispersion of aggregate particles. *KONA*, No. 12, 7–16.
2. Many of the modern physics textbooks have dropped discussions of surface tension and capillary action of fluids in channels in favor of more glamorous atomic physics etc. A good classical text discussing surface tension, capillary action and the force of adhesion between two wetted plates held together by liquid, is Fender, B.H. (1967) *General Physics and Sound*, English University Press, Aylesbury.
3. Pietsch, W. (1991) *Size Enlargement by Agglomeration*, John Wiley & Sons, New York.
4. MacDonald, E.M. (1967) *Chambers Etymological English Dictionary*, Published by W. and R. Chambers Ltd, Edinburgh.
5. For a basic introduction to the subject of non-Newtonian fluids see Walker, J. (1978) The amateur scientist. *Scientific American*, (Nov), 186–96.
6. Clarke, B. (1967) Rheology of coarse settling suspensions. *Transactions of the Institution of Chemical Engineers*, **45**, T251–T256.
7. Grazino, F.R., Cohen, R.E. and Medalia, A.I. (1979) Rheology of concentrated suspensions of carbonblack in low molecular weight vehicles. *Rheological Acta*, **18**, 641–56.
8. Smith, T.L. (1972) Rheological properties of dispersion of particulate solids in liquid media. Journal of *Paint Technology*, **44**(75), 71–79.
9. Sweeny K.H. and Geckler, R.D. (1954) The rheology of suspensions. *Journal of Applied Physics*, **25**(9), 1135–44.
10. Darby, R. and Melson, J. (1981) How to predict the friction factor for flow of Bingham plastics. *Chemical Engineering* (28 Dec), 59–62.
11. Akhter, S.K. (1982) Fineparticle morphology and the rheology of suspensions and powder systems. *M.Sc. Thesis*, Laurentian University, Sudbury, Ontario, Canada.
12. Starling, S.G. and Woodall, A.J. (1964) *Physics*, Longman, Green and Company, London.
13. Green, H. (1949) *Industrial Rheology and Rheological Structures*, John Wiley & Sons Inc., New York.
14. For a discussion of the effect of the fractal structure of powder grains from the viscosity of suspension of irregularly shaped powder grains, see Kaye, B.H. (1994) *A Randomwalk Through Fractal Dimensions*, 2nd edn, VCH Publishers, Weinheim, Germany, p. 82.
15. A discussion of the properties of viscous suspensions of fibres is given in Gordon, J.E. (1984) *The New Science of Strong Materials*, 2nd edn Princton University Press.
16. For a good introduction to spray drying of slurried materials see Shore, F.W. (1994) Fresh options in drying. *Chemical Engineering*, (Jul), 76–84.
17. Broadhead, J. *et al.* (1992) Spray drying of pharmaceuticals. *Drug Development and Industrial Pharmacy*, **18**(11 and 12), 1169–206.
18. Masters, K. (1991) *Spray Drying Handbook*, John Wiley & Sons, New York.

Author index

Page numbers appearing in **bold** refer to **figures**.

Subject index

Page numbers appearing in **bold** refer to **figures** and page numbers appearing in *italic* refer to *tables*.